D1646327

General Geography in Diag

R.B. Bunnett, B.Sc.

This book belongs to

Mairi b Rose

Longman

LONGMAN GROUP LTD
London, England

*Associated companies, branches and representatives
throughout the world*

© R.B. Bunnett 1973

All rights reserved. No part of this publication may be
reproduced, stored in a retrieval system, or transmitted
in any form or by any means, electronic, mechanical,
photocopying, recording, or otherwise, without
the prior permission of the Copyright owner.

First published †1973
Third Edition †1976
New impression †1977
New impression †1978

ISBN 0 582 69969 X

*Printed in Hong Kong by
Sheck Wah Tong Printing Press Ltd*

Acknowledgements

For permission to reproduce the photographs on the pages mentioned below, the Publishers are indebted to the
following:

Aerofilms, 71 right, 82 top, 170, 192 top, 215, 218, 324, 342 top, 337; J. Allen Cash, 49 bottom, 53, 57, 65 bottom,
82 bottom; 116, 117 top left, 167, 171, 189, 219 left, 226, 227, 341; American Museum of Natural History 87; Ansel
Adams, Magnum 175; Argyll Oil Field Press Service 285; Australian News & Information Bureau, 69, 168 top;
174 left, 218 bottom, 348; Barnaby's 168 bottom, 174 right, 190 top and bottom, 304; H.H. Bennett, F.A.O., 180
centre; Victor Bianchi, 182 top; B.O.A.C. 19 top, 203, 230, 236 top and bottom, 247, 331; British Airways 338;
British Information Service, 65 top, 193, 320, 347; British Petroleum 199 top; British Tourist Authroity 344;
R.B. Bunnett 19 bottom, 21, 44 right lower, 76 left, 78, 79; Camera Press 182 bottom; Canadian National Film
Board, 263, 264, 275, 287; H.E. Dale, U.S. Department of Commerce Weather Bureau, 137; Department of
Highways, South Dakota, 44 bottom left; Ewing Galloway 57 bottom right; Exclusive News Agency 70; F.A.O.
220, 226, 244, 253, 357; Fairchild Aerial Surveys 59; French Government Tourist Office, 35 bottom left; Geological
Survey Photograph 76 right, 84; Government of British Colombia 270; Hans Wild, Reed International 266; Harrison
and Crossfield 236 top and bottom; R. Hertzog, 68; Hong Kong Government Information Service 182 centre;
W. Heerbrogg 35 top; I.P.S. (U.S.I.S.) 31, 51 centre; Institute Geographique National 96; International Society
for Educational Information Press Inc., 256 left and right; Japan National Tourist Organisation 204; Japan
Information Centre 317; J.K. Joseph 94 bottom; J. Launois, Black Star, 54; Middle East Archives 40; Mustograph
117; Nouosti Press Agency 192 bottom, 197; Ontario Ministry of Industry & Tourism 332; Paul Popper 29, 71 top,
97, 172, 180 bottom, 199 bottom, 200, 219 right, 227, 251 bottom; Port of Singapore Authority, 335 top and bottom;
Quebec Government 77; Radio Times, 2 (from Hale Observatory), 34 top right, 72 left; Rubber Research Institute,
Malaya, 258, 259; Sky Foto 34 left, 336; S.A.S. 82 top; Singapore Ministry of Culture 324 top; Society
Encyclopedique Universelle, 180 top; Society for Cultural Relations with U.S.S.R. 183 bottom; Swissair Photo
93 bottom; Swiss Embassy 327; Swiss National Tourist Office 44 top, 91, 93 top, 94 top; Tin Industry & Development
Board, 298; U.S. Forest Service 173, 245, 251 top; U.S.I.S. 95, 181, 183 top, 184 top and bottom, 185 top
and bottom, 209, 287, 333; Zambia Information Dept. 51 left; Wong Siew Luen 269.

Contents

Preface to the Third Edition

This book, which was first published in 1973, has been considerably revised by expanding and updating the text where relevant, and by introducing new and more acceptable theories to explain the origin of geographical features, and geographical processes. In addition, multiple-choice questions have been introduced because examining authorities in many countries are now using this type of question in place of the traditional type.

Whilst the author admits the limitations of textbooks, and accepts that no textbook can be a substitute for the good teacher, he believes that a study of geography can be enriched and made more meaningful by the use of well-annotated diagrams, especially where these illustrate the stages in the development of features of the physical landscape, and in the presentation of basic ideas and concepts fundamental to geography.

Note to the Teacher

The aims of this book are to encourage and enable students to develop a knowledge of the basic concepts, and of the terminology, processes, trends and sequences, and patterns of generalisation used in geography. By doing so, it is hoped that students will obtain an appreciation of the environment, an understanding of spatial patterns, and the processes contributing to them, and develop abilities to cope with the changing environment. To achieve these aims, teachers should use teaching techniques which enable their students to exercise skills in enquiring into, understanding, communicating and applying knowledge. Questions of various types are given in the book which will help to develop these skills, but the resourceful teacher will be able to add to these appreciably.

Introduction

This book contains over 400 multiple-choice questions, most of which are contained at the ends of the chapters. In addition, there is a selective range of traditional-type questions which are also given at the end of each chapter. The multiple-choice questions have been most carefully chosen to ensure that all the basic types of such questions are adequately covered. If the student covers and understands the work contained in this book, and if he works through all the multiple-choice questions, he will be able to answer, with confidence, most of the questions set in the examination at the end of his geography course.

1 The Meaning of Objective Testing

The values and opinions of an examiner may affect the way in which marks are given to the answers to essay-type questions. This certainly happens when the candidate's values and opinions are different to those of the examiner, and as a result there is often more than one correct answer to an essay-type question. Such questions are said to be 'subjective'. In objective testing, questions are set which prevent the values and opinions of the examiner from affecting the way marks are given. The removal of the 'subjective' element in the marking means that each question in objective testing can have only one correct answer.

2 The Importance of Objective Testing

An objective test consists of multiple-choice questions of which there are several types. Such questions are not easy to set because they have to be written in a way to ensure that the whole syllabus for a specific examination is effectively tested. Also, the questions have to be devised so that the alternative answers to each question are of equal value. But an objective test is easy to mark because each question has only one correct answer and the marking is therefore objective. In contrast, a test consisting of essay-type questions is easy to set but difficult to mark because of the subjective nature of the questions.

In an objective test the candidate must answer all the questions which means that the candidate is examined on the whole syllabus. In an essay-type test, the candidate is given a choice of questions which means that the candidate is not examined on the whole syllabus.

Types of Multiple-Choice Questions

Multiple-choice questions are of several types. Some are factual, some are deductive, and some are easier than others. This book contains three types of multiple-choice questions: (i) **completion type,** (ii) **question type,** and (iii) **negative type.**

(i) Completion Type

This type of question consists of a 'stem' which is the problem, which you, the candidate has to resolve, together with four 'responses' lettered A, B, C, D, from which you must choose the correct solution. The stem, which consists of an incomplete statement, can be correctly completed by one only of the responses offered. Here are two examples:

Example 1

STEM   Minimum and maximum temperatures are obtained from an instrument called:

RESPONSES   **A**   a Barometer
               B   a Six's Thermometer

C an Anemometer

D a Clinical Thermometer

Explanation

Of these four responses, two (**A** and **C**) are not correct because the instruments are not used for measuring temperature, and one (**D**) is not correct because the instrument is used for measuring a person's temperature. Only response **B** is correct because this instrument is a thermometer which is used for measuring maximum and minimum temperatures. This correct response is recorded by you by underlining the letter **B** as follows:

A B C D

Example 2

STEM   A polder is

RESPONSES   A   an area of drained land

B   an area of the sea in Holland which has been drained

C   an area of drained land used for agriculture

D   an area of land reclaimed from the sea

Explanation

All four responses refer to land which has been drained and/or reclaimed, but a polder is the name given to land in Holland which has been obtained by enclosing a part of the sea which is subsequently drained. The correct response is **B** and you record this as follows:

A B C D

(ii) **Question Type**

This type of question also consists of a 'stem' together with four 'responses'. The stem consists of a question to which one or two of the responses may be a partly correct answer. But only one of the responses will be fully correct. Here are two examples:

Example 1

STEM   Which one of the following groups of terms is applicable to some parts of the ocean floor?

RESPONSES   A   basin, deep, cirque, plateau

B   trench, deep, basin, plateau

C   plateau, basin, dune, ridge

D   ridge, deep, basin, waterfall

Explanation

Only response **B** is fully correct. Each of the other responses contains three correct features and one incorrect feature. The three incorrect features are cirque, dune and waterfall, all of which are features which occur on land surfaces only. This correct response is recorded by you by underlining **B** as follows:

A B C D

Example 2

STEM Which one of the following statements most clearly defines the meaning of the term *hinterland*?

RESPONSES
A the area around a port
B the area served by a port
C the port area
D the area connected to a port by road and rail communications

Explanation

Each of these responses is partially true because the hinterland of a port comprises the region in which the port is located and perhaps adjacent regions as well. However, the word hinterland implies more than just physical area. It also implies the service which a port gives to its surrounding region. The correct response is therefore **B** and you record this by underlining **B** as follows:

A **B** C D
.. ‒ .. ..

(iii) Negative Type

This type of question may be of the completion type or the question type. All of the responses given for the stem are true except one, and it is this one that you must recognise. Here are two examples:

Example 1

STEM Which of the following statements is **not** true in respect of sedimentary rocks?

RESPONSES
A the particles of rock are sometimes completely of organic origin
B the rocks are non-crystalline
C they are rocks whose structure is determined by great pressure or heat
D the rocks have been deposited in layers

Explanation

Responses **A**, **C**, and **D** are all true of sedimentary rocks but response **B** is not true because there are many sedimentary rocks which are composed of crystalline mineral particles. Response **B** which is not true is recorded by you by underlining the letter **B** as follows:

A **B** C D
.. ‒ .. ..

Example 2

STEM The successful cultivation of coffee requires all of the following natural conditions **except**

RESPONSES
A a rich, deep, well-drained soil
B an early morning mist followed by cloudy skies
C an annual rainfall of about 1800 mm evenly distributed throughout the year
D an average monthly temperature of between 21°C and 26°C.

Explanation

The only response which is not essential for the successful cultivation of coffee is response **B**. Whilst this condition will not harm the cultivation of coffee, it is not essential for its cultivation, whereas all of the conditions of responses **A**, **C**, and **D** are essential. The correct answer is therefore response **B** and you record this as follows:

A   B   C   D

Note 1

These are the three basic types of multiple-choice questions and most of the questions that you will have to answer will be of these types. However, there are some topics in Geography for which multiple-choice questions present a difficulty. The topics which are of this type are those for which it is difficult to state categorically whether a response is absolutely right or wrong. Here is an example:

STEM   A region which has high average temperatures and low average rainfall will probably have a vegetation of

RESPONSES   A   mosses on peat bogs
            B   plants with bulbous roots
            C   shallow rooted trees
            D   trees with thin bark

Explanation

Of these four types of plants, those which have bulbous roots are the most likely plants to be growing in a region which has high average temperatures and low average rainfall, and **B** is the correct response. However, this does not mean that any or all of the other three types of plant will not be growing in such a region. You record the correct response as follows:

A   B   C   D

Note 2

Multiple-choice questions can also be set to test your ability in geographical reasoning. These items test such skills as comprehension, analysis, synthesis and application. Here is an example of a multiple-choice question designed to test your ability to understand the ideal conditions for the formation of a delta.

STEM   A delta can form under several similar, though not identical combinations of natural conditions. Which one of the following combinations would result in the formation of a large delta in the mouth of a river carrying a very heavy load, where there is no subsidence?

RESPONSES   A   a saline sea with strong currents
            B   a saline sea with weak currents
            C   a non-saline sea with strong currents
            D   a non-saline sea with weak currents

Explanation

Deltas will develop under conditions stated in responses **B** and **D**, and they may develop under conditions stated in responses **A** and **C**, but it is the conditions stated in response **B** that will best promote the development of a large delta. The correct response is therefore **B**. You record the correct response as follows:

A   B   C   D
·· ‒   ·· ··

6   The Marking of Objective Tests

All objective tests aim to cover the syllabus as evenly as possible and at the same time to maintain a balance between questions which test your factual knowledge of geography, and those which test your skills in the use of geographical data. The abilities for which you are tested are of four basic types. These are:

(i)   Factual knowledge

This refers to the remembering of specific geographical facts, geographical terminology, conventions and definitions.
Here is an example:

STEM   Which one of the following features is the produce of river deposition?

RESPONSES   A   a barchan
            B   a fan
            C   a drumlin
            D   a beach

The correct response is **B** and you record this as follows:

A   B   C   D
·· ‒   ·· ··

(ii)   Comprehension

(a)   This refers to the ability to understand and use geographical data which is presented in visual, that is, map or diagram or photographic form. Here is an example:

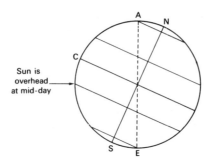

STEM   Which of the following is true in respect of this diagram?
RESPONSES   A   there are 24 hours of daylight at A
            B   there are 12 hours of daylight at C
            C   there are 24 hours of daylight at E
            D   there are 12 hours of daylight for all latitudes

The correct response is **B** and you record this as follows:

A   B   C   D
·· ‒   ·· ··

(b) The ability to understand and explain common geographical phenomena. Here is an example:

STEM   There are very few low-growing plants in a tropical humid forest because there is

RESPONSES   A   lack of moisture at ground level because of the covering of tall trees

            B   little sunlight at ground level because of the canopy of tall trees

            C   a thick covering of leaves and decaying vegetation on the forest floor which prevents plant growth

            D   a high evaporation rate on the forest floor which causes the plants to wither

The correct response is **B** and you record this as follows:

A   B   C   D

(iii) Application

This refers to the ability to select and apply theories and principles to geographical problems. Here is an example:

STEM   The windward slope of a mountain which is at right angles to winds blowing from the sea is wetter than the leeward slope. This is because

RESPONSES   A   the windward slope is nearer the sea

            B   the winds have to rise to cross the mountain

            C   descending winds are warm

            D   the sea is warmer than the land

The correct response is **B** and you record this as follows:

A   B   C   D

(iv) Analysis and Evaluation

This refers to the ability to recognise and analyse individual elements in a complicated situation and the inter-relation between them. Here is an example:

STEM   Which one of the following combinations of natural conditions would best promote the formation of a spit?

RESPONSES   A   longshore drift carrying a load along a straight coast

            B   longshore drift carrying a load along an indented coast

            C   waves breaking close inshore where the backwash is more powerful than the swash

            D   waves breaking around a headland

The correct response is **B** and you record this as follows:

A   B   C   D

Metric Conversion

LENGTH

| 1 metre (m) | = 100 centimetres (cm) |
| | = 1000 millimetres (mm) |

1 kilometre (km) = 1000 metres (m)

1 international nautical mile = 1852 metres (m)

From metric

1 millimetre (mm)	= 0·03937 inch
1 metre (m)	= 1·094 yards
	= 3·281 feet
1 kilometre (km)	= 0·621 mile

To metric

1 inch	= 25·4 millimetres (mm)
1 foot	= 304·8 millimetres (mm)
	= 0·3048 metre (m)
1 yard	= 0·9144 metre (m)
1 mile	= 1·609 kilometres (km)

AREA

1 hectare (ha)	= 10 000 square metres (m²)
1 square kilometre (km²)	= 1 000 000 square metres (m²)
1 square kilometre (km²)	= 100 hectares (ha)

From metric

1 square metre (m²)	= 1·196 square yards
	= 10·76 square feet
1 square kilometre (km²)	= 247·105 acres
	= 0·386 square mile
1 hectare (ha)	= 2·471 acres

To metric

1 square foot	= 0·0929 square metre (m²)
1 square yard	= 0·836 square metre (m²)
1 acre	= 4046·86 square metres(m²)
	= 0·4047 hectare (ha)
1 square mile	= 2·590 square kilometres (km²)
	= 259 hectares (ha)

MASS

| 1 kilogramme (kg) | = 1000 grammes (g) |
| 1 metric tonne (t) | = 1000 kilogrammes (kg) |

From metric

1 gramme (g)	= 0·0353 ounce
1 kilogramme (kg)	= 2·205 pounds (lb)
1 tonne (t)	= 0·984 ton
	= 2204·2 pounds (lb)

To metric

1 ounce	= 28·35 grammes (g)
1 pound	= 0·4536 kilogramme (kg)
1 ton	= 1·016 tonnes (t)

VOLUME AND CAPACITY

| 1 litre (l) | = 1 cubic decimetre (dm³) |
| 1 cubic metre (m³) | = 1000 litres |

From metric

| 1 litre | = 0·220 gallon = 1·76 pints |
| 1 cubic metre | = 220 gallons = 1·308 cubic yards |

To metric

| 1 gallon | = 4·546 litres |
| 1 pint | = 0·568 litre |

TEMPERATURE

From °C

$$t = \tfrac{9}{5}\theta + 32$$

Where θ = temperature in degrees Celsius (Centigrade) °C

t = temperature in degrees Fahrenheit (°F)

To °C

$$\theta = \tfrac{5}{9}(t - 32)$$

Where θ = temperature in degrees Celsius (°C)

t = temperature in degrees Fahrenheit (°F)

For a more complete list of conversions see 'Metric and Other Conversion Tables' by G.E.D. Lewis (Longman 1973).

GEOLOGICAL TIME SCALE

Era	Period or System	Epoch or Series	Important Physical Events and Fauna	Time in Millions of Years
CENOZOIC	QUATERNARY	RECENT	Glaciers melted; many mammals disappeared; warmer climates.	
CENOZOIC	QUATERNARY	PLEISTOCENE	Glaciation. Invertebrates; large mammals and Man.	1
CENOZOIC	TERTIARY	PLIOCENE	Mountain building. Large mammals.	10
CENOZOIC	TERTIARY	MIOCENE	Uplift of Rockies. Grazing animals.	25
CENOZOIC	TERTIARY	OLIGOCENE	Lands generally low: Alps and Himalayan Systems develop; Rockies area had volcanoes. Sabre-toothed cats appeared.	40
CENOZOIC	TERTIARY	EOCENE	Erosion, lakes in North America. Tropical/mild climate. All modern mammals.	60
CENOZOIC	TERTIARY	PALAEOCENE	High mountains; cool climates. Birds and primitive mammals.	70
MESOZOIC	CRETACEOUS		Lowlands widespread. Mild climates. Flowering plants and insects; extinction of giant reptiles.	135
MESOZOIC	JURASSIC		Lowlands widespread; Europe under seas; mild climates. Mountains rise in W. North America and eruptions widespread. Dinosaurs.	180
MESOZOIC	TRIASSIC		Continents mountainous; deserts widespread. Eruptions in W. North America	220
PALAEOZOIC	PERMIAN		First mammal-like reptiles. Appalachians formed	270
PALAEOZOIC	CARBONIFEROUS		Lowlands emerged from seas; tropical coal swamps formed. Large reptiles and amphibians. Mountain building in North America.	350
PALAEOZOIC	DEVONIAN		N. America low and flat but mountains and volcanoes in E. North America; Europe arid and mountainous. Fishes dominant.	400
PALAEOZOIC	SILURIAN		Flat continents; mild climates; slate deposits.	440
PALAEOZOIC	ORDOVICIAN		Low continents; mild climates; shallow seas. Some mountains.	500
PALAEOZOIC	CAMBRIAN		Seas in Geosynclines. Mild climates. Algae and trilobites.	600
PRE-CAMBRIAN	PROTEROZOIC (Algonician)		Seas in Geosynclines. Mild to cold. Few Fossils. Lake Superior iron deposits formed.	1000
PRE-CAMBRIAN	ARCHAEOZOIC (Primitive life)		Extensive mountain building. Graphite and carbon. Earliest known life.	3000
PRE-CAMBRIAN	AZOIC (Without life)		Formation of the Earth's crust. No rocks have been found.	4500 – 6000

1 Our Home in the Universe

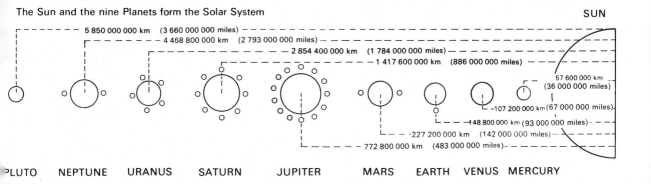

The Sun and the nine Planets form the Solar System

SUN

5 850 000 000 km (3 660 000 000 miles)
4 468 800 000 km (2 793 000 000 miles)
2 854 400 000 km (1 784 000 000 miles)
1 417 600 000 km (886 000 000 miles)

57 600 000 km
(36 000 000 miles)
107 200 000 km (67 000 000 miles)
148 800 000 km (93 000 000 miles)
227 200 000 km (142 000 000 miles)
772 800 000 km (483 000 000 miles)

PLUTO   NEPTUNE   URANUS   SATURN   JUPITER   MARS   EARTH   VENUS   MERCURY

INTRODUCTION TO THE UNIVERSE

At night the sky often appears to be full of stars each of which seems to be no bigger than a twinkling speck. But it comes as a surprise to learn that every star is much bigger than the earth: indeed some are several millions of times bigger. Again, the distances between stars in the night sky do not appear to be very great, but astronomers have calculated that despite the millions of stars in the universe, they are so scattered in space that together they occupy only a very small part of space. Some idea of the immensity of the universe can be obtained by considering speed, distance and size. Light and radio waves travel 298 000 kilometres or 186 000 miles per second which means that if we transmitted a radio signal from earth it would take about $1\frac{1}{4}$ seconds to reach the moon, 8 minutes to reach the sun and almost 4 years to reach Proxima Centauri, our nearest star, and just under 20 years to reach Delta Pavonis, a slightly more distant star. Stars tend to form clusters, which are known as *galaxies*, and galaxies form *groups*. Our Local Group, that in which the earth is located, contains 27 galaxies. In this Local Group is a distant galaxy, just visible with the naked eye, which is called Andromeda Spiral. The radio wave transmitted from earth would take about 2·2 million years to reach this galaxy. In other words, when we look at Andromeda Spiral we are seeing it as it was 2·2 million years ago. Distances are so immense that it is impossible for us to draw a scaled map of the universe. Let us try doing this by using an infinitesimally small scale. On this scale let us regard the sun as being the size of a hydrogen atom, that is, one 100 millionth of a centimetre in diameter. Our galaxy would then have a diameter of 69 metres (225 feet) and the sun would be about 25 metres (82 feet) from its centre. Andromeda Spiral would be about 1·6 kilometres (1 mile) away and the edge of the known universe would be almost 3200 kilometres (2000 miles) away.

THE SOLAR SYSTEM

This system contains the sun and its nine planets which revolve around the sun in elliptical orbits. The light of the sun falls on each of the planets and it is in turn reflected by them. This is the only light that the planets reflect.

All the energy of the Solar System is derived from the sun whose surface is covered with burning gases and whose temperature is about 6000°C. Mercury, which is the smallest planet, is nearest to the sun while Pluto, which is smaller than Earth, is the farthest away from the sun. Some of the planets, eg. Earth, Jupiter and Saturn have *satellites*: the moon is the satellite of the Earth.

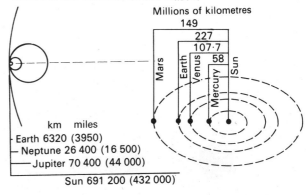

Millions of kilometres
149
227
107·7
58

Mars   Earth   Venus   Mercury   Sun

km   miles
Earth 6320 (3950)
Neptune 26 400 (16 500)
Jupiter 70 400 (44 000)
Sun 691 200 (432 000)

1

The Andromeda Spiral Nebula

Because the planets are at varying distances from the sun, and because they revolve around the sun, they each take a different time in which to complete one orbit. Mercury completes its orbit in 88 days, that is, a year on Mercury lasts 88 days. The earth completes its orbit in $365\frac{1}{4}$ days, which is the length of one year on earth. The moon, which revolves around the earth, takes approximately 27 days to do so.

PHASES OF THE MOON

As the moon revolves around the earth the illuminated part of it apparently varies in size. In the diagram the two circles represent moon positions. The outer circle clearly shows that exactly half the moon is illuminated all the time. The inner circle shows what the moon looks like to us on earth during its different positions. The illuminated part of the moon takes different shapes – at full moon it is a circle. Look at the moon on different nights in any one month and find out whether that part of the moon which is not illuminated can be seen.

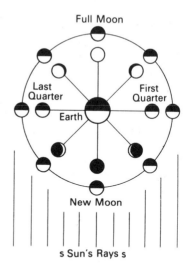

Eclipse of the Moon
(earth comes between moon and sun)

Eclipse of the Sun
(moon comes between earth and sun)

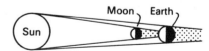

Eclipse of the Sun as seen from the Earth

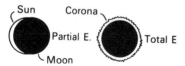

2

THE SHAPE OF THE EARTH

At one time the earth was thought to be flat, but today it is a proven fact that it is round. This is accepted by all. There is much evidence to show that the earth is round but before looking at this it is as well to remember that it is not quite a perfect sphere. The polar circumference is about 130 kilometres (80 miles) shorter than the equatorial circumference and the polar diameter is almost 40 kilometres (25 miles) shorter than the equatorial diameter. The earth is, therefore, slightly flattened at the poles and its spherical shape is called a *geoid*.

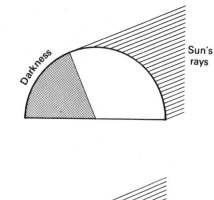

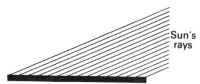

The evidence

Aerial photographs. Numerous photographs have now been taken by astronauts from satellites at great distances from the earth, and all of these show that the earth is spherical.

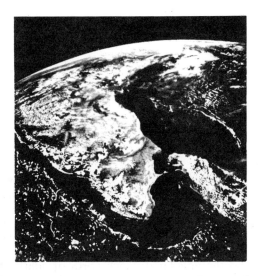

Satellite photograph showing Southern India and Sri Lanka

2 The moon's eclipse. When there is an eclipse of the moon, the shadow of the earth which is thrown on the moon is always round. Again, only a sphere can cast a shadow which is circular.

3 Circumnavigation of the earth. The earth has been circumnavigated innumerable times by land, sea and air.

4 Sunrise and sunset. The earth rotates from west to east which means that places in the east see the sun before places in the west. This is a proven fact. If the earth was flat, all places would see the sun at the same time.

5 The earth's curved horizon. The earth's horizon, when seen from a ship, a plane, or a high cliff appears curved. The curved horizon widens as the observer's altitude increases until it becomes circular. This is how astronauts see the horizon from their space ships. If the earth was not spherical, there would be no circular horizon.

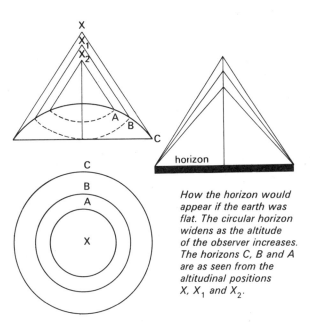

How the horizon would appear if the earth was flat. The circular horizon widens as the altitude of the observer increases. The horizons C, B and A are as seen from the altitudinal positions X, X_1 and X_2.

6 A ship's visibility. The diagram on page 4 shows two ships but you can see that the observer is only able to see one of them. If the earth was flat, the observer would see both ships.

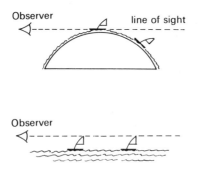

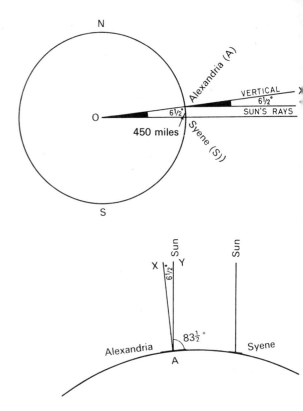

THE SIZE OF THE EARTH

About 200 B.C. Eratosthenes measured the altitude of the mid-day sun at Alexandria and found that it was $83\frac{1}{2}°$. He knew that at the same time on this particular day the altitude of the sun at Syene, which was about 720 kilometres (450 miles) to the south, was 90°. Eratosthenes knew that angle XAY was $6\frac{1}{2}°$ and that angle AOS was also $6\frac{1}{2}°$. He then calculated that if an angle of $6\frac{1}{2}°$ is subtended by an arc of 720 kilometres (450 miles) then an angle of 360° would be subtended by an arc of 40 000 kilometres (25 000 miles).

THE POSITION OF A PLACE ON THE EARTH'S SURFACE

Let us take a large ball and mark two points on it so that they are exactly opposite to each other. Now draw a line right round the ball so that it is midway between the points all the way. The line will now divide the ball into two equal parts, and, because the ball is a sphere, each part can be called a *hemisphere*. We will now call the line the *equator* and you will see that it is a *circle*. One point we will call the *North Pole* and the other the *South Pole* (*fig. a*).

We can now draw more circles parallel to and to the north and south of the equator. These can be called *parallels* or lines of *latitude* (latitude referring to the angular distance north or south of the equator). This idea is applied to the earth. The equator is given a value of 0°, and, as you can see

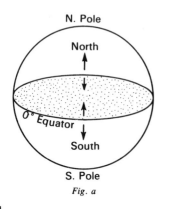

Fig. a

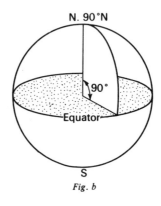

Fig. b

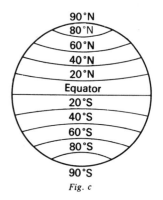

Fig. c

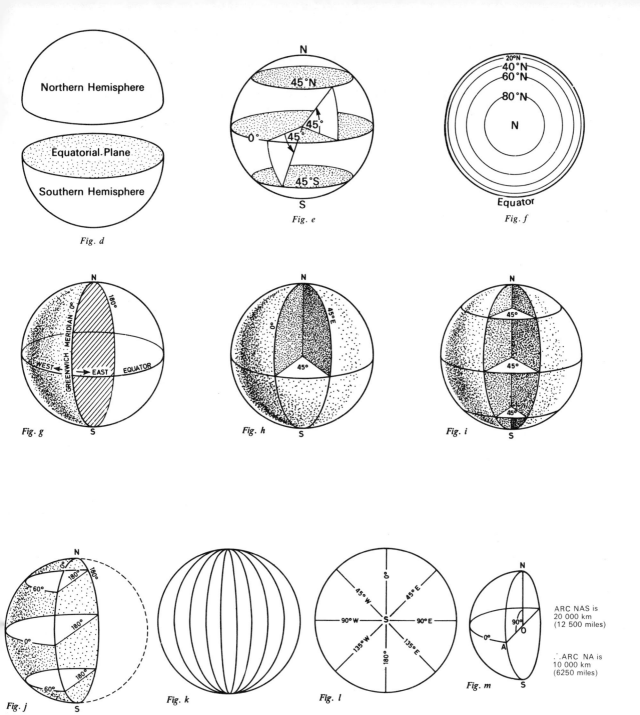

Fig. d

Fig. e

Fig. f

Fig. g

Fig. h

Fig. i

Fig. j

Fig. k

Fig. l

Fig. m

ARC NAS is
20 000 km
(12 500 miles)

∴ ARC NA is
10 000 km
(6250 miles)

from *fig. b*, the North Pole has a latitude of 90°N. The South Pole has a value of 90°S. and every other place on the earth's surface has a latitude of so many degrees north or south (of the equator). You will also notice that the equator is the longest parallel. *Figs. c* and *f* show what the parallels look like on a globe from the side and from the pole respectively.

We can now draw on the ball another set of circles all of which pass through the two poles (*fig. k*). That part of each circle between the poles can be called a *meridian* or line of *longitude*. This idea is also applied to the earth and the meridian which passes through Greenwich is given a value of 0°; the meridian that is opposite to it will therefore have a value of 180° (*fig. g*). *Longitude refers to the*

5

angular distance east or west of the Greenwich Meridian, and all places except those on meridian 180° will therefore have a longitude of so many degrees east or west (of Greenwich). *Figs. k* and *l* show what the meridians look like from the side and from the pole respectively.

How long is 1° of Latitude?
Fig. m is a diagram of a hemisphere and N and S stand for the North and South Poles. Angle NOA is 90° and this is the latitude of N or the angular distance of N from the equator (0°). This angle is subtended by arc NA whose length is one half of a meridian. On the earth arc NA has a length of 10 000 kilometres approx.

If an arc of 10 000 kilometres subtends 90° then an arc of $\frac{10\,000}{90}$ kilometres subtends 1° i.e. *1° of latitude represents 111 kilometres approx.*

How long is 1° of Longitude?
Every parallel has an angle of 360° at its centre and every half-parallel is subtended by an angle of 180° (*fig. j*). If the length of the parallel or half-parallel is known then the length of the arc subtended by 1° can be calculated. For the equator this is 111 kilometres, but for other parallels it is less than this because parallels decrease in size away from the equator. *1° of longitude represents 111 kilometres along the equator.*

Great Circles
Any circle which divides a globe into hemispheres is a *great circle*. The equator is a great circle and Greenwich Meridian together with Meridian 180° make another great circle. Likewise Meridian 10°E. and 170°W., and 20°E. and 160°W., make two more great circles. The number of great circles is limitless. Great circles can extend in any direction: east to west, north to south, north-east to south-west, and so on. Great circles are of equal length.

MOVEMENTS OF THE EARTH

I It rotates   II It revolves

Rotation of the Earth
The earth rotates once in 24 hours and this results in:
(i) Day and night
(ii) A difference of 1 hour between two meridians 15° apart

(iii) The deflection of winds and ocean currents
(iv) The daily rising and falling of the tides.

Day and Night
These four diagrams show what is happening along Greenwich Meridian during one rotation of the earth on March 21st.

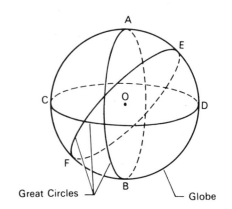

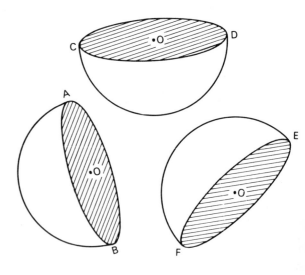

Great Circles   Globe

The sun is rising along Greenwich Meridian. People here see it 'rising' over the Eastern Horizon.

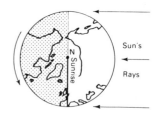

he earth has turned through $\frac{1}{4}$ of a rotation and is noon along the Meridian. The sun has reached s highest position in the sky.

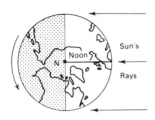

The earth has passed through $\frac{3}{4}$ of a rotation and it is midnight along Greenwich Meridian.

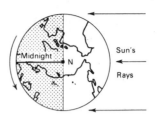

he earth has now turned through $\frac{1}{2}$ of a rotation nd the sun is setting along the **Meridian**. People ere see the sun 'sinking' below the Western Ho-izon.

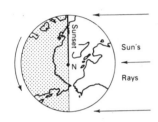

Day and Night as seen from Space

The 6 photographs below, all taken in one day from an altitude of approximately 36 000 kilometres (22 500 miles), show the progress of sunlight from early morning (a) to late afternoon (f). The photographs were taken from an American satellite in synchronous orbit above the intersection of the equator and the International Date Line. Parts of the U.S.A., Mexico and South America are visible at the right of photographs (*a*) and (*b*).

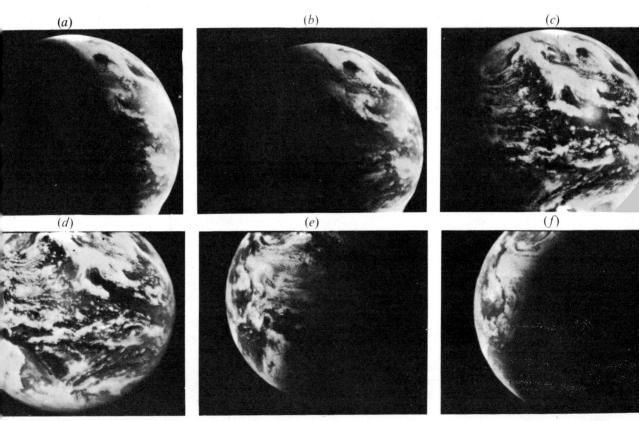

(a)   (b)   (c)

(d)   (e)   (f)

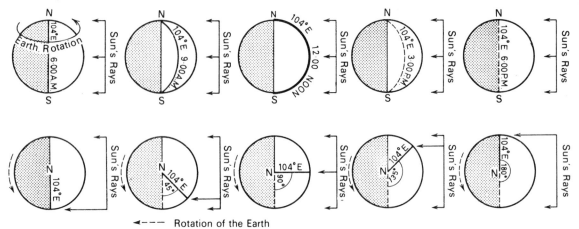

At the Equinox

◄--- Rotation of the Earth

Rotation and time

The diagram above shows the positions of meridian 104°E. at intervals of 3 hours. The top row of diagrams shows the appearance of the meridian from above the equator and the bottom row of diagrams the appearance from above the North Pole.

The sun reaches its highest position in the sky for this meridian when it lies under the sun. At this time it is said to be *1200 noon Local Time* along this meridian. Local time is sometimes called *Sun Time*. The highest position of the sun for any place can be observed from a study of the lengths of the shadows cast by a vertical stick. The shortest of these is cast by the sun when it is in its highest position in the sky (study the Sun Path Diagram for Singapore).

The diagram above also shows that all places on meridian 104°E. have noon at the same time. This means that *all the places on the same meridian will have the same local time.*

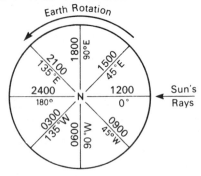

Local times for selected meridians when it is 1200 noon G.M.T.

When Greenwich Meridian lies under the sun the local time along this meridian is of course 1200 noon, but this local time is 1200 noon *Greenwich Mean Time* or G.M.T.

Sun Path Diagram for Singapore on June 21

The vertical stick indicates the position of Singapore. The shortest shadow points due south and occurs at *noon*, i.e. when the sun reaches the highest point in its 'path' across the sky.

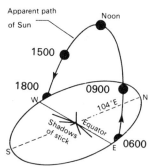

Behind and Ahead of G.M.T.

All meridians to the east of Greenwich Meridian have sunrise before that meridian. Local times along these meridians are therefore *ahead* of G.M.T. Meridians to the west of Greenwich Meridian have sunrise after this meridian and therefore their local times are *behind* G.M.T.

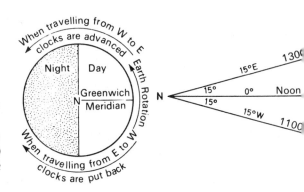

Longitude can be calculated from Local Time and G.M.T.

The local time at X is 1600 and G.M.T. is 1400. The difference in time between X and Greenwich is therefore 2 hours. This represents a difference of 30° of longitude between the two places (15° of longitude represents 1 hour). Since the local time at X is *ahead* of that at Greenwich then X is *east of Greenwich. The longitude of X is 30°E.* Similarly, if the local time at Y is 0800 and G.M.T. is 1400, then Y is 6 hours *behind* Greenwich, that is, Y is 90° to the *west* of Greenwich and its longitude is 90°W. *Note* If any two of the above three facts are given, the third can always be calculated.

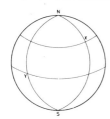

The Significance of the International Date Line

The diagram on the right shows what happens when two travellers set off at the same time (1600) on a Monday from a place A (long. 0°). One traveller goes westwards and the other eastwards to a place B (long. 180°). The traveller going west calculates the local times at 90°W. and 180° to be 1000 Monday and 0400 Monday respectively. The traveller going east calculates the local times at 90°E. and 180° to be 2200 Monday and 0400 Tuesday respectively.

In theory along meridian 180° it is both 0400 Monday and 0400 Tuesday. When the traveller going west crosses this meridian he finds it is 0400 *Tuesday*, i.e. he has *lost one day*. When the traveller going east crosses this meridian he finds it is 0400 *Monday*, i.e. he has *gained one day*. The line at which a day is lost or gained is called the *international date line*. This line follows meridian 180° except where this crosses land surfaces. To avoid confusion to the peoples of these regions the line bends round them so passing over a sea surface.

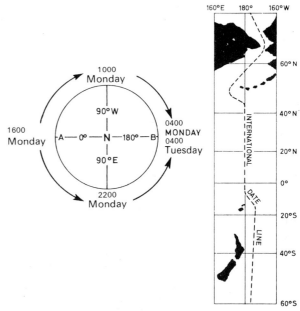

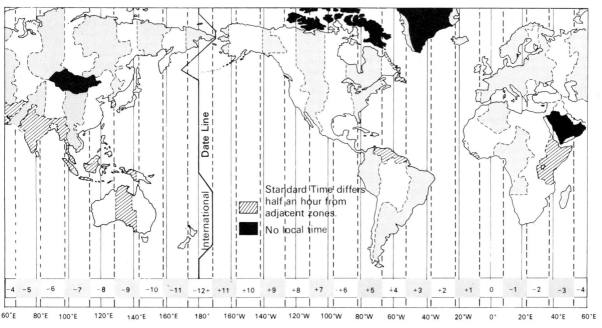

9

Standard Time and Time Zones

Each meridian has its own local time. Thus when it is 1200 noon local time in Georgetown (Penang) which is 100° 20′E. it is 1214 local time in Singapore whose longitude is 103° 50′E. Great confusion would arise if all places used local time. To avoid this the world is divided into 24 belts, each 15° of longitude wide. The local time of the central meridian of each belt is applied to that belt which is called a *time zone*. The local time of the central meridian for each time zone is called *standard time*. Neighbouring time zones have a difference of 1 hour. The boundaries of time zones are frequently adjusted to conform to political boundaries. (Diagram on page 9, bottom).

Deflection of Winds and Ocean Currents

All places on the earth's surface make 1 revolution in 24 hours. A place on the equator moves eastwards at a greater velocity than a place on say parallel 60°N. because the equator is longer than this parallel. A mass of air or water on parallel 60°N. will have an eastward speed equal to that of the parallel. If this mass moves towards the equator it will cross over parallels whose eastward speeds increase with decreasing latitude. The path of the mass when plotted appears as a curve which bends to the *right* (from the starting point) of the path it would have taken if the earth had not been rotating. If a similar air or water mass had moved from a high latitude to a low latitude in the Southern Hemisphere its path would appear as a curve to the *left* of the path it would have taken if the earth had not been rotating. This can be summarised by stating that in the N. Hemisphere winds and currents are deflected to the *right* whilst in the S. Hemisphere they are deflected to the *left*.

The figure (below, left) shows deflection to the right of an air or water mass moving (i) towards the pole, and (ii) towards the equator in the Northern Hemisphere.

The figure (below, right) shows deflection of an air or water mass moving (i) towards the pole, and (ii) towards the equator in the Southern Hemisphere.

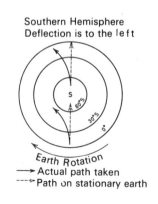

Northern Hemisphere
Deflection is to the right

Southern Hemisphere
Deflection is to the left

→ Actual path taken
---→ Path on stationary earth

Revolution of the Earth

The earth takes $365\frac{1}{4}$ days to revolve once round the sun. Every fourth year is given 366 days and this is called a *leap year*. All other years have 365 days. The earth's axis always points in the same direction in the sky, i.e. to the Pole Star. It is also permanently tilted at an angle of $66\frac{1}{2}°$ to the earth's Orbital Plane. The revolution of the earth and the inclination of its axis result in:

(i) Changes in the altitude of the mid-day sun at different times of the year

(ii) Varying lengths of day and night at different times of the year

(iii) The four seasons.

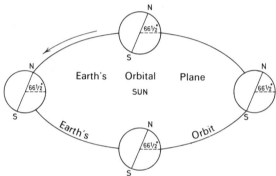

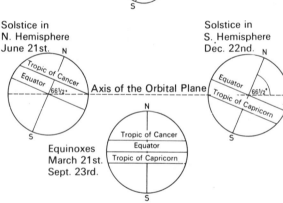

Solstice in N. Hemisphere June 21st.

Equinoxes March 21st. Sept. 23rd.

Solstice in S. Hemisphere Dec. 22nd.

Changing Altitudes of the Mid-day Sun at Different Times of the Year

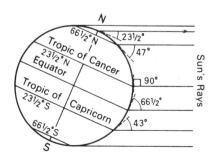

Summer solstice June 21

Equinoxes Mar. 21, Sept. 23

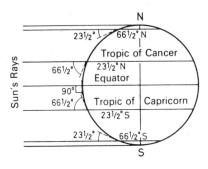

Winter solstice Dec. 22

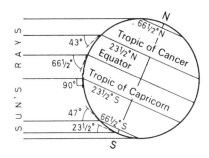

The apparent path of the sun between sunrise and sunset for selected latitudes during the solstices and the equinoxes

(I) North Pole

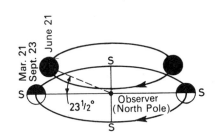

March 21   Sun circles the Pole, one half of it being visible above the horizon.

June 21   After March 21 sun rises higher in sky and is visible 24 hours each day. Highest altitude of sun is on June 21.

Sept. 23   Sun's path the same as for March 21. Sun is visible from March 21 to Sept. 23.

Dec. 22   After Sept. 23 sun is not visible above the horizon until March 21.

(II) Arctic Circle

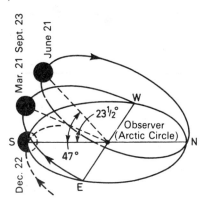

March 21   Sun rises due east and sets due west. It is visible for 12 hours.

June 21   Sun is visible for 24 hours.

Sept. 23   Sun rises and sets as for March 21.

Dec. 22   Sun is only visible for a few minutes when it appears above the southern horizon.

(III) Tropic of Cancer

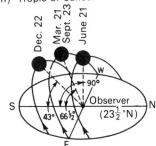

March 21   Sun rises due east and sets due west. It is visible for 12 hours.

June 21   Sun rises north of east and sets north of west. At noon its altitude is 90°.

Sept. 23   Sun rises and sets as for March 21.

Dec. 22   Sun rises south of east and sets south of west. It is visible for less than 12 hours.

(IV) Equator

11

The range of mid-day altitudes of the sun for selected latitudes.

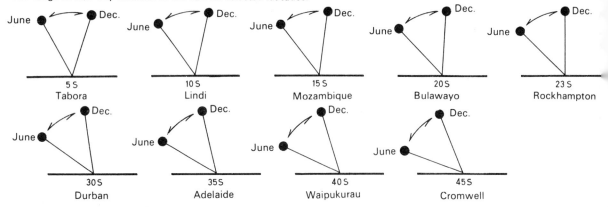

5 S Tabora	10 S Lindi

(Diagrams labelled: 5 S Tabora, 10 S Lindi, 15 S Mozambique, 20 S Bulawayo, 23 S Rockhampton, 30 S Durban, 35 S Adelaide, 40 S Waipukurau, 45 S Cromwell)

March 21 Sun rises due east and sets due west. Mid-day altitude is 90°.

June 21 Sun rises north of east and sets north of west.

Sept. 23 Sun rises and sets as for March 21.

Dec. 22 Sun rises south of east and sets south of west.

Note The sun is visible for 12 hours every day of the year.

The Varying Lengths of Day and Night at Different Times of the Year

The shaded part of each diagram represents night. The lengths of day and night for a selected parallel can be found by comparing that part of the parallel in the shaded zone with that part of it in the non-shaded zone.

In each diagram one half of the equator has night while the other half has day, i.e. DAY = NIGHT along the equator throughout the year.

Throughout the year one half of the earth has day while the other half has night. Only during the equinoxes does the dividing line between day and night coincide with meridians (see diagram below). During the equinoxes the sun is overhead at noon along the equator and at these times (March 21 and September 23) DAY = NIGHT along every parallel.

On June 21 the sun is overhead at noon along the Tropic of Cancer and all parallels in the Northern Hemisphere have their longest day of the year. At this time the length of the day increases as latitude increases north of the equator until there is continuous day north of the Arctic Circle. South of the equator the length of the day decreases with increasing latitude until there is continuous night south of the Antarctic Circle.

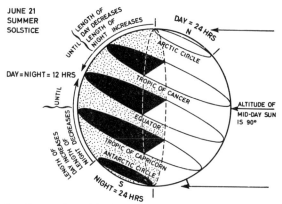

On December 22 the reverse takes place, i.e. the length of the day increases with increasing latitude south of the equator but decreases with increasing latitude north of the equator.

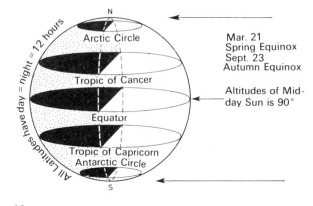

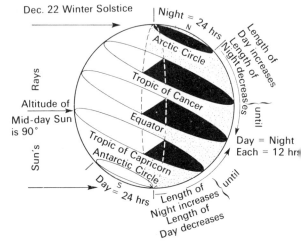

The Seasons

All part of the earth's surface except the equatorial latitudes experience a definite rise in temperature during one part of the year and a corresponding fall in temperature during another part of the year. This rise and fall in temperature is chiefly caused by the varying altitude of the mid-day sun and the number of hours of daylight. High mid-day sun altitudes cause high temperatures whereas low mid-day sun altitudes cause low temperatures.

The diagram below shows three bands of light A, B, C) each containing the same amount of sun energy (this is indicated by equal diameters shown in dotted lines). Band A has its energy spread over Surface A; Band B has its energy spread over Surface B, and Band C has its energy spread over Surface C. Clearly the temperature will be highest at A and lowest at C.

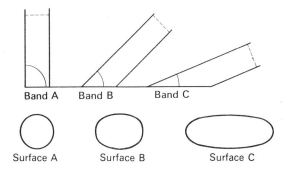

Sun's altitude in relation to heating

The monthly positions of the earth in its revolution round the sun are shown in the diagram below. The parallel at which the sun is overhead at noon is shown for each position. The overhead position of the sun 'moves' from the equator on March 21 northward to the Tropic of Cancer on June 21, then back to the equator on September 23 and then 'moves' southwards to the Tropic of Capricorn on December 22 and finally returns to the equator on March 21.

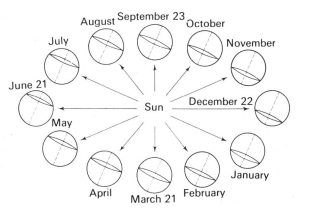

Associated with the overhead sun is a belt of heat and it is the movement of this belt between the Tropics which results in the alternation of the seasons. The Northern Hemisphere receives its maximum amount of solar radiation during June and its minimum amount during December (diagrams on page 12 and below). Between March 21 and September 23 this hemisphere has its summer while between September 23 and March 21 it has its winter. Spring and Autumn are two shorter seasons which occur between the two main seasons and which represent a transition from one to the other.

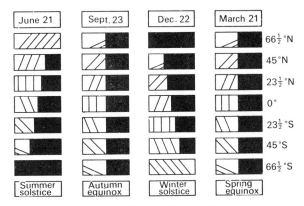

Each rectangle represents 24 hours. Night is shaded and day is left white.

The lines represent the altitude of the sun at mid-day.

The relationship between length of day, latitude and time of year is shown in the diagram above. Also shown is the altitude of the mid-day sun for each latitude. Notice how the length of day decreases from a maximum at $66\frac{1}{2}°$N to a minimum at $66\frac{1}{2}°$S on June 21, and how it increases from a minimum at $66\frac{1}{2}°$N to a maximum at $66\frac{1}{2}°$S on December 22. Finally notice how all latitudes have day equal to night at the equinoxes and how day always equals night along the equator.

EXERCISES

Draw a diagram to illustrate the revolution of the earth around the sun by showing the positions of the earth at the two equinoxes and the two solstices. On your diagram insert the following:
  (i) the path in orbit
  (ii) an arrow showing the direction of the earth's movement along the orbit
  (iii) an arrow to show the direction of the earth's rotation at one position of the earth
  (iv) the equator and the two tropics
  (v) the date at each position.

Objective Exercises

1 The distance of the Earth from the Sun is about
   A 1500 million km
   B 300 million km
   C 227 million km
   D 149 million km
   E 120 million km

   A B C D E

2 What is the name of the planet which takes 88 days to make one revolution of the Sun?
   A Venus
   B Mercury
   C Earth
   D Pluto
   E Neptune

   A B C D E

3 What is the longitude of a place when Greenwich time is 3 p.m. and the time by the Sun is noon?
   A 50° West
   B 45° East
   C 15° East
   D 15° West
   E 45° West

   A B C D E

4 Which of the following is true of lines of latitude?
   A they are great circles
   B they are numbered from 0° to 180°
   C they are circles on a globe which are parallel to the Equator and which are to the north and south of the Equator
   D they are concentric circles numbered from 0° to 90°
   E they are semi-circles on a globe

   A B C D E

5. Which of the following statements best describes longitude?
   A An imaginary line on the Earth's surface joining the North and South Poles.
   B The angular distance east or west of the Greenwich Meridian.
   C The distance of a place east or west of the Greenwich Meridian.
   D The position of a place on the Earth's surface with reference to the Prime Meridian.
   E A line on a map that cuts the Equator at right angles.

   A B C D E

6 All of the following are caused by the Earth's rotation **except**
   A the seasons
   B day and night
   C the deflection of winds and ocean currents
   D the daily rising and falling of the tides
   E the increasing altitude of the Sun from sunrise to midday

   A B C D E

7 The Earth makes one complete revolution of the Sun in
   A 365 days
   B 360 days
   C 365 ¼ days
   D 1 day

   A B C D

8 This question is based on the diagram given below. The position of the overhead Sun at noon indicates that this diagram shows the Earth's orbital position on
   A 22nd December
   B 21st March
   C 21st June
   D 23rd September

   A B C D

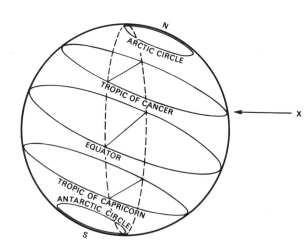

9 An eclipse of the Sun, takes place
   A when the Moon passes between the Sun and the Earth
   B once every five years
   C when the Moon is full
   D when the Earth comes between the Sun and the Moon

   A B C D

0 The permanent tilt of the Earth's axis and the revolution of the Earth in its orbit together cause
A day and night
B varying lengths of day and night at different times of the year
C differences in time between places on different meridians
D the deflection of winds

A B C D

1 At the summer solstice in the Northern Hemisphere, which of the following latitudes will have the longest night?
A 45°N
B 23½°S
C 66°N
D 66°S

A B C D

2 When the Sun is vertically overhead along the Tropic of Capricorn at midday
A days and nights are of equal length in the Northern Hemisphere
B nights are longer than days in the Southern Hemisphere
C days and nights are of equal length at the Poles
D night is equal to 24 hours at the North Pole

A B C D

3 Which one of the following is **not** connected with proofs of the Earth's shape?
A rotation and revolution
B circumnavigation
C the Earth's shadow on the moon during an eclipse
D the Bedford Level Canal Experiment

A B C D

4 What is meant by the eclipse of the Moon?
A It is the path along which the moon revolves.
B When the Moon comes between the Sun and the Earth it causes the shadows of the Moon to fall on the Earth.
C It occurs when the Earth comes between the Sun and the Moon and the centres of all three are on the same straight line.
D For any place, it is the average angle made by a line drawn from the Moon to a place and the horizon at midnight.

A B C D

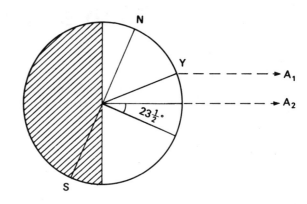

15 The above diagram shows the Sun's rays A_1 and A_2 shining on the Earth on 21st June. The latitude of Y, a point on the Earth's surface, is 45°N. What is the elevation of the Sun at Y?
A $23\frac{1}{2}°$
B $21\frac{1}{2}°$
C $68\frac{1}{2}°$
D 45°

A B C D

16 Which of the following statements can be taken as evidence to show that the Earth is spherical?
A The rotation of the Earth from west to east.
B Some parts of the Earth have day when other parts have night.
C The Earth's horizon is seen to be curved when seen from an aeroplane.
D The Earth's revolution around the Sun.

A B C D

2   The Earth's Crust

STRUCTURE OF THE EARTH

Although man has not been able to sink boreholes more than a few kilometres into the earth's crust, he is able to obtain information on the nature of the earth's interior by studying lavas emitted from volcanoes and by studying the behaviour of earthquake waves. From these studies it appears that the earth is composed of three parts, the core, the mantle and the crust as shown in this diagram. Our knowledge of the earth's interior increases year by year and this knowledge is used to modify our theories about the earth's interior. Until recently it was thought that the earth's outer crustal layer was composed of rafts of SIAL (light rocks rich in silica and alumina) which 'floated' on a 'sea' of SIMA (heavier rocks rich in silica and magnesia). But recent studies have revealed that large areas of the outer crustal layer are made of basaltic rocks similar to the SIMA. For example, there are vast areas of newly formed basaltic rocks which form the ocean bed in the central parts of the Atlantic and Indian Oceans. Similar rocks form the ocean bed of the waters around Antarctica. It is thought that these basaltic rocks emerge from the earth's interior near to the mid-oceanic ridges and slowly push outwards away from these ridges. The earth's crust is now regarded as a series of *plates* which are gradually being pushed apart, away from the zone where they formed. The sial rocks carried on some of the plates form the continents.

The plates eventually collide and when this happens one of them is drawn underneath another to make room for the newly forming plates. There is evidence which suggests that this is happening in *zones of subduction* along the edges of the continental plates. Active zones of subduction occur off the coasts of Japan, California and South America, and in the Caribbean area and around New Zealand. The map on page 17 shows the locations of the earth's main plates in relation to the oceanic ridges.

The earth, which is thought to have been formed about 6000 million years ago, has experienced a great many changes. Although these changes take place very slowly, there is evidence that considerable changes in the shape and character of the land masses have taken place since man first appeared.

Distribution of Land and Water

The water surface of the earth accounts for just over 70% of the earth's surface. This distribution is not the same for both Northern and Southern Hemispheres. In the latter it is as much as 80% of the total surface area of that hemisphere. The

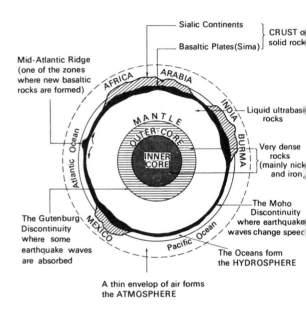

Section through the Earth at 20 °N

average height of the land is just under 900 metres (2950 feet) while the average depth of the oceans is 3800 metres (12 400 feet). See the diagram below. The greatest height is Mt. Everest (8875 metres or 29 120 feet) on the northern border of Nepal and the greatest depth is the Mindanao Deep (10 490 metres or 34 400 feet) off the east coast of the Philippines.

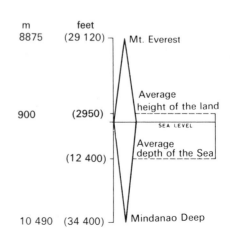

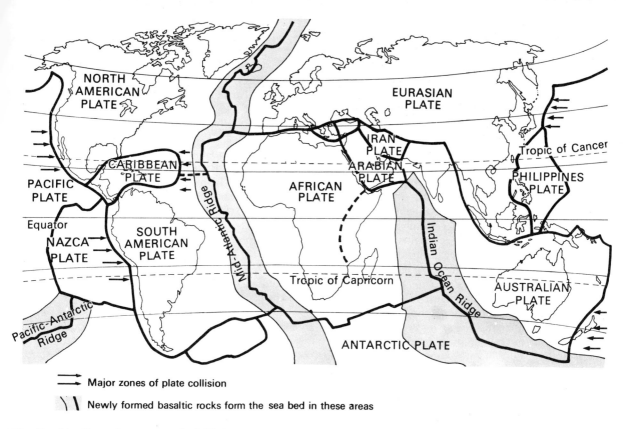

Major zones of plate collision

Newly formed basaltic rocks form the sea bed in these areas

The Earth's Crust is composed of Plates

The relationship between land and water surfaces is shown by the diagram below. Two main levels can be recognised: (i) the *ocean floor* level, and (ii) the *continental level*. The two are connected by the *continental slope*. The edge of the continental level is submerged to a depth of about 200 metres (650 feet) and this zone is called the *continental shelf*. The seas on this shelf are called *epicontinental* or *shelf seas* (the importance of these is discussed in a later section). The more important of these shelf seas in the Tropics are (i) between N. Australia and New Guinea, (ii) between Borneo and the Malay Peninsula and Thailand, and (iii) along the Gulf Coast of North America.

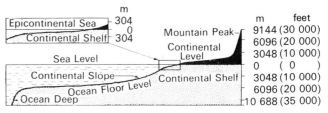

The continental surface is broken by mountain ranges, plateaus, and plains, whilst the ocean floors are far from level. Basins and deeps cause the floor to plunge to great depths. Extending across most of the oceans are ridges, some of which rise above the level of the sea to form chains of islands. Extensive plateaus also occur in some oceans.

ROCKS AND MINERALS

The earth's crust is composed of rocks each of which is made up of minerals. Most minerals are compounds of several elements, e.g. silica (SiO). A few minerals are themselves elements, e.g. carbon (diamond), gold and sulphur. Silica often combines with other oxides to form *silicates*, the most common of which are *felspars*. *Mica* is another common silicate.

Minerals are frequently crystalline, i.e. the atoms forming the crystals are arranged in a definite manner. Some minerals are non-crystalline, i.e. the atoms forming the mineral are not arranged in any definite order.

Felspars are silicates of aluminium (Al), potassium (K), sodium (Na) and calcium (Ca).

Augite, hornblende and *olivine* are silicates of iron (Fe), magnesium (Mg), calcium (Ca) and aluminium (Al).

Clay minerals are complex silicates derived from

weathered minerals such as felspars. They are silicates of aluminium (Al).

Felspars which weather under tropical humid conditions lose their silica. The residue is chiefly oxides of aluminium and these are called *bauxite*.

Classification of Rocks according to origin

IGNEOUS — These rocks have been formed inside the earth, under great pressure and heat. They do not occur in layers and most of them are crystalline. Some rocks, e.g. granite, have cooled slowly and contain large crystals: others, e.g. basalt, have cooled quickly and contain small crystals. Some rocks contain a high percentage of silica and are called *acid rocks*. Granite is a good example. Other rocks, such as basalt, contain a high percentage of iron, or aluminium or magnesium oxides and are called *basic rocks*. Igneous rocks do not contain fossils.

SEDIMENTARY — The most common sedimentary rocks are composed of particles of rocks which have been deposited, usually in layers, by water, wind or moving ice. These are called *mechanically-formed*, or *elastic sedimentary rocks* which are largely or completely made of particles of organic matter, or they are made of minerals which have been chemically deposited. All sedimentary rocks are non-crystalline. They contain fossils.

METAMORPHIC — These are rocks whose structure and appearance have been changed by great heat, or great pressure, or both. Any rock can be changed into a metamorphic rock.

Igneous Rocks

There are two main groups.

I *Volcanic* (these have been poured out onto the earth's surface, and they are called *lavas*) e.g. basalt

II *Plutonic* (these have solidified deep in the earth's crust and they reach the surface only by being exposed by erosion) e.g. granite

Sedimentary Rocks

There are three main groups.

I *Mechanically-formed*
   (i) Wind-deposited e.g. loess
   (ii) River-deposited e.g. clays, gravels and alluviums
   (iii) Glacier-deposited e.g. moraines, sands and gravels and boulder clay
   (iv) Sea-deposited (very similar to those of (ii)

II *Organically-formed*
   (i) From animals e.g. chalk and coral
   (ii) From plants e.g. peat, lignite, coal

III *Chemically-formed*
   e.g. rock salt, borax, gypsum, nitrates, potash and certain limestones

Metamorphic Rocks

e.g. marble (from limestone)
   slate (from clay)
   gneiss (from granite)
   quartzite (from sand)
   graphite (from coal)

Grouping of sedimentary rocks according to their texture

CLAYS — Composed of microscopically fine particles

SILTS — Composed of particles not quite so fine

SANDS — Composed of coarser particles (easily seen with the naked eye). When cemented together they form *sandstones* (these are called *grits* when the sand grains are angular)

GRAVELS — Composed of rounded and large particles. When cemented together they form *conglomerates*

BRECCIA — Composed of coarse and angular particles which have been cemented together

JOINTS

Rocks very often develop cracks when they are subjected to strain produced by compression or tension. The strain may be caused by earth movements, or by contraction when molten rocks solidify, or by the shrinking of sedimentary rocks on drying. The cracks so formed are called *joints*. In sedimentary rocks joints are often at right angles to the bedding plane. Sometimes more than one set of joints develops. When this happens the rock becomes broken into blocks, e.g. limestone and sandstone, or into columns as in some lavas, e.g. basalt.

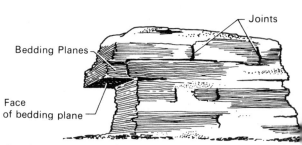

Sandstone cliff showing bedding planes

Note Jointing does not result in the displacement of rocks as does faulting (see pages 21 and 22).

Horizontal bedding planes in cliffs on the coast of South-West England

Joints in granite rocks in South-West England

THE FORCES WHICH PRODUCE PHYSICAL FEATURES

A. Internal (operate within the earth's crust)

I Earth Movements

(i) *Vertical* (up and down) movements cause faulting of the crustal rocks. Features produced: *plateaus*; *block mountains* (horsts); *basins*; *some types of escarpment*.

(ii) *Lateral* (sideways) movements cause folding of the crustal rocks. Features produced: *fold mountains*; *rift valleys*; *horsts, block mountains*

II Volcanic Eruptions

(i) *External* (lavas reach the earth's surface). Features produced: *lava plains and plateaus*; *volcanic cones*; *geysers*

(ii) *Internal* (lavas solidify in the crust). Features produced: *dykes*; *sills*; *batholiths*; *laccoliths*

B. External (operate on the earth's surface)

I Denudation

(i) *Weathering* (the break-up of rocks by alternate heating and cooling; chemical actions; and the action of living organisms). Features produced: *soil*; *earth pillars* (by rain action); *screes*

(ii) *Erosion* (the break-up of rocks by the action of rock particles being moved over the earth's surface by water, wind and ice). Features produced: *valleys*; *peneplains*; *cliffs*; *river* and *coastal terraces*; *escarpments*

(iii) *Transport* (the movement of rock particles over the earth's surface by water, wind and ice)

II Deposition

(i) *By Water* (river or sea). Features produced: *flood plains*; *levées*; *alluvial fans*; *deltas*; *beaches*; *lake plains*; *marine alluvial plains*

(ii) *By Ice* (ice sheets and valley glaciers). Features produced: *boulder clay plains*; *outwash plains*; *moraines*; *drumlins*; *eskers*

(iii) *By Wind* Features produced: *loess plains*; *sand dunes*

(iv) *By Living Organisms* (e.g. coral) Features produced: *coral formations*

(v) *By Evaporation and Precipitation* Features pro-

(vi) *Of Organic Matter*

duced: *salt deposits*

Features produced: *coal deposits*

Internal Forces
Earth movements

The face of the earth is full of features, some small and some large, all of which are being slowly but constantly changed by the agents of denudation – rivers, waves, wind, moving ice, frost, rain and sun. The major features such as mountains, plateaus and plains have been formed by earth movements. These movements, which are lateral and vertical, exert great forces of tension and compression, and although they usually take place very slowly, they eventually produce very impressive features. Before studying the effects of earth movements it is as well to get to know the meanings of some common terms. Sedimentary rocks are formed from sediments which have been laid down in horizontal layers. This layering is called *stratification*, and sedimentary rocks are therefore *stratified rocks*. The face of each layer is called the *bedding plane*.

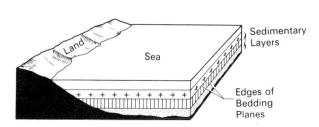

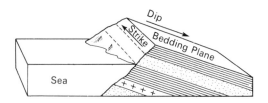

Earth movements cause sedimentary rocks to be displaced, i.e. to be pushed out of the horizontal plane so that the rocks are *tilted* or *inclined*. The inclination of the rocks is called the *dip*. The direction parallel to the bedding plane and at right angles to the dip is called the *strike*. The bottom diagram on page 17 has had the upper sedimentary layer removed so as to show the bedding plane, dip and strike.

Earth movements can also cause *folding* and *faulting* of the sedimentary rocks. Folding results from *lateral forces of compression*.

The Nature of Folds

The diagrams show a *simple fold*. The layers of rock which bend up form an *upfold* or *anticline*. Those which bend down form a *downfold* or *syncline*. The sides of a fold are called the *limbs*.

If compression continues then simple folds are changed first into *asymmetrical folds*, then into *overfolds*, and finally into *overthrust folds*.

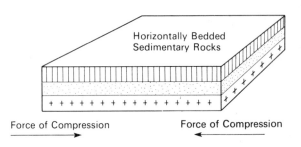

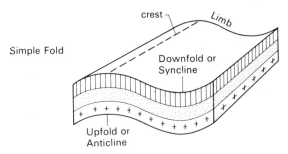

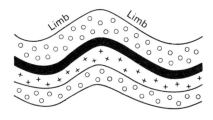

Simple Fold

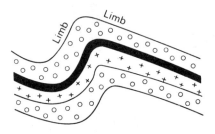

Asymmetrical Fold

Asymmetrical fold: one limb steeper than the other

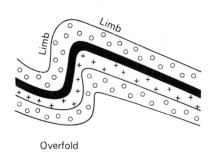

Overfold

Overfold: one limb is pushed over the other limb

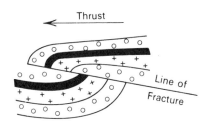

Overthrust Fold

rthrust fold (nappe) when pressure is very great a fracture occurs in the fold and one limb is pushed forward over the other limb.

Folded rocks in South-West England

The Nature of Faults

Faulting can be caused by either *lateral* or *vertical* forces of either *compression* or *tension*. Tension causes a *normal fault* (right, top) and compression causes a *reverse fault* (right, bottom).

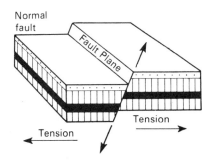

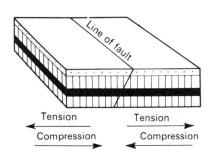

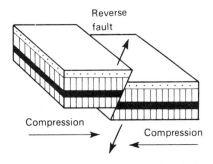

Forces of Tension and Compression

Rocks of the earth's crust are subjected to tension and compression when vertical or lateral earth movements take place. If one part of the crust is compressed then clearly another part must be stretched, i.e. be under tension. Rocks when under tension usually fault, but when under compression they may fold or fault depending on whether they are brittle or flexible under stress.

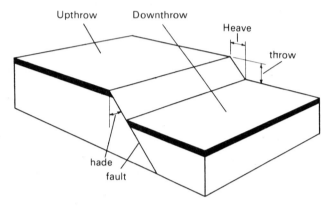

The nature of Faults

Upthrow	upward displacement
Downthrow	downward displacement
Throw	vertical displacement
Heave	lateral displacement
Hade	inclination of the fault to the vertical

Normal Fault (caused by tension)

(i)

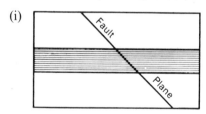

Fault plane develops

(ii)

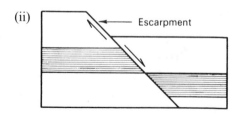

Rocks on each side of the fault plane are usually displaced as shown by the arrows.

(iii)

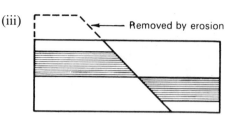

Faulting sometimes produces an escarpment. This is sometimes removed by erosion.

Note Surface area is increased.

Reverse Fault (caused by compression)

(i)

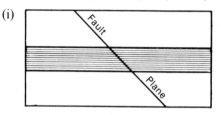

Fault plane develops.

(ii)

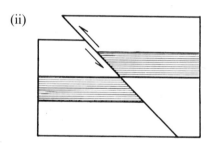

Rocks are displaced as shown by the arrows. Those on one side of the fault plane ride up over those on the other side.

(iii)

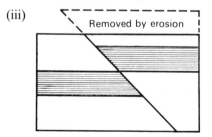

An escarpment may mark the fault. Erosion may later remove this.

Note Surface area is reduced.

Tear Fault

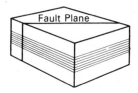

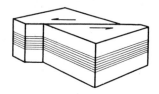

The fault plane is often almost vertical. The rocks are displaced horizontally as shown by the arrows. There is no vertical displacement.

Tear faults usually occur during earthquakes. The earthquakes which wrecked San Francisco produced tear faults e.g. the San Andreas Fault.

Note:
Earthquakes sometimes produce vertical movements or a combination of vertical and horizontal movements e.g. Midori, Japan.

Vertical Movements

There is evidence in many countries to show that these movements take place. There are beaches in Scotland and Norway which now lie several feet above the sea and there are submerged forests around the coasts of Britain and other countries.

Horizontal Movements

There is considerable evidence which suggests that the surface of the earth has been moving, and that it is still moving horizontally. We have seen that new rocks appear to be forming around the mid-oceanic ridges and as these move outwards, older rocks are drawn down into the mantle in the zones of subduction, which is where crustal plates collide. Wegener was the first person to suggest that the positions of the continents had changed, and he suggested that originally there was only one continent, which he called PANGEA. He postulated that this was located around the South Pole during the Carboniferous Period (see the Geoglogical Time Scale on page xiii), and that during the early Tertiary Period, this ancient continent started to split up into several parts. Nobody has yet explained why this should have happened, or what forces could have been powerful enough to set such an enormous change in motion. It may have been caused by strong convection currents forming in the mantle, but no reason can be found to explain why such currents, powerful enough to start what is known as *Continental Drift,* should have started. The Himalayas, the Rockies and the Andes were formed at about this time, and Wegener believed that they originated through 'crumpling' at the edges of the continents as they drifted away from the South Pole, northwards, eastwards and westwards.

There is a considerable variety of evidence in all the continents which suggests that at one time the continents may have all been joined together. This evidence includes types of fossils, both plant and animal, and rock types and rock structures. The shapes of the continents suggest that they can be fitted together. This is especially true of the Americas, Europe and Africa. A group of scientists recently studied the rocks in the Gulf of Guinea in West Africa, and by using a computer map they decided to visit a specific area on the coast of Brazil to see whether the rocks there had any similarity with those of Guinea. On visiting that part of the Brazilian coast they discovered that the rocks were identical.

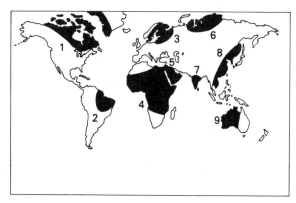

1 Laurentian Shield	6 Siberian Shield
2 Brazilian Shield	7 Deccan Shield
3 Baltic Shield	8 China Shield
4 African Shield	9 Australian Shield
5 Arabian Shield	

Pangea in the late Palaeozoic Period

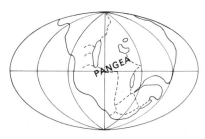

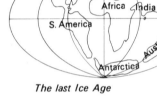

The early Tertiary Period (the continents begin to drift apart)

Note   Europe and N. America were partly joined, forming *Laurasia,* while Africa was the ancient continent of *Gondwanaland*

The last Ice Age

The Continents Today (position of similar rocks across the Atlantic Ocean is shown)

_ _ Main mid-ocean ridges where new rocks are being formed

INTERNAL AND EXTERNAL FORCES OPERATE TOGETHER

The theory of *isostacy* has been accepted for a long time to explain why vertical and horizontal movements take place in the earth's crust. This theory compares the continental blocks with corks floating on a liquid. If a thin layer is cut from the top of the corks then they will rise and the corks will not project as far down into the liquid as before. In a similar way the theory states that the light sial rocks of the continents are 'floating' on the denser sima rocks which are in a semi-liquid state. As the continents are worn down by denudation they gradually rise up. This concept may be applied to the theory of Plate Tectonics, which attempts to explain the destruction and renewal of crustal rocks, as follows. The zone of subduction is said to be a low-angle fault at the junction of two colliding plates. To relieve the pressure between the plates, the advancing plate slides down below the stronger plate, as shown in this diagram. As the rocks in the subsiding plate melt and mix with the magma at the base of the crust, the pressure frequently causes earthquakes, and volcanic mountains and islands often develop 150 to 400 kilometres (about 250 to 650 miles) away from the inclined fault. The lava emitted from these volcanoes may be a mixture of the old rocks, which were drawn down into the zone of subduction, and completely new rocks (magma) from the earth's interior. Many of the world's volcanic zones, for example, those of Japan, the Philippines, the Caribbean Islands, the Andes and the Rockies are near to zones of subduction. All of these volcanic zones experience earthquakes.

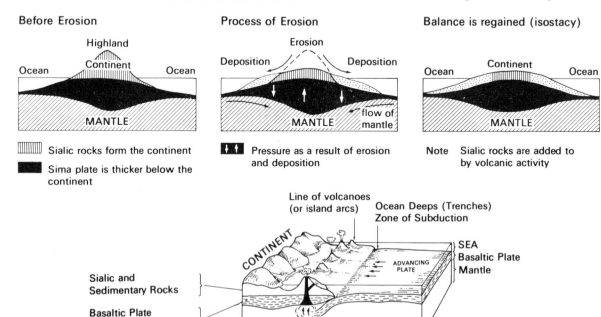

Before Erosion

Highland
Continent
Ocean   Ocean
MANTLE

▤ Sialic rocks form the continent

■ Sima plate is thicker below the continent

Process of Erosion

Erosion
Deposition   Deposition
flow of mantle
MANTLE

◆▮▲ Pressure as a result of erosion and deposition

Balance is regained (isostacy)

Continent
Ocean   Ocean
MANTLE

Note   Sialic rocks are added to by volcanic activity

Line of volcanoes (or island arcs)
Ocean Deeps (Trenches) Zone of Subduction
SEA
Basaltic Plate
Mantle
CONTINENT
ADVANCING PLATE

Sialic and Sedimentary Rocks

Basaltic Plate carrying the Continent

Old crustal rocks are consumed and mixed with mantle to form **new magma**

Block diagram showing the Zone of Subduction and related Volcanic Mountains

Ice Sheets can cause the earth's crust to sag

1 The northern parts of North America and Europe lay under extensive ice sheets of vast thickness during the *Ice Age*. The weight of the ice caused the crust to be depressed.

2 Since the melting of these ice sheets, the crust has been slowly rising. The Scandinavian coasts have many sea beaches which lie from 8 metres to 30 metres (26 feet to 100 feet) above the present-day sea beaches. Evidence suggests that these *old sea beaches* have been raised because the land has been uplifted.

FEATURES PRODUCED BY EARTH MOVEMENTS

1 Fold Mountains

Fold mountains consist of great masses of folded sedimentary rocks whose thickness is often as much as 12 000 metres (40 000 feet) or more. Originally the sedimentary rocks must have been laid down in horizontal layers which later became folded through compression. To begin with, earth movements must have caused a part of the earth's crust to warp downwards into large depressions, called *geosynclines* which became the sites of seas and in which the sediments collected. In time the sediments caused further subsidence by their own weight. Also, the process of folding caused the width of the sedimentary rock zone to decrease and its thickness to increase further.

At one time many people thought that fold mountains developed through the contraction of the earth's crust on cooling but this explanation is not accepted today. The origin of fold mountains and other large landforms is much more complex than this. The explanation given today, though largely dependent on Holme's theory, takes into account the theory of Plate Tectonics. The following diagrams suggest how fold mountains may have been formed. During the later stages of their formation, the fold mountains were faulted and igneous intrusions and volcanic features were added to the pattern of fold sedimentary rocks. Magma intrusions into the base of fold mountains formed batholiths (see Chapter 3), and some of these now lie exposed at the surface as a result of prolonged denudation which has removed the overlying rocks.

Types of Fold Mountains

Sometimes the folding is never intense and this simple folding gives rise to mountains and valleys. The anticlines become the mountains and the synclines the valleys as in the Jura of France. Simple fold mountains are rare.

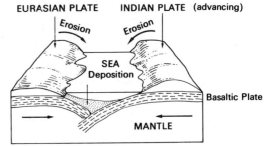

Colliding plates cause depression of the land and the formation of a shallow sea. Layers of sediments accumulate on the bed of this sea.

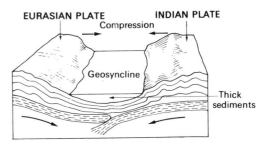

Sedimentary layers occupy a large depression or geosyncline and they become folded as the plates move together

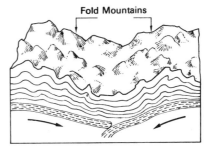

The sea has drained away and the rocks have been more intensely folded

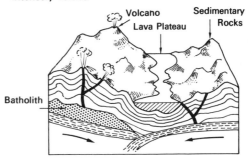

Batholiths, volcanoes and lava plateaus form at a later stage

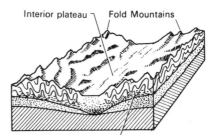

Volcanic rocks form the base of the fold mountains.

Complex folding is much more common. When this happens there is little or no relationship between the anticlines and the mountains, and the synclines and the valleys.

The peaks and valleys which occur in complex fold mountains have been formed by glacial and, or, river erosion (this will be dealt with later).

Some complex fold mountains do not appear to have been formed between two continents, e.g. the mountains of Java and Sumatra. The floor of the ocean probably acted as the 'missing' continent.

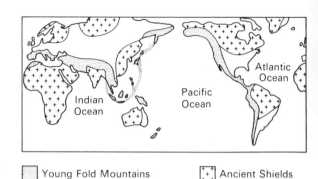

☐ Young Fold Mountains   ⊡ Ancient Shields

World Distribution of Young Fold Mountains

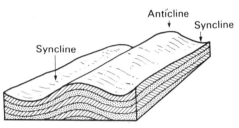

1 *Anticline and synclines*

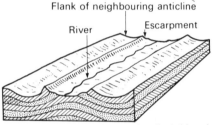

2 *Anticline opened by erosion (neighbouring anticlines will be opened up later by same process)*

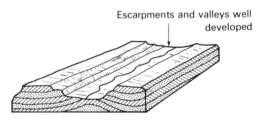

3 *Anticlinal valley develops rapidly as river excavates weak strata*

4 *Synclines stand up as mountains*

The Fate of Fold Mountains

Weathering and erosion attack fold mountains as they begin to form. The building of mountains like the building of all other major features of relief takes vast periods of time to complete. From the beginning of their formation weathering and erosion, operating together, attack and wear away the mountains. To begin with, earth movements are more powerful than weathering and erosion and the mountains thus reach heights of thousands of feet, but as earth movements weaken the wearing away process becomes dominant. In time they are reduced to an almost level surface not far above sea level. This surface is called a *peneplain*.

In this process of mountain reduction, strange things can happen. Rocks which were contorted into synclines are sometimes turned into mountains by river erosion. This is achieved by rivers opening up the weakened rocks of neighbouring anticlines which in time are completely removed leaving the adjoining synclines as mountains. The diagrams on the left explain this.

How many times have Fold Mountains formed on the earth's surface?

There is evidence to show that there have been three main mountain-building periods during the last 400 million years. The remnants of the first of these can be seen in the Altai Mountains, the Scandinavian Mountains and the Lake District of England. Remnants of the second mountain-building period occur in the Tien Shan Mountains and in the plateaus of central France and Spain. As far as present-day mountain systems go it is the last mountain-building period which is the most important. Mountains belonging to this period include the Himalayas, Andes, Rockies and Atlas Mountains. They are all called *Young Fold Mountains* (see diagram at top of page).

Is Mountain building still going on?

We know that as fold mountains are worn away the deposition of sediments in neighbouring oceans gives rise to new sedimentary rocks. Scientific surveys show that there is an extensive geosyncline forming from south of the Atlas Mountains to south of Indonesia. The geosyncline is represented by the Shotts of Tunisia and Algeria, the Mesopotamian Alluvial Plain, the Persian Gulf, the Indo-Gangetic Plain and the ocean deep south of Java. The stage of mountain building reached in the geosyncline varies from one part of it to another. In the eastern part there is frequent volcanic activity.

Influence of Fold Mountains on Human Activities

1 They often act as climatic barriers. Regions on one side of a mountain range may have an entirely different climate from that of the region on the other side. The coastlands of British Columbia have mild winters, warm summers and rain throughout the year. To the east of the Rockies the prairies have cold winters, hot summers and there is a maximum of rain in the summer.

2 They often receive heavy rain and/or snow which may give rise to important rivers. Most of the rivers of Asia rise in the mountain ramparts of central Asia. These rivers may be used for irrigation, e.g. the Ganges and Indus, or they may be used for developing hydro-electric power (H.E.P.), e.g. the Colorado and Columbia and the rivers of Switzerland and Japan, or they may be used for both irrigation and developing H.E.P., e.g. Murray River in S.E. Australia.

3 Some mountains and their plateaus may contain minerals, e.g. Nevada (copper and gold), Bolivia (tin).

4 They may act as barriers to communications or they may make the construction of communications difficult.

Some mountain ranges have valuable timber resources, e.g. coast ranges of Western America (coniferous soft woods), foothills of the Himalayas (teak).

2 Basins and plateaus

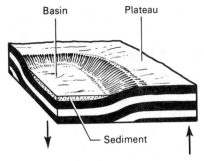

Vertical earth movements can cause the crust to warp, and sometimes large areas of it are uplifted whilst others are depressed. The uplifted areas form *plateaus*, sometimes called *tectonic plateaus*, and the depressed areas *basins*. There are two types of tectonic plateaus. Some plateaus slope down to surrounding lower land. The Deccan Plateau of India is an example. Other plateaus slope up to surrounding mountains. These are called *intermont plateaus* and the Tibetan and Bolivian Plateaus are examples.

Basins, too, are of different types. Some are *sea-basins*, e.g. the Celebes Sea (Indonesia) and the Black Sea (U.S.S.R.); others are high above sea level and rimmed by mountains, e.g. Lake Victoria Basin (East Africa), the Great Basin of Nevada (U.S.A.) and the Tarim Basin (central Asia); and others are filled with sediments and either have an external drainage, as in the Zaire Basin (central Africa), or are *basins of inland drainage*, as in the Chad Basin (north-central Africa) and the Tarim Basin (central Asia).

On page 28 is a diagrammatic section across the Western Highlands of the U.S.A. The Colorado Plateau is an intermontane plateau and lies between the Rockies and the Wasatch Mountains. The Great Basin is really a plateau which has been block-faulted (see second diagram). Block sections of the Plateau now form block mountains. Some of the depressions of the Great Basin have no

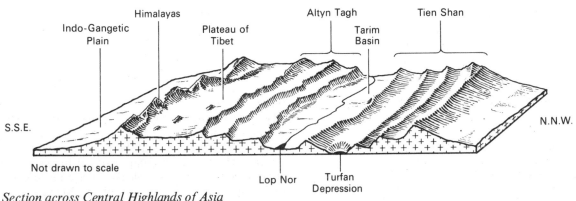

Section across Central Highlands of Asia

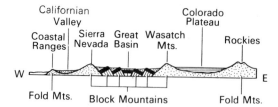

Californian Valley — Coastal Ranges, Sierra Nevada, Great Basin, Wasatch Mts., Colorado Plateau, Rockies — W / E — Fold Mts., Block Mountains, Fold Mts.

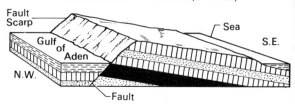

Tilt-Block Mountain of Somalia (E. Africa) — Fault Scarp, Gulf of Aden, Sea, S.E., N.W., Fault

external drainage and because of a high rate of evaporation salt lakes frequently form. Some of these dry up and form salt flats or *playas*.

Central Asia contains numerous plateaus and basins. Tibet is the highest plateau in the world and is an intermontane plateau. The Turfan Depression lies below sea level while the Tarim Basin is between 600 and 2000 metres (1950 and 6550 feet) above sea level.

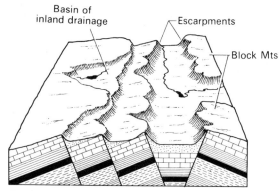

Block Faulting in the Great Basin (U.S.A.)

Basin of inland drainage — Escarpments — Block Mts

Horsts, Tilt Blocks and Rift Valleys

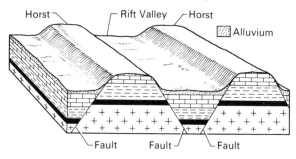

Horst — Rift Valley — Horst — Alluvium — Fault — Fault — Fault

Earth movements sometimes cause the crust to be divided into rectangular-shaped blocks some of which are uplifted and others are depressed. This is called *block-faulting*. Uplifted blocks may either be tilted when they form *tilt blocks* or they may be horizontal to form *horsts*. The tilt block usually has one pronounced scarp, e.g. the Western Ghats of the Deccan, while horsts have two pronounced scarps.

Tilt Blocks	*Horsts*
Deccan Plateau	Korea
Arabian Plateau	Sinai
Brazilian Plateau	Black Forest

When blocks of the crust are depressed between parallel faults they often form *rift valleys*.

East African Rift Valleys

The most outstanding belt of rift valleys extends for 4800 kilometres (3000 miles) from Syria to the River Zambesi in East Africa. The belt contains a number of well-developed rift valleys some of which contain lakes whose beds are below sea level. Many of the valleys have precipitous sides which are fault scarps and which are often very straight for many miles. The broken line on the map indicates the main belt of rift valleys.

The River Rhine in Europe flows through an impressive rift valley between the fault blocks of the Black Forest and the Vosges.

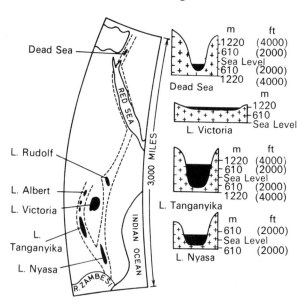

Dead Sea, RED SEA, 3,000 MILES, INDIAN OCEAN, L. Rudolf, L. Albert, L. Victoria, L. Tanganyika, L. Nyasa, R. ZAMBESI

Dead Sea: 1220 (4000), 610 (2000), Sea Level, 610 (2000), 1220 (4000)
L. Victoria: 1220, 610, Sea Level
L. Tanganyika: 1220 (4000), 610 (2000), Sea Level, 610 (2000), 1220 (4000)
L. Nyasa: 610 (2000), Sea Level, 610 (2000)

Note The level of the Dead Sea is 393 m (1290 ft) below sea level, and its floor is 819 m (2680 ft) below sea level.

Origin of Rift Valleys

They are thought to have developed either from the action of tensional forces in the crust which caused fault blocks to sink between parallel faults, or from the action of compressional forces in the crust which caused fault blocks to rise up towards each other and over a central block. Many people think that compression has been responsible for most rift valleys. They argue that it would not be possible for blocks of the crust to sink into the heavier rocks of the sima below the crust, unless they were trapped there by later pressure.

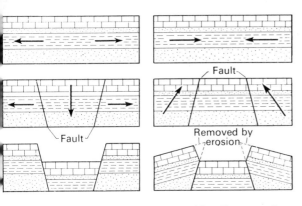

Formed by Tension
Layers of rocks are subject to tension.
Faults develop and the centre block begins to subside.

Formed by Compression
Layers of rocks are subject to compression.
Faults develop and the outer blocks begin to thrust up over the centre block.
The over-hanging sides of the rift valley are worn back by erosion.

After subsidence a depression with steep fault scarp sides, i.e. a rift valley, is formed. It is trapped in position by later pressure.

Rift Valley Formation

Recent research also suggests that some rift valleys, e.g. the Red Sea Rift and the rift in the Mid-Atlantic Ridge, were formed from one fault only, and not from two parallel faults. Yet their structure, with a depressed central portion and high horsts on either side cause them to be classified as rift valleys. The horsts of these rifts are maintained at a high level by the divergence of two plates at the rift and the subsequent replacement of the diverging rocks by newly formed basaltic rocks from the rift. It can be seen that rift valleys may form in different ways, and even within one rift valley the structure may vary in different parts of its length. The East African Rift has definite reverse faults in some places, while in other places the faults appear to be normal. This rift extends north to the Red Sea, which is thought to have been formed from a single fault in a zone of divergence. In addition to the very large rift valleys discussed here, there are many other, though smaller, rift valleys.

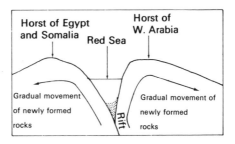

Influence of Plateaus on Human Activities

1 Some plateaus are rich in mineral deposits, e.g. the Highlands of Brazil (iron ore and manganese); the African Plateaus (copper and gold); Western Australia (gold).
2 High plateaus in tropical latitudes are sometimes of agricultural value, e.g. the Kenya Plateau (coffee, sisal and maize); the Brazilian Plateau (coffee).

Jordan Rift Valley

EXERCISES

1 With the aid of well-labelled diagrams, explain the differences in origin and appearance of the following:
   (i) faults and folds
   (ii) block mountain and rift valley
   (iii) plateau and basin.

2 For each of the following features: young fold mountain, block mountain, and rift valley:
   (i) draw a clear diagram to show its main features
   (ii) explain its possible origin
   (iii) name and locate a region where a good example may be found.

3 Explain the meanings of the following terms by using well-labelled diagrams:
   (i) syncline and anticline
   (ii) dip slope and scarp slope
   (iii) tensional force and compressional force.

Objective Exercises

1 Which one of the following groups of terms is applicable to some parts of the ocean floor?
   A basin, deep, cirque, plateau
   B trench, ridge, drumlin, plateau
   C plateau, basin, dune, ridge
   D ridge, deep, basin, waterfall
   E trench, ridge, basin, plateau

   A B C D E

2 Which of the following statements is **not** true in respect of sedimentary rocks?
   A The particles of rock are sometimes completely of organic origin.
   B The rocks are non-crystalline.
   C They are rocks whose structure is determined by great pressure or heat.
   D The rocks have been deposited in layers.
   E The rocks may be formed under water.

   A B C D E

3 Which statement is **not** true of fold mountains?
   A They are often adjacent to a stable area of old crystalline rocks.
   B They contain a core composed of metamorphic and igneous rocks.
   C They form rugged peaks.
   D They are caused by contraction of the Earth's crust.
   E Their tops are often buried beneath snow and ice.

   A B C D E

4 Questions 4 and 5 are based on the map given below which shows some important structural features.
   Which of the following numbers marks the position of an ancient tilted plateau
   A 1
   B 2
   C 3
   D 4
   E 5

   A B C D E

5 A well-developed rift valley is represented by the number
   A 1
   B 2
   C 3
   D 4
   E 5

   A B C D E

6 Which of the following statements is **irrelevant** when explaining factors involved in the formation of young fold mountains?
   A Thick layers of sediment are crushed between continental blocks.
   B sediments are deposited in a geosyncline.
   C As sediments in the geosyncline accumulate the floor of the geosyncline subsides.
   D Continental blocks move in opposite directions.

   A B C D

7 The stratification of sedimentary rocks is mainly caused by
   A lateral pressure being applied from two sides
   B the deposition of rock particles in layers
   C the removal by erosion of rocks exposed at the surface
   D the compression of rock particles by earth movements

   A B C D

8 A rift valley is formed mainly by
   A forces of tension in the Earth's crust
   B the subsidence of the floor of a river valley
   C the formation of fold mountains
   D the over-deepening of a valley by ice action

   A B C D

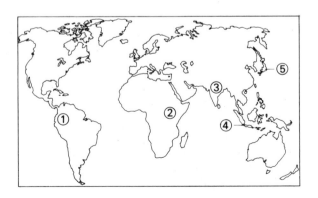

3 Earthquakes and Vulcanicity

Destruction caused in an Alaskan city by an earthquake in 1964

EARTHQUAKES

These are sudden earth movements or vibrations in the earth's crust. They are caused by:
(i) The development of faults (cracks) in the crust which result from collision between plates.
(ii) Movements of molten rock below, or within the crust, or the sudden release of stress which has slowly built up along the fault plane.

Nature of Earthquakes

Waves passing out from the point of origin (focus) set up vibrations which may reach up to 200 per minute. The vibrations cause both vertical and lateral movements and this violent shaking causes great destruction to buildings. Earthquakes are frequently associated with faults.

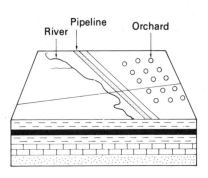

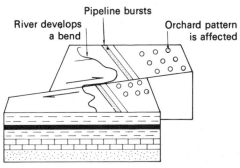

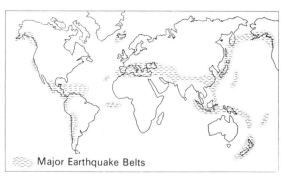

Major Earthquake Belts

Effects of Earthquakes

1 They can cause vertical and lateral displacement of parts of the crust.
2 They can cause the raising or lowering of parts of the sea floor as in Sagami Bay (Japan) in 1923. Parts of the Bay were uplifted by 215 metres (700 feet). This causes *tsunamis* or tidal waves.
3 They can cause the raising or lowering of coastal regions as in Alaska in 1899 when some coastal rocks were uplifted by 16 metres (50 feet).
4 They can cause landslides as in the loess country of North China in 1920 and 1927.
5 They can cause the devastation of cities, fires and diseases.

Some Catastrophic Earthquakes

1755	Portugal	Caused depression of the sea floor near Lisbon
1868	Peru	30 000 people killed
1899	Alaska	Coast of Disenchantment Bay uplifted
1906	California	San Francisco destroyed
1906	Chile	3000 people killed
1920	Japan	Level of Sagami Bay changed and 200 000 people killed
1927	China	Landslides killed 100 000 people
1931	New Zealand	Napier destroyed
1931	Nicaragua	Managua (the capital) destroyed
1960	Agadir (Morocco)	Town destroyed and 10 000 people killed
1962	Iran	Over 20 000 people killed
1970	Peru	Earthquake on 31st May killed 50 000 people and made 1 000 000 people homeless
1972	Nicaragua	Managua, the capital, devasted and 50 000 people killed

VOLCANIC ACTION AND THE FEATURES IT PRODUCES

Rocks below the crust have a very high temperature, but the great pressure upon these keeps them in a semi-solid state. If the pressure weakens (as happens when faulting or folding takes place) then some of the rocks become liquid. This liquid is called *magma*. The magma forces its way into cracks of the crust and may either reach the surface where it forms *volcanoes* or *lava flows*, or it may collect in the crust where it forms *batholiths*, *sills* and *dykes*. Magma reaching the surface may do so quietly or with great violence. Whichever happens it eventually cools and solidifies.

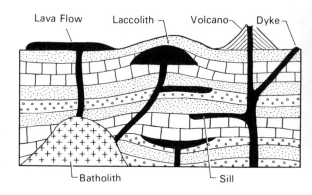

This diagram shows the chief types of volcanic forms. They do not occur together like this.

Volcanic Features formed in the Crust

BATHOLITH This is a very large mass of magma which often forms the root of a mountain. It is made of granite and can become exposed on the surface by the removal of the overlying rocks by erosion.

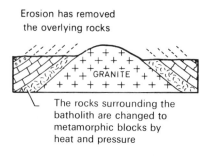

Erosion has removed the overlying rocks

The rocks surrounding the batholith are changed to metamorphic blocks by heat and pressure

SILL When a sheet of magma lies along the bedding plane it is called a sill. Some sills form ridge-like escarpments when exposed by erosion, e.g. Great Whin Sill in Northern England. Others may give rise to waterfalls and rapids where they are crossed by rivers.

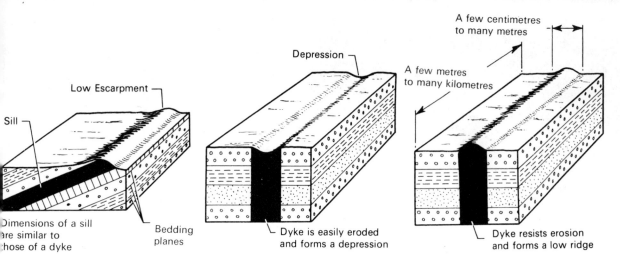

Sill

Low Escarpment

Depression

A few centimetres to many metres

A few metres to many kilometres

Dimensions of a sill are similar to those of a dyke

Bedding planes

Dyke is easily eroded and forms a depression

Dyke resists erosion and forms a low ridge

DYKE When a mass of magma cuts across the bedding planes and forms a wall-like feature it is called a dyke. Dykes may be vertical or inclined. Some dykes when exposed on the surface resist erosion and stand up as ridges or escarpments. Others are easily eroded and form shallow depressions. Examine the diagrams above. Like sills they sometimes give rise to waterfalls or rapids.

Volcanic Features formed on the Surface

Sometimes magma reaches the surface through a *vent* (hole) or a *fissure* (crack) in the surface rocks. When magma emerges on the surface it is called *lava*. If lava emerges via a vent it builds up a *volcano* (cone-shaped mound) and if it emerges via a fissure it builds up a *lava platform* or *lava flow*.

Vent Eruptions and Types of Volcanoes

Structure of a Volcano
The cone is made of either lava, or a mixture of lava and rocks torn from the crust, or ash and cinders (small fragments of lava).

Ash and Cinder Cone
When lava is violently ejected it is blown to great heights and it breaks into small fragments. These fall back to earth and build up a cone. Examples: Vulcano de Fuego (Guatemala), Paracutin (Mexico).

2 Lava Cones
The slope of the cones depends upon whether the lava was fluid or viscous when it was molten.
Fluid lavas give rise to gently sloping cones, e.g. Mauna Loa (Hawaii).

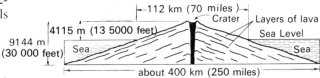

112 km (70 miles)

Crater

Layers of lava

Sea Level

4115 m (13 5000 feet)

9144 m (30 000 feet)

Sea

Sea

about 400 km (250 miles)

Viscous lavas give rise to steeply sloping cones. Sometimes the lavas are so viscous that when they are forced out of the volcano they form a *spine* or *plug*. Spines are rare because they often rapidly break up on cooling. This was the fate of Mont Pelée.

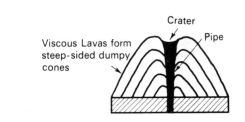

Crater

Pipe

Viscous Lavas form steep-sided dumpy cones

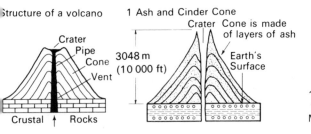

Structure of a volcano

Crater
Pipe
Cone
Vent

Crustal Rocks

1 Ash and Cinder Cone

Crater Cone is made of layers of ash

3048 m (10 000 ft)

Crater

Earth's Surface

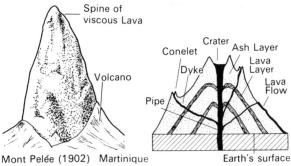

Spine of viscous Lava

Volcano

Mont Pelée (1902) Martinique

Conelet
Dyke
Pipe

Crater Ash Layer
Lava Layer
Lava Flow

Earth's surface

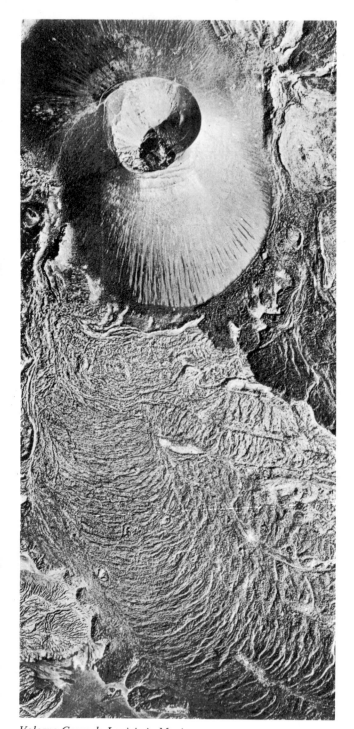

Volcano Cerro de Lopizio in Mexico

In this aerial view the cone and its crater stand out extremely well. This volcano is only 48 kilometres (30 miles) south of Paracutin which first erupted in 1943 and which is now 460 metres (1500 feet) high.

Bromo and Batok in Java

These cones are made of ash and cinders. The sides are deeply grooved by rain erosion. The flat land around the volcanoes consists of fine black volcanic dust.

3 Composite Cone

This is made of alternate layers of lava and ash.

This type of volcano begins each eruption with great violence which accounts for the layers of ash. As the eruption gets under way the violence ceases and the lava pours out forming layers on top of the ash. Lava often escapes from the sides of the cone where it builds up small conelets. The best examples of this type of volcano are Vesuvius (Italy), Etna (Sicily) and Stromboli (Italy). Sometimes explosive eruptions are so violent that the whole top of the volcano sinks into the magma beneath the vent. A huge crater called a *caldera* later marks the site of the volcano.

Formation of a Caldera

A caldera may become the site of a lake, e.g. Lake Toba (Northern Sumatra) and Crater Lake (U.S.A.). Sometimes eruptions begin again and new cones form in the caldera, e.g. Krakatoa (Sunda Strait). The town of Crater near Aden is built in a caldera.

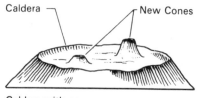

Caldera with new cones

Formation of a Caldera

Before violent
eruptions take place

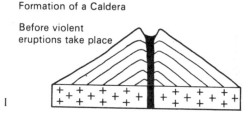

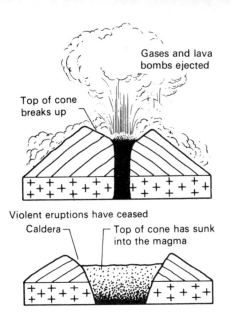

Gases and lava bombs ejected

Top of cone breaks up

Violent eruptions have ceased

Caldera — Top of cone has sunk into the magma

Mt. Meru in Tanganyika

The caldera of this volcano contains a conelet, the vent of which is clearly visible. Note the steep walls of the caldera.

Volcanic Activity

Are Volcanoes always active?

Volcanoes usually pass through three stages in their life cycle. In the beginning eruptions are frequent and the volcano is *active*. Later eruptions become so infrequent that the volcano is said to be *dormant* (sleeping). This is followed by a long period of inactivity. Volcanoes which have not erupted in historic times are said to be *extinct*.

Like all land forms a volcano is attacked by weathering and erosion and by the time it is extinct most if not all the volcano may have been removed. Sometimes the neck of lava remains and this stands up as a steep-sided plug. There are many plugs in central France and some of them became the sites of castles in the Middle Ages (see photograph below).

Fissure Eruptions and the Features they form

Large quantities of lava quietly well up from fissures and spread out over the surrounding countryside. Successive lava flows result in the growth of a lava platform which may be extensive and high enough to be called a plateau. The thickness of such lava plateaus may be as much as 1800 metres (5900 feet) (Deccan lava plateau near Bombay).

Examples of lava plateaus are: Columbia and Snake Plateau of the U.S.A. (520 000 sq km or 325 000 sq miles); north-west Deccan of India (648 000 sq km or 405 000 sq miles); parts of South Africa where the edge of the plateau forms the Drakensberg Mountains; Victoria and Kimberley Districts of Australia.

Rivers crossing lava plateaus often carve out deep gorges, e.g. Snake River of Oregon (North America). Sometimes the lavas weather to give fertile soils, as in the north-west Deccan which is used for cotton cultivation.

Stages in the formation of a Lava Plateau

Original relief

Le Puy in Central France

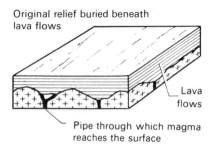

Original relief buried beneath lava flows

Lava flows

Pipe through which magma reaches the surface

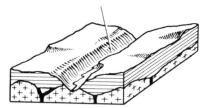

River valley cut through lava into rocks below

Stages in the formation of a lava plateau

Why do some volcanoes erupt violently?

Most magma contains gases which are under great pressure. In some instances there is a sudden decrease in pressure in the rising magma which causes the gases in it to expand very rapidly. This sudden expansion can cause violent explosions. Water vapour is often one of the gases and it may have originated from the magma or from water in the crust with which the magma came into contact. Many of the gases burn with a fierce heat and some of them like sulphur gases can form a dense cloud which rolls down the side of the volcano killing everything in its path. When eruptions are violent the lava explodes into small pieces which are blown to great heights. The sizes of the pieces vary from grains to small chunks of rock. The latter are called volcanic bombs. If the explosions are particularly violent the fine dust can reach such great heights that it gets carried along by the air currents of the upper atmosphere. When Krakatoa exploded in 1883 some of its dust passed right round the world causing vivid sunsets in many countries.

Other forms of Volcanic Activity

Emissions of gases and steam periodically take place from dormant volcanoes. Similar emissions of gases and steam take place in some volcanic regions where active lava eruptions have long since ceased. Super-heated water may flow quietly as in *hot springs*, or it may be thrown out with great force and accom-

panied by steam as in *geysers*. Thus a geyser differs from a hot spring in that its water is ejected explosively. Geysers often form *natural fountains*. Hot springs and geysers are common in Iceland, North Island of New Zealand and the Yellowstone National Park of U.S.A.

Formation of a Geyser

1 *Based on underground cavern*

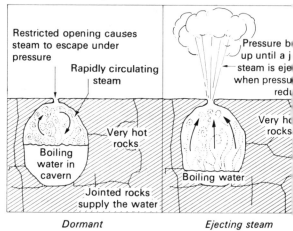

Restricted opening causes steam to escape under pressure

Rapidly circulating steam

Very hot rocks

Boiling water in cavern

Jointed rocks supply the water

Dormant

Pressure builds up until a jet of steam is ejected when pressure reduced

Very hot rocks

Boiling water

Ejecting steam

When the jet of steam has been ejected the geyser returns to the dormant phase. These geysers are often very regular, e.g. *Old Faithful* in the Yellowstone National Park of the U.S.A., ejects steam every half an hour.

2 *Based on underground cavern and sump*

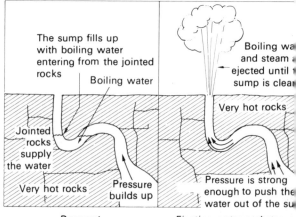

The sump fills up with boiling water entering from the jointed rocks

Boiling water

Jointed rocks supply the water

Very hot rocks

Pressure builds up

Dormant

Boiling water and steam are ejected until the sump is cleared

Very hot rocks

Pressure is strong enough to push the water out of the sump

Ejecting water and steam

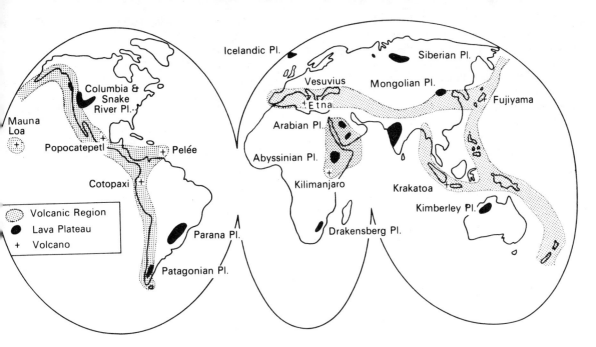

World Distribution of Volcanoes and Lava Plateaus

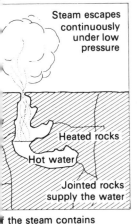

Formation of a Fumerole

Steam escapes continuously under low pressure

Heated rocks

Hot water

Jointed rocks supply the water

If the steam contains much dissolved sulphurous gas, the vent becomes surrounded by yellow sulphur deposits. This type is known as a *solfatara*.

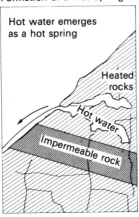

Formation of a Hot Spring

Hot water emerges as a hot spring

Heated rocks

Hot water

Impermeable rock

The hot water contains minerals dissolved from the rocks. The water is believed to be good for rheumatism and similar ailments which has resulted in Spas (health resorts) developing at many hot springs.

Influence of Volcanic Eruptions on Man
Destructive Influences
1 Some eruptions cause great loss of life, e.g. Vesuvius in A.D. 79 (out-pourings of gases and ash); Krakatoa in 1883 (caused great sea waves which drowned 40 000 people in neighbouring islands); Mont Pelée in 1902 (out-pourings of gases killed 30 000 people).
2 Some eruptions cause great damage to property, e.g. Vesuvius buried Herculaneum and Pompeii with ash; Mont Pelée caused the destruction of St. Pierre.

Constructive Influences
1 Some lava out-pourings have weathered to give fertile soils, e.g. in Java, the north-western part of the Deccan Plateau, and the plains around Etna. These regions are of important agricultural value.
2 Volcanic activity sometimes results in the formation of precious stones and minerals. These occur in some igneous and metamorphic rocks, e.g. diamonds of Kimberley; copper deposits of Butte (U.S.A.); and the nickel deposits of Sudbury in Canada.
3 Some hot springs are utilised for heating and supplying hot water to buildings in New Zealand and Iceland.

37

EXERCISES

1 Explain the main differences in appearance and origin between the members of each of the following pairs of features:
   (i) batholith and lava flow;
   (ii) sill and dyke;
   (iii) crater and caldera;
   (iv) hot spring and geyser.
   Illustrate your answer with diagrams.

2 Carefully explain the differences between the appearance and the origin of (i) a lava volcano, and (ii) a composite volcano. Draw a large diagram for each type of volcano and on this mark and name the main parts.

3 Locate by shading on a map of the world, *two* important volcanic and earthquake regions. Briefly describe one major volcanic eruption and one major earthquake which have occurred in the last twenty years. Your description should indicate the causes and effects of these natural catastrophes.

4 (a) Carefully explain the differences between:
      (i) igneous and sedimentary rocks;
      (ii) sedimentary and metamorphic rocks;
      (iii) sial and sima.
   (b) Carefully explain how a sedimentary rock originates and name two common examples of this rock.

Objective Exercises

1 Which one of the following best describes the world distribution of active and recently active volcanoes?
   A they are found in association with young fold mountain chains
   B they occur in river flood plains
   C they are associated with old eroded mountain chains
   D they are located on the western sides of continents
   E they tend to form chains around ocean basins

   A B C D E

2 Which of the following features is the product of vulcanicity?
   A geosyncline
   B escarpment
   C atoll
   D fold mountain
   E caldera

   A B C D E

3 A volcanic eruption is most likely to be violent when
   A the volcano is near to the sea
   B the neck of the volcano is sealed by a plug
   C the lava is viscous
   D the lava reaches the surface through a fissure
   E the volcano is on the ocean floor

   A B C D E

4 Which of the following features may occur when lava cools at the surface?
   A basalt plateau
   B sill
   C batholith
   D dyke

   A B C

5 An intrusion of magma along a bedding plane is called
   A dyke
   B sill
   C batholith
   D laccolith

   A B C

6 Which of the following features is **not** an aspect of vulcanicity?
   A geyser
   B batholith
   C dyke
   D fold

   A B C

7 Which of the following statements best describes the usefulness of vulcanicity?
   A A violent volcanic eruption represents the release of an enormous amount of energy.
   B The high density of population in parts of Java is dependent on agriculture.
   C Some volcanoes are dormant.
   D Fissure eruptions are usually non-violent.

   A B C

38

4 Features Produced by Running Water

External forces lower the level of the land by wearing it away and this process is called *denudation*. They also raise the level of the land by *deposition*. Denudation consists of (i) Weathering, (ii) Erosion. *Weathering* refers to the gradual disintegration of rocks which lie exposed to the weather. The effect of weathering can be seen on stone monuments and buildings where pieces of stone have flaked off, and on iron railings which have rusted.

Erosion refers to the disintegration of rocks which lie exposed to what are called the agents of erosion, i.e. running water, wind and moving ice.

Deposition refers to the laying down of rock particles by the agents of erosion.

1. WEATHERING AND THE FEATURES IT PRODUCES

It is effected by physical forces and chemical forces.

Physical Weathering

I By Temperature Changes

In arid regions, such as hot deserts, rock surfaces heat up rapidly when exposed to the sun and the surface layers expand and break away. At night when the temperature falls rapidly the same layers contract and more cracks develop. In time the layers of rock peel off and fall to the ground. Rock break-up of this type is called *exfoliation*. Exfoliation is best seen in rocks of uniform structure. This process ultimately changes rocky masses into rounded boulders.

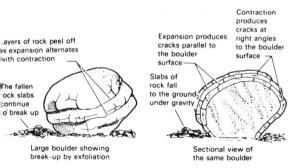

Layers of rock peel off as expansion alternates with contraction

The fallen rock slabs continue to break up

Large boulder showing break-up by exfoliation

Expansion produces cracks parallel to the boulder surface

Contraction produces cracks at right angles to the boulder surface

Slabs of rock fall to the ground under gravity

Sectional view of the same boulder

Exfoliation domes are common in the Kalahari, Egyptian and Sinai Deserts. Physical weathering on steep slopes often produces *screes* which collect at the bottom of the slopes.

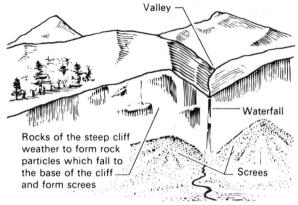

Valley

Waterfall

Rocks of the steep cliff weather to form rock particles which fall to the base of the cliff and form screes

Screes

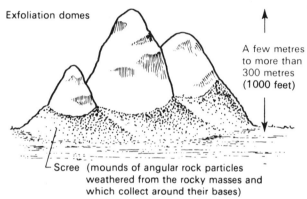

Exfoliation domes

A few metres to more than 300 metres (1000 feet)

Scree (mounds of angular rock particles weathered from the rocky masses and which collect around their bases)

II By Frost Action

When water freezes its volume increases. If water in the cracks of rocks freezes a tremendous power is applied to the sides of the cracks and the cracks widen and deepen.

Frost action in time breaks up rocky outcrops into angular blocks which later break up into small fragments.

Frost action is very common in the winter season in temperate regions and in some high mountains all the year, e.g. the Himalayas. It usually involves the freezing of water in the cracks of rocks during the night and the thawing of the ice during the following day. The angular fragments of rocks which break off the main rocky masses through frost action form screes around the lower slopes of the rocky outcrops.

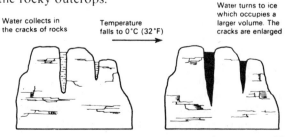

Water collects in the cracks of rocks

Temperature falls to 0°C (32°F)

Water turns to ice which occupies a larger volume. The cracks are enlarged

Desert Screes near St. Catherine's Monastery in Sinai

Note Screes formed by frost action contain *angular* rock particles, those formed by other types of weathering contain *rounded* rock particles.

Some rocks break up into large rectangular-shaped blocks under the action of mechanical weathering. This may be partly frost action and partly expansion and contraction through temperature changes. This is called *block disintegration*.

Joints are opened by both frost action and expansion and contraction

Block disintegration

III By Plant and Animal Action

The roots of plants, especially trees, can force joints and cracks apart in rocks. Some animals by burrowing also help to break up rocks.

Note A covering of vegetation often protects rocks from weathering. It binds the soil together and reduces changes in temperature. The removal of a covering of vegetation can result in *soil erosion*.

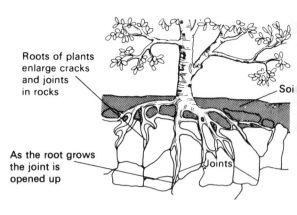

Roots of plants enlarge cracks and joints in rocks

Soil

As the root grows the joint is opened up

Joints

Chemical Weathering
I By Rain Action

Rain is really a weak acid because it dissolves oxygen and carbon dioxide as it falls through the air. Some minerals especially carbonates are dissolved out of rocks by rainwater. These rocks are therefore weakened and begin to break up. Chemical weathering is most active in limestone rocks the surface of which become weathered into deep narrow grooves called *grikes* which are separated by flat or round-topped ridges called *clints*. The rainwater enters the limestone via the joints which become enlarged to form clints.

In humid tropical countries chemical weathering is very active in many types of rocks.

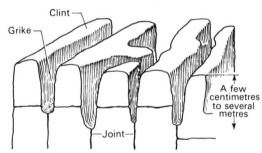

Clint

Grike

A few centimetres to several metres

Joint

Weathering of limestone rocks

II By Plants and Animals

Bacteria in the presence of water break down certain minerals in the soil, and all plants absorb minerals from the soil. Decaying vegetation produces organic acids which cause a further break-down of minerals. All these actions help to weaken and break up the rocks.

Note Chemical weathering takes place in all regions where there is rain, but it is most marked in wet regions which have high temperatures. Physical weathering takes place in all regions where there are changes in temperature but it is most marked in the hot deserts which have a large daily temperature range.

2. UNDERGROUND WATER AND THE FEATURES IT PRODUCES

When rain falls some of it runs off the surface forming streams and rivers; some of it evaporates directly, or indirectly via plants; some of it soaks into the surface rocks.

The amount of run-off, evaporation and percolation depends upon the nature of the rocks, the slope of the land and the climate. Run-off on steep slopes is greater than on gentle slopes; evaporation in dry climates is greater than in humid climates, and water percolates into sands more easily than into granites.

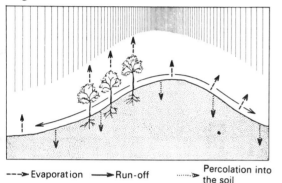

---→ Evaporation ——→ Run-off ·······> Percolation into the soil

How Water enters into the Rocks

Water will enter rocks which are *porous* (i.e. rocks having small air spaces, e.g. sandstone) or rocks which are *pervious* (i.e. rocks having joints or cracks, e.g. granite). Rocks which allow water to pass through them are said to be *permeable*. Sandstone is a permeable rock. Rocks which do not allow water to pass through them are said to be *impermeable*. Clay is an impermeable rock.

Note Clay is porous (water enters it), but it is impermeable (water will not pass through it).

The Water-table or the Level of Saturation

Water entering the surface rocks moves downward until it comes to a layer of impermeable rock when further downward movement ceases. There are three water zones below the surface:

(i) *The zone of permanent saturation.*
  The pores of the rocks of this zone are always filled with water.

(ii) *The zone of intermittent saturation.*
  The pores of the rocks of this zone contain water only after heavy rain.

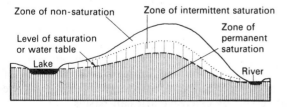

(iii) *The zone of non-saturation.*
  This lies immediately below the surface. Water passes through but never remains in the pores of the rocks of this zone.

Springs

When water flows naturally out of the ground it is called a *spring*. There are many types of springs. Here are some of the more common ones.

I A permeable rock lying on top of an impermeable rock in a hill.

This diagram shows two lines of springs which occur where the junction of the two rock layers meets the surface. Notice that one line of springs is temporary.

I Intermittent spring occurring in wet season only
P Permanent springs
--- Water Table during the wet season
····· Water Table during the dry season which is too low to supply the intermittent springs

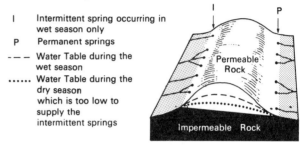

II Well-jointed rocks forming hilly country produce springs.

Water enters the rocks via the joints. Springs frequently occur where the water-table meets the surface.

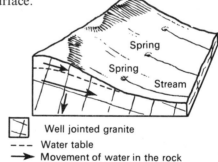

Well jointed granite
--- Water table
——→ Movement of water in the rock

III The impounding of water by a dyke can give rise to springs.

If a dyke cuts across a layer of permeable rock then the water on the up-slope side of the dyke is impounded. The water-table here rises and it gives rise to springs where it meets the surface.

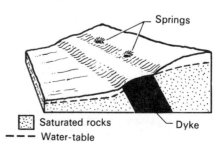

Saturated rocks
--- Water-table

IV Chalk or limestone escarpments which overlie impermeable rocks give rise to springs.

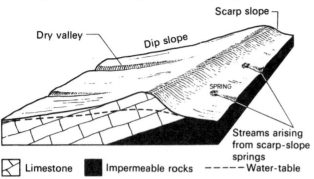

Limestone ▨ Impermeable rocks ▪ − − − − Water-table

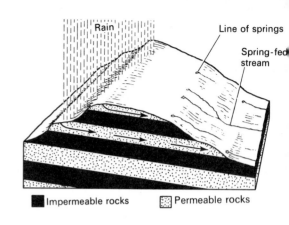

Impermeable rocks ▪ Permeable rocks ▨

The dip slope with the scarp slope (sometimes called escarpment) together form a feature known as a *cuesta*. On the dip slope of cuestas such as the North Downs in Southern England, there are many dry valleys along which streams once flowed. These valleys are now dry because the level of the water-table has fallen. The valleys are called *dry valleys*. After a prolonged period of heavy rain the water-table sometimes rises high enough to allow a seasonal flow of water. This is called a *bourne* and an example of where it occurs is in the Caterham Valley near Croydon.

Usually two lines of springs occur, one at the foot of the scarp slope and the other on the dip slope. Since there is little or no surface drainage in limestone or chalk regions, settlements often become located near to the springs.

V Gently sloping alternate layers of permeable and impermeable rocks often produce springs.
Rain falling on the exposed ends of the permeable rock layers soaks down the sloping bedding planes and finally comes out as springs. The springs are sometimes in lines.

VI Numerous springs occur where the junction of limestone rocks and underlying impermeable rocks meets the surface.
Limestone regions rarely have surface drainage.

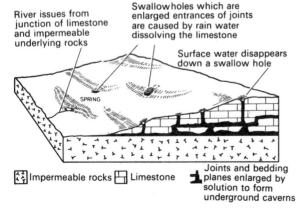

River issues from junction of limestone and impermeable underlying rocks

Swallow holes which are enlarged entrances of joints are caused by rain water dissolving the limestone

Surface water disappears down a swallow hole

SPRING

Impermeable rocks ▨ Limestone ▨ Joints and bedding planes enlarged by solution to form underground caverns

Both the rain and streams entering the region work their way by solution into the limestone rocks. Some joints become enlarged to form *swallow holes*. If the limestone rocks rest on impermeable rocks then this water will reach the surface again where the two rock types meet the surface. The water may issue out as streams or springs. Inside the limestone underground streams dissolve the rocks and form huge caverns.

Wells

A well is a hole sunk in the ground to below the water-table. Water then seeps out of the rocks into the well.

Wells which are sunk far below the water-table always contain water (A, B and C). Wells sunk only just below the water-table often go dry in periods of drought when the water-table falls (D). If a well does not reach the water-table it will be dry.

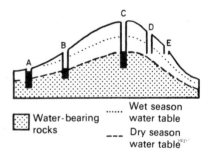

Water-bearing rocks ▨

...... Wet season water table

− − − Dry season water table

Artesian Basins and Artesian Wells
Artesian Basin

An artesian basin consists of a layer of permeable rock lying between two layers of impermeable rock such that the whole forms a shallow syncline with one or both ends of the permeable rock layer exposed on the surface. Rain water enters the permeable layer at its exposed ends. This layer becomes saturated with water and is called an *aquifer*. Western Australia, the Sahara Desert and parts of North America from Saskatchewan to Kansas are underlain by extensive artesian basins.

The diagram on the right shows a part of the artesian basin of the Sahara Desert. In places the aquifer bends up towards the surface and wind erosion sometimes exposes it. When this happens pools of water occur and these are called *oases* (sing. *oasis*). If the aquifer is near to the surface wells can be sunk. This is often done. Notice some typical hot desert erosional and depositional features, e.g. rock platforms, wadis and sand dunes. Also notice that the exposed part of the aquifer which receives the rain is called the *catchment area*. The London Basin consists of a shallow syncline formed of chalk which lies between layers of clay. In some parts the water-table has fallen by as much as 30 metres (100 feet) during the last 50 years. One reason for this is that Waterboards and industries have taken so much water from the numerous wells sunk in the chalk aquifer.

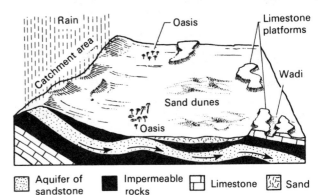

Section across part of the Sahara Desert

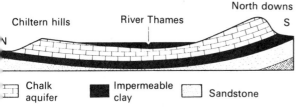

Section across the London Basin

Artesian Well

If a well is sunk in the aquifer of an artesian basin and the pressure of water is sufficient to cause the water to flow out of it, then the well is called an *artesian well.*

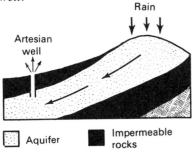

Value of Ground Water to Man

1 Springs and wells have played an important part in the siting of settlements in many regions all over the world.

2 In some regions, e.g. hot deserts and semi-arid plains, settlement is only possible by utilising the ground water. When this is too deep to be tapped, settlement does not take place.

3 In southern Algeria there is a limestone plateau which is 600 metres (1950 feet) above sea level. The Chebka people who live there have dug wells to tap the underground water. The water is used for irrigating the land and the wells have given rise to numerous oases.

4 The Soafas (people of the Suf) who live on the borders of Tunis and Algeria use artesian water which lies near to the surface. This water maintains many oases which cultivate date palms.

5 The great artesian basins which underlie Queensland, New South Wales and parts of South Australia have an area of 1 500 000 sq km (58 600 sq miles). Water taken from the many wells which tap the aquifer of the basins is too salty for irrigation but is used for watering large herds of cattle.

6 Similar artesian basins underlie parts of North America from Kansas to Saskatchewan and wells here raise water for use in cattle ranching.

7 Wells play an important part in agriculture in the Indo-Gangetic Plain of the Indian subcontinent in the dry season. The water is used for irrigation.

3. RAIN ACTION AND THE FEATURES IT PRODUCES.

Rain action is an aspect of erosion because it involves movement. It is most marked in semi-arid regions because these have little or no vegetation and the rains, though infrequent, are torrential. Rain action produces many types of features of which the following are the most common: *gully, earth pillar, soil creep,* and *landslide.*

Gully

Rain which falls on gently sloping land which has little or no vegetation moves downhill as a sheet of water. The slope is quickly eroded into deep grooves called *gullies*. They develop on a small scale on embankments and cuttings and also tip heaps. They develop on a large scale in semi-arid regions where the landscape becomes cut up into gullies and ridges of all shapes and sizes, as in the Badlands of the Dakotas (U.S.A.). See page 44.

The influence of rain-wash on soil erosion will be discussed in a later section.

Earth Pillar

When rain falls on slopes made of clay and boulders, the clay is rapidly removed except where boulders

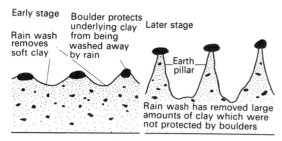

Early stage

Rain wash removes soft clay

Boulder protects underlying clay from being washed away by rain

Later stage

Earth pillar

Rain wash has removed large amounts of clay which were not protected by boulders

Note Earth pillars range from a few centimetres to several metres in height.

Earth Pillars in Upper Rhône Basin

form a protection. When this happens, columns of clay, capped by boulders, develop. These are called *earth pillars*. However, the pillars are only temporary. In time they will be removed by erosion.

Soil Creep (slow movement)
On all sloping land there is a steady movement of the soil down the slope. Rain water soaking into the soil slowly trickles down the slope and this causes a movement in the soil. Bulging fences and walls and outward bending tree trunks reflect this movement which is called *soil creep*.

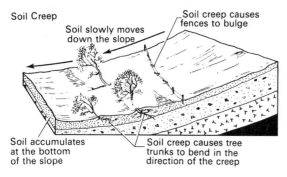

Soil Creep

Soil slowly moves down the slope

Soil creep causes fences to bulge

Soil accumulates at the bottom of the slope

Soil creep causes tree trunks to bend in the direction of the creep

Gully Erosion in a road-side laterite bank in Kuala Lumpur

Landslide (sudden movement)
A *landslide* takes place when large quantities of loosened surface rocks slide down steep slopes which may be cliff faces, embankments, valley sides or railway cuttings. Landslides are caused by the lubricating action of rain water and the pull of gravity which result in slumping or sliding.

Badlands of Dakota in U.S.A. The hill sides are deeply gullied

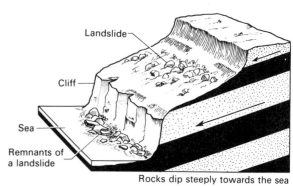

Landslide

Cliff

Sea

Remnants of a landslide

Rocks dip steeply towards the sea

Landslide caused by sliding The black arrows show both the place of lubrication and the direction of movement.

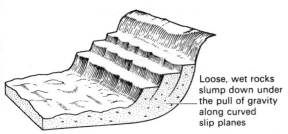

Loose, wet rocks slump down under the pull of gravity along curved slip planes

Landslide caused by slumping Slumping of this type takes place on steep slopes made of clay. It is especially common in cliffs of clay which are under wave attack.

Landslide on a Railway Cutting
Landslides are very common on the sides of railway and road cuttings, especially in mountainous regions which have heavy falls of rain. In some regions frost action speeds up the process. Frozen soils and subsoils on steep slopes become unstable when they thaw and the movement of water down the slope together with the pull of gravity triggers off a landslide.

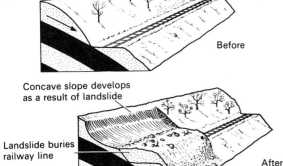

Before

Concave slope develops as a result of landslide

Landslide buries railway line

After

4. RIVER ACTION AND THE FEATURES IT PRODUCES
A river's *source* is the place at which it begins to flow. It may be in the melt waters of a glacier, e.g. the Rhône (France), or in a lake, e.g. the Nile (Africa), or in a spring, e.g. the Thames (England), or in a region of steady rainfall, e.g. the Zaire (Africa). A river's *mouth* is the place where the river ends. This is usually in the sea, e.g. the Amazon (Atlantic), the Niger (Gulf of Guinea) and the Indus (Arabian Sea), although it may be in a lake, e.g. the Volga (Caspian), or in a salt swamp, e.g. the Chari River (Lake Chad) and the Tarim River (Lop Nor).

Rivers are one of the greatest sculpturing agents at work in humid regions. They carve out valleys in the highlands and as they do so they produce peaks, ridges and hills.

The material so removed is transported from the highlands and is deposited around them as gently sloping plains. A river thus does three types of work: it *erodes*; it *transports*; and it *deposits*.

In process of time river erosion, transport and deposition turn the original surface into an almost level plain which is called a *peneplain*.

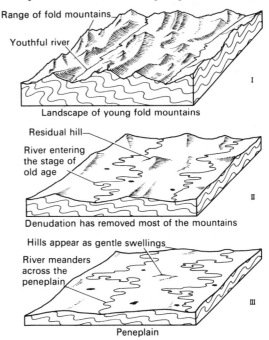

Range of fold mountains

Youthful river

Landscape of young fold mountains

Residual hill

River entering the stage of old age

Denudation has removed most of the mountains

Hills appear as gentle swellings

River meanders across the peneplain

Peneplain

The River as an Energy System
Energy is the ability to do work, and the amount of energy which a river has determines whether it can effectively erode its valley and transport the material it is carrying, or whether it drops the material in the form of deposition. We often say that a mature river has three sections: the upper course where most of the erosion takes place, the middle course where transport is the dominant process, and the lower course where the main process is deposition. This is a very simplified view to take as we shall now see. The amount of erosion that a river can achieve depends on its energy. A river's energy increases with its volume and with its velocity, and with its regime, that is, seasonal flow. This means that a large, fast-flowing river will have more power to affect erosion in time of flood than the same river will have in time of drought when it flows sluggishly with little water in its channel. However, not all the energy in a river is used for erosion. Some is needed to overcome frictional resistance both externally along the bed and banks and internally when currents, caused by turbulence, splash and roll against each other. Energy is also needed to transport the pebbles, sand, silt and the dissolved minerals in the water. The following diagram summarises these different uses of energy.

45

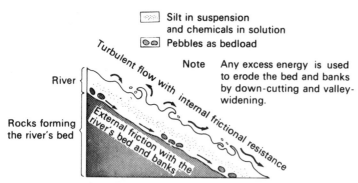

Silt in suspension and chemicals in solution

Pebbles as bedload

Note Any excess energy is used to erode the bed and banks by down-cutting and valley-widening.

River

Rocks forming the river's bed

Long Section of a steeply-flowing River

Energy is used to overcome friction

The shape of a river's channel is also important in determining how much energy a river will have for erosion. A flat, wide channel is very inefficient for transporting water, while a narrow, deep channel is much more efficient because the smaller surface area results in less frictional drag. The diagram below shows two river channels which have the same cross-sectional area and which therefore can carry the same volume of water. But the wide channel has a greater surface area of bed and banks, and therefore frictional drag will be greater than in the narrower channel.

In the upper part of a river's course the gradient of the channel is steeper but the volume of water is less. It used to be thought that a river flowed fastest at this point. This is true, in that the maximum velocity occurs here, but the water splashes and eddies so much that the average speed of the river at this point is not very great. In the middle and lower sections of a river's course the average flow of the river is at least equal to, and in many cases, greater than the speed in the upper course. This is because of the greater loss of energy to overcome internal friction in the turbulent upper section.

Cross Section of an inefficient channel shape

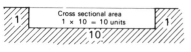

Cross sectional area
1 x 10 = 10 units

Total distance across the bed and banks is 12 units and this river has to overcome more friction and it therefore has less energy for erosion

Diagram to explain the influence of the shape of a river's channel

Cross Section of an efficient channel shape

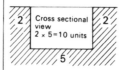

Cross sectional view
2 x 5 = 10 units

Total distance across the bed and banks is 9 units and this river has to overcome less friction and it therefore has more energy for erosion

Energy is used to transport sediments

A river transports its load in three ways: by *traction* or the dragging of the *bedload* of pebbles along its

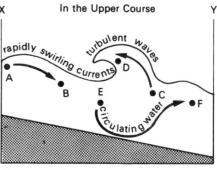

X   In the Upper Course   Y

Greater **maximum** velocity

An object is tracked in its movement from X to Y and the time is recorded. It follows an indirect path (A to B, C to D and E to F) which is greater than the direct path. The distance travelled in the time recorded is greater than it is in the Middle Course.

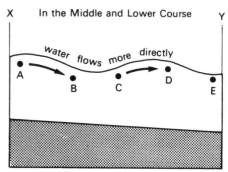

X   In the Middle and Lower Course   Y

Greater **average** velocity

The object is tracked in its movement from X to Y and the time is recorded. It follows a more direct path (A to B, C to D and E to F) but it takes about the same time to do so. This shows that the velocity in the Middle Course is lower than it is in the Upper Course

Diagram to explain the effects of turbulent flow

bed, by *suspension* of light sediments such as silt and mud in the water, and by *solution* of chemicals which are dissolved in the water. Sediments are transported by a river until it has insufficient energy to move them farther. It then deposits them. A river may lose energy where there is either a decrease in gradient, or a widening or meandering of its channel, e.g. in its lower course, or where there is a decrease in volume, e.g. after a flood.

Energy used in erosion

Erosion in a river is caused by attrition, corrasion, hydraulic action and chemical solution. Attrition is the process whereby pebbles are eroded by striking together as they are rolled along a river's bed. Corrasion is the wearing away of the bed and the banks, by a river's load. Hydraulic action is the wearing away of the bed and the banks of a river by the sheer weight of water hurled against them. This is particularly effective in fast-flowing rivers. Chemical solution is the dissolving of minerals from he rocks, and it is particularly effective with soluble minerals.

In parts of a river's course the main action of the river is erosional, and in these parts the river's surplus energy keeps the channel clear of debris and sediment. In other parts of a river's course, deposition is dominant. This happens when most of a river's energy is used up to overcome frictional drag. The channel then becomes choked with boulders, or banks of sand and shingle if the river has a large load. Between these two, there is a part of a river's course where there is a balance between erosion and deposition, where both processes occur but without one being dominant. The erosional areas more commonly occur in the upper parts of a river's course and it is here that a river appears to have more energy. The depositional areas more commonly occur near to a river's mouth, and it is here that a river has little or no surplus energy for erosion. Based on these general observations a river valley can be divided into an upper section, a middle section and a lower section, and by studying each of these we can see what landscape features are produced in each of the three parts.

A river's volume decreases:
1 When it enters an arid region (especially a hot one)
2 When it crosses a region composed of porous rocks, e.g. sand and limestone
3 In the dry season or in a period of drought.

A river's speed decreases when it:
1 Enters a lake

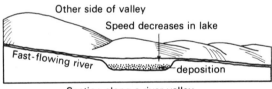

Section along a river valley

2 Enters the sea

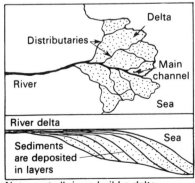

Note: not all rivers build a delta when they enter the sea

3 Enters a flat or gently sloping plain such as a valley bottom.

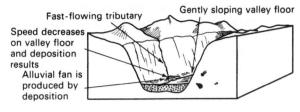

THE DEVELOPMENT OF A RIVER VALLEY

The force of a river partly depends upon its size and partly upon its *gradient*, that is the distance it has to fall before it reaches *base level*. Base level is the surface of a lake or a river or the sea into which it flows. The level of a river at its confluence with a tributary is the base level for that tributary. The upper part of a river, because it is high above the base level, will be able to deepen its channel rapidly. A river cannot, of course, erode below its base level. A river, like an animal or a plant, has a *life-cycle*. In the beginning, when it is in the *stage of youth*, it flows turbulently in a narrow, steep-sided valley whose floor is broken by pot holes and waterfalls. As time passes denudation widens the valley and lowers its floor. Now that the gradient is reduced the river has less energy to erode and the initial bends that it had, because of the nature of its valley floor, become more pronounced. It is now in the *stage of maturity*. As denudation continues the valley is opened out more and more. The gradient is further reduced and deposition now becomes active. Layers of sediments are dropped by the river and these ultimately extend over the entire floor of its valley where they build up a gently sloping plain called a *flood plain*. The river wanders in great *meanders* or loops across this plain and often it becomes divided into many channels by its own deposition. The river is now in the *stage of old age*. Deposition within the mouth of an old river sometimes builds up a triangular-shaped piece of land called a *delta*.

Many river valleys such as those of the Nile, Indus and Irrawaddy contain all three stages. The torrent or *upper course* represents the stage of youth; the valley or *middle course* represents the stage of maturity, and the plain or *lower course* represents the stage of old age.

Note Youthful Stage is often called Torrent Stage. Mature Stage is often called Valley Stage. Old Age Stage is often called Plain Stage.

A river valley grows in length by *headward erosion*. Rain wash, soil creep and undercutting at the head of a river combine to extend the valley up the slope. A river's valley is deepened by *vertical erosion* and widened by *lateral erosion*. The former is entirely a river process; the latter is effected by weathering on the valley sides and by the river on the river banks. When a river's gradient is steep, i.e. when it is in the stage of youth, vertical erosion is dominant. When a river's gradient is very gentle, i.e. when it is in the stage of old age, there is little erosion and deposition is dominant. In a mature valley lateral erosion is dominant.

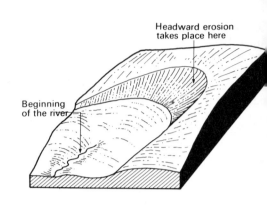

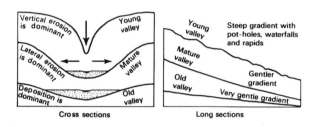

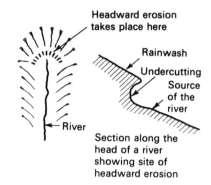

Over 1 in 10 Youthful (Torrent) Stage.
Between 1 in 10 and 1 in 100 Mature (Valley) Stage.
Under 1 in 100 Old Age (Plain) Stage.

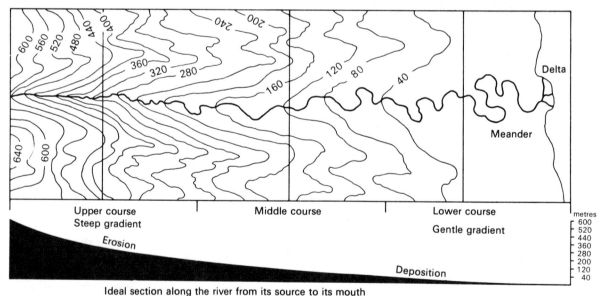

Ideal section along the river from its source to its mouth

THE CHARACTERISTIC FEATURES OF A YOUTHFUL VALLEY

- Deep, narrow valley (V-shaped)
- Valley has a steep gradient (river is fast-flowing)
- Pot-holes
- Interlocking spurs
- Waterfalls and rapids

Pot-holes

The water of a fast-flowing river swirls if the bed is uneven. The pebbles carried by a swirling river cut circular depressions in the river bed. These gradually deepen and are called *pot-holes*. Much larger but similar depressions form at the base of a waterfall. These are called *plunge pools*.

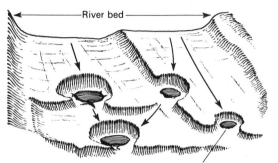

Swirling water falling into a slight depression turns it into a cylindrical hole called a pot hole

Pot hole in a river's bed

Interlocking Spurs

Vertical erosion rapidly deepens the valley. The river twists and turns around obstacles of hard rock. Erosion is pronounced on the concave banks of the bends and this ultimately causes spurs which alternate on each side of the river to *interlock*. The undercut concave banks often stand up as *river cliffs*. On the opposite convex banks there is little or no erosion. The banks form gentle *slip-off slopes*.

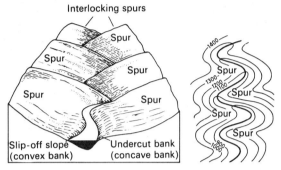

V-shaped valley with interlocking spurs

Waterfalls and Rapids

These occur where the bed of a river becomes suddenly steepened. Waterfalls are of two types:
 (i) Those caused by differences in rock hardness into which the river is cutting.
 (ii) Those caused by uplift of the land, lava flows and landslides, etc.

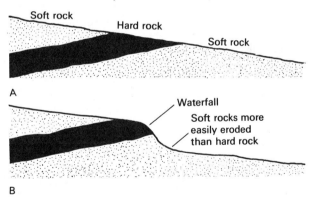

Waterfalls caused by Differences in Rock Hardness

When a layer of hard rock (rock which resists erosion) lies across a river's course, the soft rocks on the down-stream side are more quickly eaten into than is the hard rock. The river bed is thus steepened where it crosses the hard rock and a *waterfall* or a *rapid* develops. A waterfall arises when the hard rock layer is: (i) horizontal, (ii) dips gently upstream, or (iii) is vertical.

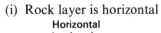

(i) Rock layer is horizontal

Horizontal hard rock layer

Undercutting weakens overlying hard rock

Soft rock

Plunge pool

(ii) Rock layer dips up-stream

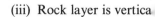 (iii) Rock layer is vertical

Hard rock layer dips gently up-river

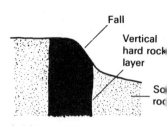

Fall

Vertical hard rock layer

Soft rock

A rapid develops when:

1 A waterfall of types (i) and (ii) above retreats up-stream.

2 A hard rock layer dips down-stream.

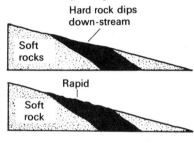

Hard rock dips down-stream

Soft rocks

Rapid

Soft rock

Some Good Examples of Waterfalls and Rapids

1 *Gersoppa Falls* (253 metres) (827 feet) — Western Ghats of India (in the wet season it is the greatest fall in the world).

2 *Kaieteur Falls* (225 metres) (736 feet) — Potaro River (British Guiana): 5° 0′N; 59° 0′W.

3 *Aughrabies Falls* (137 metres) (448 feet) — Orange River (South Africa): 28° 49′S; 20° 22′E.

4 *Victoria Falls* (110 metres) (360 feet) — 49′S; 25° 51′E. The gorge below the Falls is 96 kilometres long.

5 *Niagara Falls* (52 metres) (170 feet) — Between Lakes Erie and Ontario: 43° 7′N; 79° 1′W. The gorge below the Falls is 12 kilometres long.

6 *Livingstone Falls* (274 metres) (900 feet) — Zaire River: 5° 0′S; 14° 15′E. The Falls are formed by 32 rapids.

7 *Nile Cataracts* — Between Aswan and Khartoum.

Young Valleys of Special Interest

Some valleys have very steep sides and are both narrow and deep. These are called *gorges*. A gorge often forms when a waterfall retreats up-stream. The diagrams below show how this takes place. One of the most impressive gorges formed in this way lies below the Victoria Falls.

A gorge will also form when a river maintains its course across a belt of country which is being uplifted. Only very powerful rivers are able to do this.

The diagrams below show how this comes about. Notice that only parts of the region crossed by the river are uplifted and *not* the whole region. If the latter happened a gorge would develop along the entire length of the river.

The Indus, Brahmaputra and the headwaters of the Ganges have cut deep gorges in the Himalayas. The Indus gorge in Kashmir is 5180 metres (16940 feet) deep. The Columbia River has also cut a gorge across the Cascades in North America.

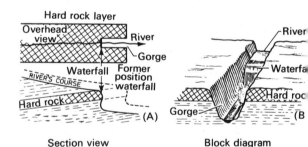

Section view

Block diagram

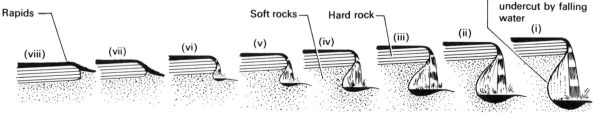

Undercutting of the hard rock layer causes waterfall to recede upstream

as waterfall moves upstream its height decreases

The gorge below the Victoria Falls on the River Zambesi (Rhodesia). It is normally obscured by spray from the waterfall

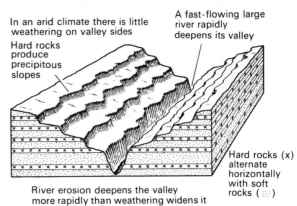

When a river flows across a plateau which is composed of horizontal and alternate layers of hard and soft rock, the valley it cuts will be deep and narrow. If the region is arid then there will be little weathering on the valley sides and the gorge will be very impressive. The Colorado River has cut a gorge 1·6 kilometres (1 mile) deep and 480 kilometres (300 miles) long into the Colorado Plateau (U.S.A.). Because of its size the gorge is called a *canyon*.

Canyons usually occur in dry regions where large rivers are actively eroding vertically and where weathering of the valley sides is at a minimum. This means that the river valleys are deepened more than they are widened. Canyons of great size develop when the land over which they flow is being uplifted but at a rate which enables rivers to maintain their courses across the area of uplift.

An impressive canyon in Utah's National Park (U.S.A.). The canyon is 792 metres (2600 feet) deep.

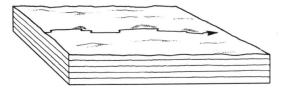

(i) Before formation of mountains

(ii) After formation of mountains

Gorge   Gorge

Vertical erosion by the river enables it to maintain its course across the rising mountains

Folded rocks

Earth movements have resulted in the formation of mountain ranges across the river's course

The Grand Canyon in U.S.A.

River action along the banks and bed of a river bend

The following two diagrams on page 52 show the erosional and depositional actions of a river as it flows round a meander. The water flows round the meander in a corkscrew manner and this causes erosion to take place on the concave bank and deposition on the convex bank.

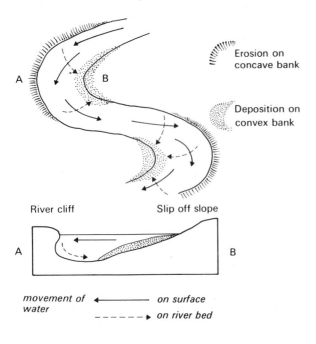

Erosion on concave bank

Deposition on convex bank

River cliff   Slip off slope

A   B

movement of water
⟶ on surface
⇢ on river bed

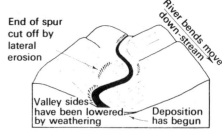

End of spur cut off by lateral erosion

River bends move down-stream

Valley sides have been lowered by weathering

Deposition has begun

3 The valley is now mature. The ends of spurs are cut right back and they stand up as *bluffs*.

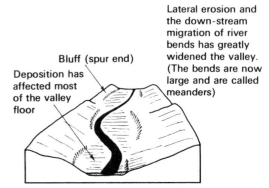

Bluff (spur end)

Deposition has affected most of the valley floor

Lateral erosion and the down-stream migration of river bends has greatly widened the valley. (The bends are now large and are called meanders)

4 The valley is now fully mature and it is approaching the stage of old age. Lateral erosion has developed a wide valley whose floor is almost completely covered with sediments. A flood plain is clearly being formed. The meander belt is as wide as the valley.

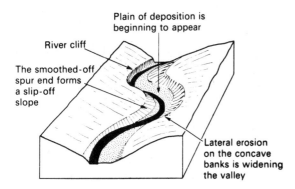

Plain of deposition is beginning to appear

River cliff

The smoothed-off spur end forms a slip-off slope

Lateral erosion on the concave banks is widening the valley

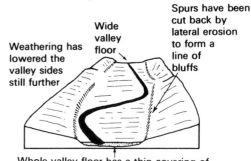

Wide valley floor

Weathering has lowered the valley sides still further

Spurs have been cut back by lateral erosion to form a line of bluffs

Whole valley floor has a thin covering of coarse sediments (gravel)

1 The young valley has well-developed interlocking spurs as shown in the diagram below. Lateral erosion on the concave bank has begun.

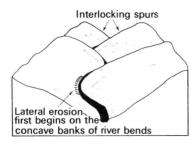

Interlocking spurs

Lateral erosion first begins on the concave banks of river bends

2 The valley is widened as the river meanders from side to side. Weathering lowers the valley sides. The river meanders migrate downstream and by doing so both widen and straighten the valley. Deposition takes place on the convex banks of meanders.

CHARACTERISTIC FEATURES OF A MATURE VALLEY

1 The valley has the shape of an open V in cross-section.
2 The gradient is more gentle than in a young valley.
3 River bends are pronounced. The concave banks stand up as river cliffs; the convex banks slope gently as slip-off slopes (smoothed ends of spurs).
4 Spurs are removed by lateral erosion. Their remains form a line of bluffs on each side of the valley floor.
5 The valley floor is wide and by the time the valley enters the stage of old age it is covered with a layer of sediments.

The building of the flood plain

1 During maturity the valley floor is widened by lateral erosion which is effected by meanders migrating downstream.

2 Active deposition begins to take place on the convex banks of meanders during maturity. Ultimately the whole valley floor is affected as meanders wander across it.

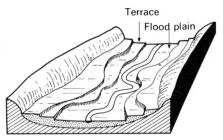

Meander terraces cut into flood plain.

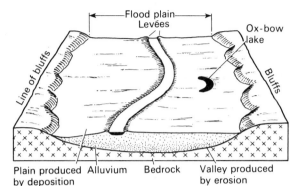

3 After the stage of maturity is reached the river begins to overflow its banks and it deposits fine silts and muds on the valley floor. This is the final stage in the growth of a flood plain.

Note Meanders still migrate downstream and effect both lateral and vertical erosion. In doing so they may remove most of the original flood plain deposits though parts of these remain as terraces above the new flood plain which is developing. These meander terraces are of various heights and those on one side of the valley do not match in height

those on the other side. They are not paired as are terraces formed through *rejuvenation* (page 61).

Characteristic features of a flood plain

1 The river carries a heavy load some of which is deposited on its bed. This may produce mounds which divide the river into several channels. When this happens the river is said to be *braided*.

2 Meanders are pronounced and cut-offs develop and produce *ox-bow lakes*.

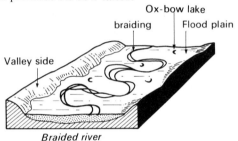

Braided river

3 The river builds up its bed and banks with alluvium (the banks are called *levées*). The river thus flows between pronounced banks and above the level of the flood plain.

Ox-bow river entering Lake Athabasca in Canada

4 The river mouth sometimes becomes blocked
   with sediments and a delta forms.

The Development of an Ox-bow Lake

(i)

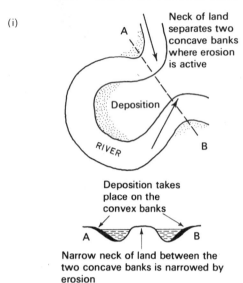

Neck of land separates two concave banks where erosion is active

Deposition takes place on the convex banks

Narrow neck of land between the two concave banks is narrowed by erosion

An acute meander where a narrow neck of land separates two concave banks which are being undercut.

(ii)

The neck is ultimately cut through. This may be accelerated by river flooding

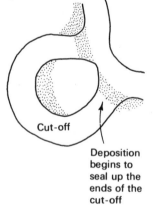

Deposition begins to seal up the ends of the cut-off

Erosion has broken through the neck of land. This often happens when the river is in flood. The meander has been cut off.

(iii)

Deposition seals the cut-off which becomes an ox-bow lake

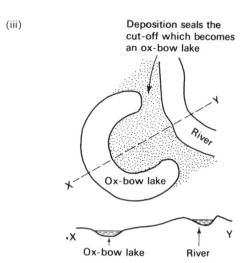

Deposition takes place along the two ends of the cut-off and it is eventually sealed off to form an ox-bow lake.

Note how the area of deposition along the convex banks is increasing. After the formation of the ox-bow lake the river bed and banks are steadily raised by deposition and ultimately the river lies above the level of the ox-bow lake.

Mekong Meander

The photograph shows two pronounced meanders and an ox-bow lake (bottom left). The flood plain is clearly shown. Natural levees occur along both banks of the river. These are fairly conspicuous in the centre of the photograph.

The Formation of Levees and the Raising of the River Bed

1 Active deposition takes place along the banks of an old river when it is in flood. Each time this happens the banks get higher and they are called natural levées.

A meandering river in Sabah (East Malaysia) which has produced numerous ox-bow lakes.

2 When the river is not in flood deposition takes place on the river's bed. The bed is thus raised.

3 In time the river flows between levées and it is above the general level of the flood plain.
The Hwang-ho and Yangtze-kiang in China; the Mississippi* in the U.S.A. and the River Po in Italy all flow above the level of the flood plain in their lower courses.

* The bottom of this river is, however, below sea level.

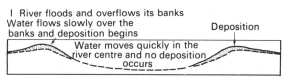

I River floods and overflows its banks
Water flows slowly over the banks and deposition begins
Deposition
Water moves quickly in the river centre and no deposition occurs

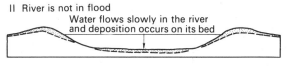

II River is not in flood
Water flows slowly in the river and deposition occurs on its bed

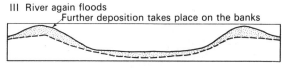

III River again floods
Further deposition takes place on the banks

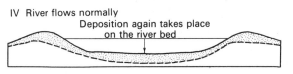

IV River flows normally
Deposition again takes place on the river bed

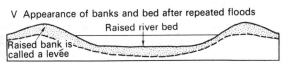

V Appearance of banks and bed after repeated floods
Raised river bed
Raised bank is called a levée

The Influence of Levels and Raised River Beds

When a river flows above the level of its flood plain, its tributaries find difficulty in joining it. They often parallel the river for many miles and in doing so they frequently meander themselves. They may cross depressions in the flood plain which then become swampy. Tributaries whose confluence with the main river is interfered with in this way are called *deferred junctions*.

Rivers which flow above the level of their flood plains are a constant menace. In times of severe floods they sometimes burst through the levées and disastrous floods spread out over the flood plains. The Hwang-ho, Yangtze-kiang and Mississippi all flow above the level of their flood plains in their lower valleys, and all periodically produce disastrous floods.

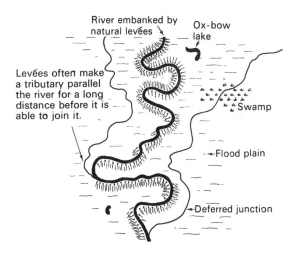

River embanked by natural levées
Ox-bow lake
Levées often make a tributary parallel the river for a long distance before it is able to join it.
Swamp
Flood plain
Deferred junction

Deltas

Most of the load carried by a river is ultimately dumped into the sea or lake into which it flows. The deposited load sometimes collects in the river mouth where it builds up into a low-lying swampy plain called a *delta*. As deposition goes on in the river mouth, the river is forced to divide into several channels each of which repeatedly divides. All these channels are called *distributaries*. Stretches of sea or lake become surrounded by deposited sediments and these are filled in with sediments when they may persist for some time as swamps. *Spits* and *bars* develop along the front of the delta. (Explanations on spits and bars are given later in the book).

There are Three Basic Types of Delta

1 *Arcuate*

This type is very common. It is composed of coarse sediments such as gravel and sand and is triangular in shape (see diagram of Nile Delta on page 56).

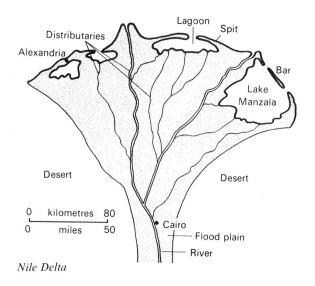

Nile Delta

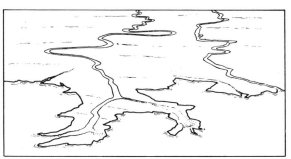

Vardar Delta

It always has a great number of distributaries. Rivers having this type of delta are: Nile, Ganges, Indus, Irrawaddy, Mekong, Hwang-ho and Niger.

2 *Bird's Foot or Digitate.*
This type is composed of very fine sediments called *silt*. The river channel divides into a few distributaries only and these maintain clearly defined channels across the delta. The Mississippi Delta is one of the best examples of this type. This is a diagram of the Vardar Delta. Two main distributaries can be seen. Both of these are flowing between levées which are another characteristic feature of this type of delta. This type occurs in seas which have few currents and tides to disturb the sediments.

3 *Estuarine*
An estuarine delta develops in the mouth of a submerged river. It takes the shape of the estuary. The deltas of the Elbe (Germany), Ob (U.S.S.R.) and the Vistula (Poland) are of this type.

Stages in the Development of a Delta
In stage 1 deposition divides the river mouth into several distributaries. Spits and bars arise and lagoons are formed. The levées of the river extend into the sea via the distributaries.

In stage 2 the lagoons begin to get filled in with sediments, and they become swampy. The delta

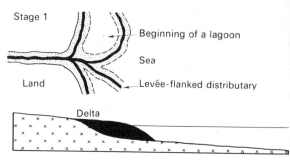

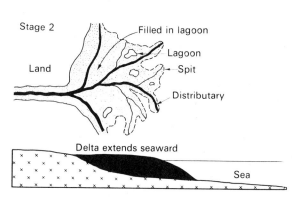

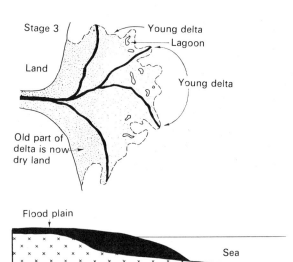

begins to assume a more solid appearance.
In stage 3 the old part of the delta becomes colonised by plants and its height is slowly raised as a result of this. Swamps gradually disappear and this part of the delta becomes dry land.
Note All three stages can often be seen in most deltas. As a delta grows larger and larger, the old parts merge imperceptibly with the flood plain, and

they no longer have the appearance of a delta. Much of the North China Plain is the deltaic plain of the Hwang-ho; the plains of Iraq are the deltaic plains of the Tigris-Euphrates Rivers, and so on.

Conditions necessary for the Formation of a Delta

1 The river must have a large load, and this will happen if there is active erosion in the upper section of its valley.
2 The river's load must be deposited faster than it can be removed by the action of currents and tides.

Note
1 Deltas can, and do form on the shores of highly tidal seas, e.g. River Colorado (Gulf of California), and River Fraser (British Columbia).
2 Any river, irrespective of its stage of development, can build a delta. The Kander, whose valley is in the stage of youth, has built a delta in Lake Thun (Switzerland).

The Rate at which a Delta Grows

The formation of a delta results in an extension of the flood plain. Some deltas grow more rapidly than others. The Mississippi Delta is being extended seawards by 75 metres (245 feet) each year. The River Po (Italy) is extending its delta seawards by 12 metres (39 feet) per year. The advance of the Hwang-ho Delta has resulted in an island becoming joined to the mainland thus forming a peninsula (Shantung).

The delta of the Niger (West Africa)

Some Common Deltas from Asia and Africa

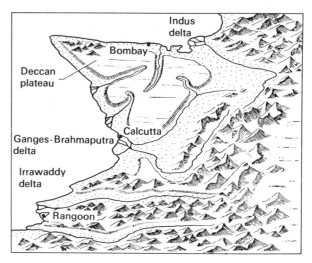

Deltas of India, Bangladesh and Burma

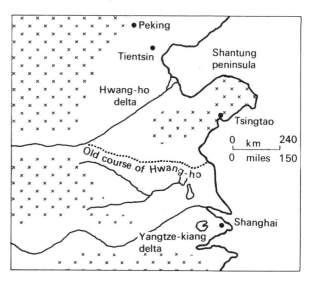

Hwang-ho Delta

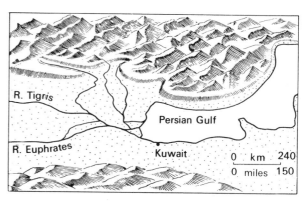

Tigris-Euphrates Delta

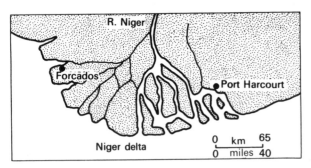

Niger Delta

The Value of Rivers and their Valleys to Man
Rivers
1 Some rivers, especially those in the stage of old age, form natural routeways which can be used for transport. The Yangtze-kiang and Mississippi and Rhine are particularly important.
2 Many rivers can be used for supplying irrigation water to agricultural regions, e.g. Nile, Tigris-Euphrates, Indus and Yangtze-kiang (irrigation will be discussed more fully in a later section).
3 For the development of hydro-electric power (H.E.P.). Young rivers which have waterfalls or which flow through gorges offer possibilities for H.E.P. development, e.g. of H.E.P. sites: *Kariba Dam* across the end of the gorge below the Victoria Falls on the River Zambesi; *Boulder Dam* near the Grand Canyon on the River Colorado; and *Owens Dam* near the Owens Falls on the River Nile where it leaves Lake Victoria. A dam is being built across the *Sanmen Gorge* on the Hwang-ho. Mature rivers can also be used for H.E.P. development, e.g. the dams at *Dnepropetrovsk* on the Dnieper, *Tsimlyanskaya* on the River Don, and at *Kuybyshey* on the River Volga all of which are in the U.S.S.R.
4 Some river mouths contain deep sheltered water and enable ports to be developed there, e.g. *Calcutta* on a branch of the Ganges, *Alexandria* on a distributary of the Nile, *Shanghai* in the delta of the Yangtze-kiang and *New Orleans* in the Mississippi Delta.

Valleys
1 Young valleys extend into highland regions by headward erosion. In doing so they often develop gaps by river capture. These gaps together with river gaps offer fairly easy passageways across the highlands. Roads and railways often take advantage of these.
2 Mature valleys by virtue of their wide floors offer gently sloping routeways to roads and railways across highland areas.
3 Mature valleys also offer good sites for settlements. Bluffs and river cliffs are frequently used as settlement sites.

4 The flood plains and deltas of old valleys contain fertile soils and provide Man with some of the best agricultural land. This is especially true of the subtropics and humid tropics where flood plains and deltas have for long been the home of large populations. It was on the riverine plains of the Nile, Tigris-Euphrates, Indus and Hwang-ho that the early civilisations grew up. All of these except the Hwang-ho have remarkably similar physical environments:
  (i) each is bordered by mountains, or deserts, or both
  (ii) each is open to the sea on one side
  (iii) each has a mild winter, a hot summer and an abundance of river flood water.
In Asia the flood plains and deltas form the most important physical landscapes, because most of the people live here. Wherever temperature and water conditions are suitable they are used for growing rice. The deltas of the Yangtze-kiang (China), Red River (N. Vietnam), Mekong (S. Vietnam), Irrawaddy (Burma) and the Indus (Pakistan) are especially important rice-growing regions. Outside Asia the Nile Delta, the Mississippi Delta and the Rhine Delta are also of great agricultural value.
Flood plains and deltas have their disadvantages. Sometimes their rivers cause serious flooding. This is particularly true when a river flows above the level of its flood plain and bursts its levées. Damage is done to crop land, settlements and communications. There may be considerable loss of life as there was in 1887 when the Hwang Ho burst its banks. Over 1 000 000 people lost their lives on that flood plain and delta in that year. The dangers of such serious flooding can be lessened by strengthening the levées and dredging the river to deepen its channel. The construction of dams across rivers also lessens the danger of flooding e.g. as in the Nile Valley, and if reafforestation is practised around the headwaters, less silt will be brought down.

DRAINAGE PATTERNS
All rivers are joined by smaller rivers or streams which are called *tributaries*. The area drained by a river and its tributaries is known as a *river basin* and its boundary is formed by the crest line of the surrounding highland. This boundary forms the main *watershed* of the basin.
A river system usually develops a pattern which is related to the general structure of its basin. Three distinct river patterns can be recognised. They are
  (i) Dendritic
  (ii) Trellis
  (iii) Radial
A *dendritic* pattern develops in a region which is made of rocks which offer the same resistance to erosion and which has a uniform structure. The

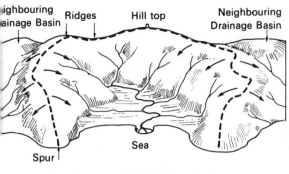

– – – Watershed around the drainage basin
in the centre of the diagram

Diagram of a Watershed

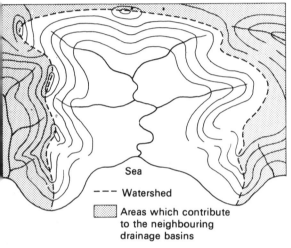

Sea

– – – Watershed

▓ Areas which contribute
to the neighbouring
drainage basins

directions of the river and its tributaries are determined by slope. The land between main valleys and
between these and tributary valleys stands up as
ridges and spurs. The crests of these form the
watersheds. The river and its tributaries make a
pattern like the veins of a leaf.

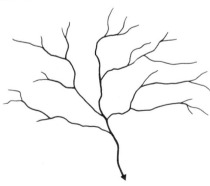

Dendritic drainage

A *trellis* drainage pattern develops in a region which
is made up of alternate belts of hard and soft rocks
which all dip in the same direction and lie

at right angles to the general slope, down which
the principal river flows. The tributaries of the
main river extend their valleys by headward erosion
into the weak rocks which are turned into wide
valleys, called *vales*, and the hard rocks stand up
as *escarpments*.

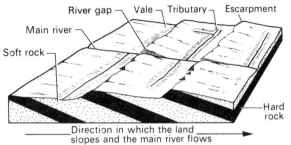

River gap   Vale   Tributary   Escarpment
Main river
Soft rock
Hard rock

Direction in which the land
slopes and the main river flows
The main river cuts through the hard rock forming
the escarpment and a river gap is produced

The principal river which flows down the slope is
called a *consequent river* (C), and the tributaries
which cut out the vales and which do not flow
down the main slope are called *subsequent rivers*
(S).

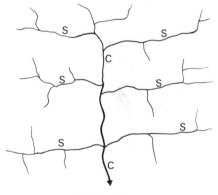

Trellis drainage

Dendritic Drainage

59

A *radial* drainage pattern develops on a dome or volcanic cone. The rivers flow outwards forming a pattern like the spokes of a wheel.

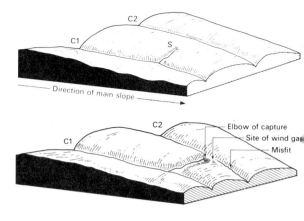

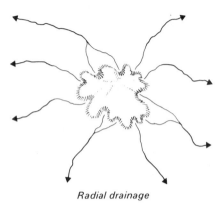

Radial drainage

River Capture

These diagrams show how river capture can take place. C_1 and C_2 are consequent rivers and S is a subsequent tributary of C_1. If S is a more powerful river than C_2 then it will lower its valley more rapidly than will C_2. This in turn will cause its tributary (S) to effect headward erosion and in time it will cut back into the valley of C_2 and effect river capture. When this happens the upper part of C_2 will flow into S and hence into C_1 thus making it an even more powerful river. The lower part of C_2 is now deprived of its headwaters and its volume decreases and it becomes too small for its valley. It is called a *misfit*. At the elbow of capture the valley now contains no river and it becomes a *wind gap*.

This process can be repeated until several parallel rivers have had their headwaters captured by which time C_1 will be the dominant river of the region.

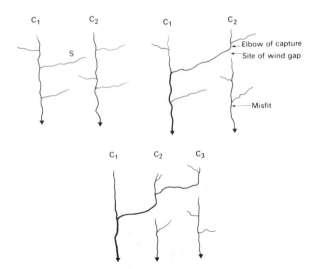

Superimposed drainage patterns

Some rivers have developed a drainage pattern which is in no way related to the structure of the region in which it occurs. This happens when the drainage pattern was developed on a surface which overlay the present one. As the drainage pattern developed it cut its way into the underlying rock surface without regard for its structure and we say that it became superimposed on it. Gradually the whole of the original surface may be removed by erosion. In time the drainage pattern, especially the tributaries become affected by structure. The following diagrams show the development of a superimposed drainage pattern.

1 *Original folded surface*

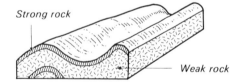

2 *Region is reduced to a peneplain*

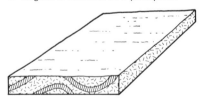

3 *Subsidence results in region being buried by newer rocks but subsequent uplift se the formation of a drainage pattern. The main rivers are draining at right ang to the axis of the original structure.*

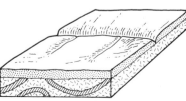

4 *Tributaries to the main river develop wide valleys in the weaker rocks. As the main river erodes vertically it cuts across the ridges of strong rock and forms gorges. The strong rock forms ridges because the weak rocks are worn away and not because the region has been uplifted.*

Rejuvenation

When the base level (page 47) of a river is lowered. the river's power to erode is increased. Most of the erosion is downwards into the valley floor and as a result new types of features are produced. A river which is given this extra power is, in effect, given new life, that is, it is *rejuvenated*.

The lowering of a river's base level may be caused by an uplift of the land or by a fall in the level of the sea. The latter happens when water is withdrawn from the seas to be locked up in ice masses when the climate of some regions gets colder, for example, as in ice ages.

A river in any stage of development from youth to old age may be rejuvenated which means that the valley eroded by a rejuvenated river, which will be a young valley, may occur in an old landscape. For example, if a river crossing a peneplain is rejuvenated, its valley will have the typical V-shape of a young valley.

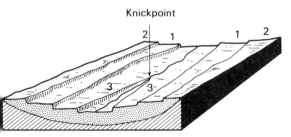

Paired rejuvenation terraces (11, 22, 33) and knickpoint.

Some features produced by rejuvenated rivers

1 If a river on a flood plain is rejuvenated, the downcutting effected by the river will produce terraces. These will be paired, that is the heights of the terraces on one side of the river will correspond with the heights of the terraces on the other side. These terraces are different from meander terraces (page 53). It sometimes happens that the point where the river crosses from the original flood plain to the new flood plain is visible and is marked by rapids or a waterfall. This point is called the *knickpoint*.

2 If a river is able to maintain its course over a region of uplift then the river will rapidly deepen its valley. If the rocks of the region are resistant and if the climate is dry, then a very impressive gorge can develop. The Grand Canyon is a good example (page 51).

EXERCISES

1 Carefully distinguish between the following:
  (i) weathering by temperature changes and weathering by frost action
  (ii) a scree and an exfoliation dome.

2 What are the main differences between each of the following:
  (i) a porous rock and a pervious rock
  (ii) a permeable rock and an impermeable rock
  Name *one* specific example of each type of rock and say which condition, if it occurs in surface rocks, is likely to promote flooding.

3 Briefly describe the manner in which a land surface can be changed by: (i) temperature changes, (ii) rain, (iii) running water. Illustrate your answer with annotated diagrams and give specific examples.

4 Draw fully labelled diagrams to show the differences between: (i) a spring, and (ii) a well. Write brief notes on each and state in what type of region either, or both, may be found.

5 By means of well-labelled diagrams, and supporting text, show how (i) an oasis, and (ii) an artesian basin, may occur.

6 Explain, by well-labelled diagrams, the main differences between the following pairs of features:
  (i) waterfall and rapid
  (ii) distributary and tributary
  (iii) radial and dendritic drainage pattern.

7 Describe the characteristic features which commonly occur in a river valley during the stages of youth, maturity and old age. Illustrate the more important features by means of well-labelled diagrams.

8 Name *two* features produced by river erosion and *two* features produced by river deposition. For each feature state:
  (i) the type of valley in which it occurs
  (ii) where it occurs in that valley.
  For *one* of the erosional and *one* of the depositional features which you have named, draw a well-labelled diagram to show its main characteristics and say how the feature may have formed.

9 Choose any *two* pairs of the following features and for *each*:
  (a) say how the features develop, (b) name *one* valley where the features occur, (c) draw well-labelled diagrams of the features.
  (i) levée and flood plain
  (ii) matched river terraces and unmatched river terraces
  (iii) ox-bow lake and cut-off

Objective Exercises

1 Which of these statements best describes the meaning of weathering?

  A the burrowing action of animals and the growth of plant roots

  B the freezing of water in cracks in rocks

  C the break-up of rocks exposed at the surface

  D the alternate heating and cooling of rocks

  E the dissolving out of mineral particles by rain water

           A B C D E

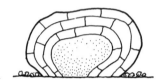

2 The diagram above shows a common type of weathering in a hot desert. This type is known as

  A attrition

  B deflation

  C extrusion

  D solution

  E exfoliation

           A B C D E

3 A well will always contain water if

  A it is sunk at the bottom of a hill

  B the bottom of the well is far below the water-table

  C it is sunk into sedimentary rocks

  D it is located in a rainy region

  E it is on a spring line

           A B C D E

4 All the following features are produced by denudation. Which one is produced by rain action?

  A gully

  B gorge

  C earth pillar

  D exfoliation dome

  E cliff

           A B C D E

5 The formation of a river delta involves the following processes. In what order do they take place?

  A transport, corrasion, deposition

  B corrasion, transport, deposition

  C deposition, corrasion, transport

  D corrasion, deposition, transport

  E deposition, transport, corrasion

           A B C D E

6 The erosive power of a river depends most upon its

  A width and depth

  B speed and depth

  C gradient and width

  D speed and width

  E speed and volume

           A B C D E

7 Deposition by a river increases when

  A the volume of water in the river increases

  B the river enters a gorge

  C the river flows more quickly

  D the river enters a lake

  E the river approaches a waterfall

           A B C D E

8 Chemical weathering takes place most effectively when it is

  A warm and wet

  B cold and dry

  C cold and wet

  D hot and wet

  E dry all the time

           A B C D E

9 Under which one of the following conditions is physical weathering most effective?

  A hot and dry all the time

  B hot and wet all the time

  C hot season alternating with dry season

  D large diurnal temperature range

  E even temperatures all the time

           A B C D E

10 Which of these features clearly indicates the presence of a flood plain?

  A seasonal floods

  B ox-bow lakes

  C meanders

  D distributaries

           A B C D

11 Most rivers flow slowly near to sea level and in consequence their main action is depositional. A river in this stage would **not** show signs of

  A a wide flat-floored valley

  B deposits of large boulders

  C river levees

  D tributaries with deferred junctions

           A B C D

12 Which one of the following combinations of natural conditions would best promote the formation of a delta at the mouth of a heavily loaded river (assuming there is no subsidence)?

A a saline sea with strong currents
B a fresh-water sea with negligible currents
C a fresh-water sea with strong currents
D a saline sea with negligible currents

A B C D

13 A drainage system which is in no way related to the structure of the region where it occurs is called a

A radial drainage pattern
B trellis drainage pattern
C superimposed drainage pattern
D dendritic drainage pattern

A B C D

14 All the following features **except** one may suggest that a drainage system has been rejuvenated

A knick-points
B paired river terraces
C flat-floored valleys
D incised meanders

A B C D

15 "The wearing away of the sides and bottom of a river's channel by the load carried by a river" is called

A corrosion
B attrition
C corrasion
D transportation

A B C D

16 Which one of the following statements best explains why the lower course of a river is sometimes choked with sediments

A The valley of a river is widest in its lower course.
B The velocity of a river in its lower course is low.
C A delta sometimes develops in a river's lower course.
D Waterfalls rarely occur in the lower course of a river.

A B C D

17 A region which has a trellis drainage pattern usually has the following features

A swallow holes and dry valleys
B wind gaps and vales
C erratics and moraines
D clints and grikes

A B C D

18 Which one of the following best explains the formation of canyons?

A a large volume of water
B greater vertical erosion by a glacier of a main valley than of the tributary valleys
C pronounced Earth movements
D vertical erosion in the valley of a river crossing an arid plateau

A B C D

19 The load of a river comes mainly from

A the banks which are undercut by the river on a meander
B the valley sides down which rock particles are moved by soil creep
C the bluffs which are undercut by the river in its upper course
D the river's bed which has been abraded by the action of 'pot-holing'

A B C D

20 The diagram above is a drawing of a meandering river. Using this diagram suggest an area where an ox-bow lake is likely to be formed.

A where the river current slows down
B where deposition is taking place along a meander
C where the meander neck is cut off
D at a meander core

A B C D

21 Which one of the following features is the product of river deposition?

A scree
B earth pillar
C landslide
D levée

A B C D

5 Limestone and Chalk Features

Limestone Landscape

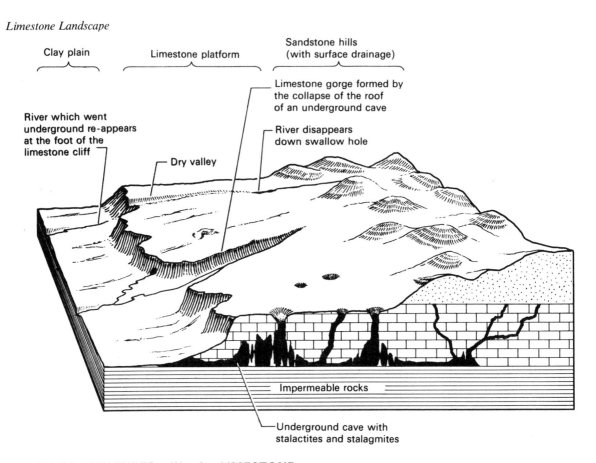

Clay plain | Limestone platform | Sandstone hills (with surface drainage)

Limestone gorge formed by the collapse of the roof of an underground cave

River which went underground re-appears at the foot of the limestone cliff

River disappears down swallow hole

Dry valley

Impermeable rocks

Underground cave with stalactites and stalagmites

DRAINAGE FEATURES IN A LIMESTONE REGION

The nature of limestone

Limestone consists chiefly of calcium carbonate which is insoluble. The carbon dioxide, which rain water absorbs from the air, turns the insoluble carbonate into soluble bicarbonate. This is the reason why rain water and rivers are able to remove limestone in solution.

$$CaCO_3 + CO_2 + H_2O = Ca(HCO_3)_2$$
carbonate $\qquad\qquad$ bicarbonate

Limestone is a well-jointed rock and its joints and bedding planes soon become opened up by rain and water, and in time the surface consists of broken and rugged rocks.

Limestone landscape

One of the most noticeable features of a limestone landscape is the almost complete absence of surface drainage. The permeability of limestone permits rain to soak into it very easily. Joints rapidly become excavated and deepened, with the result that the surface becomes criss-crossed with wide irregular gullies, known as *grikes*. The intervening blocks of limestone surface are called *clints*.

Rivers rising in a non-limestone region sometimes flow into a limestone region. When this happens the rivers disappear into vertical holes in the surface and continue to flow as *underground rivers* inside the limestone. The vertical holes, called *swallow holes* or *sink holes*, are formed by rivers and they are usually widened vertical joints. Gaping Ghyll in Yorkshire, England, is a particularly good example. Swallow holes may join together to give a very large opening, called a *doline*. Likewise, dolines may join up to give even larger openings. These are called *uvala*.

Rivers which flow inside limestone develop underground caves and caverns as they flow along joints and bedding planes. Some caves are of great size. e.g. Carlbad Cave (New Mexico – U.S.A.) is 1200 metres (3950 feet) long, 183 metres (600 feet) wide and 90 metres (295 feet) high. Batu Caves, near

Gaping Gill near Ingleborough, Yorkshire, England.

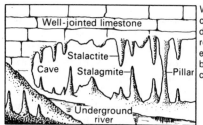

Well-jointed limestone

Stalactite

Cave — Stalagmite — — Pillar

Underground river

Water containing calcium bicarbonate drips from the cave roof. When the water evaporates it leaves behind calcium carbonate

Interior view of a limestone cave

Kuala Lumpur and the caves near Ipoh are further examples. *Stalagmites* and *stalactites* develop in these caves and sometimes they join together to form natural columns or pillars.
Sometimes the roof of an underground cave or cavern collapses and a gorge, with almost vertical sides, then develops. Cheddar Gorge (the U.K.) was formed in this way.

Rivers which disappear underground when entering a limestone region reappear on the surface again where the junction of the limestone and the underlying impermeable rocks meet the surface. Dry, gorge-like valleys often mark the former courses of such rivers and these occur between the point of disappearance and the point of emergence (diagram. page 64). The former course over the limestone in European regions was probably made possible by the frozen subsoils in Glacial Times. Dry waterfalls also occur in these valleys, especially where the rivers once crossed limestone escarpments.

The surface of a limestone region is not only broken, it is also stony. Any soil which may occur is usually

Interior of Cango Caves (South Africa)

in small shallow patches which support only a few shrubs, grasses and in some regions sweet-smelling herbs. Larger plants, such as trees, only occur in the bottom of large valleys which have been excavated down to the rocks underlying the limestone. Although the limited plant life in limestone regions varies from region to region, it being dependent upon the nature of the climate, the general appearance of all limestone regions is very much the same. The limestone region around Ipoh, in Perak (Peninsular Malaysia), is well-covered with vegetation because of the fairly deep soils which have formed under humid tropical weathering.

Limestone landscapes are called *karst* landscapes and good examples occur in north-west Yugoslavia, the Pennines of the U.K., the Yucatan Peninsula of Mexico, the Kentucky region of the U.S.A. and parts of Perak and Perlis in Peninsular Malaysia.

Value of Karst regions to Man

Because of their barren nature karst regions contain few settlements. The dryness of the surface and the limited amounts of poor soils prevent the growth of a continuous plant cover. In some regions there is sufficient grass to support sheep or goats and animal grazing takes place. Occasionally areas of good soils do occur. These are usually confined to basins which have been formed by the collapse of roofs of underground caverns. In Yugoslavia and other parts of the Mediterranean region, these soils are usually red and are called *terra rossa*. They are valuable for farming.

Limestone is quarried as a building stone and for making cement, and usually there are stone and cement works near to limestone regions, e.g. near to Ipoh in Peninsular Malaysia.

Features of a chalk landscape

Chalk, like limestone, is made of calcium carbonate but it is much softer than limestone. Its surface is not marked by outcrops of hard rock. Instead it is usually gently undulating with rounded hills, called *downs* in England, and wide open valleys, which are usually without rivers. Chalk is a porous rock and rain falling on its surface rapidly soaks into the ground. There is, therefore, very little surface run-off, that is, there are very few streams. Because the valleys are without streams, they are called *dry valleys* or *coombs*.

Good examples of chalk landscapes occur in England in the Chiltern Hills and the Downs, and in these regions dry valleys are very common. These valleys were obviously formed when the water-table was higher than it is at present. Possibly, towards the end of the last glacial period, vast quantities of melt water from the retreating ice sheets were able to flow as rivers across these chalk regions, because

the subsoils were frozen, thus presenting an im permeable zone.

EXERCISES

1 Briefly distinguish between the following:
   (i) a dry valley and an underground river
   (ii) a limestone gorge and a swallow hole.
   (iii) a clint and a grike
   Name *one* region where these types of features may be seen.
2 Carefully explain why (i) some underground river produce varied underground scenery, (ii) most limestone areas have little agriculture and few people, and (iii) there is almost no surface drainage in a limestone region. Illustrate your answer with relevant diagrams.

Objective Exercises

1 Which one of the following is characteristic of a limestone region?
A dry valleys
B meandering streams
C deep soils
D good cover of natural vegetation
E salt marshes

A   B   C   D   E

2 All of the following features are likely to occur in a limestone region **except**
A underground rivers
B ox-bow lakes
C dry valleys
D clints
E caves

A   B   C   D   E

3 A karst landscape is most likely to develop in a region whose rocks are
A porous
B impermeable
C chiefly made of calcium carbonate
D well jointed

A   B   C   D

4 A doline is
A a large crack produced by erosion on the surface of a limestone plateau
B a large swallow hole produced by 'solution'
C an underground cave
D the opening through which an underground river re-emerges at the surface.

A   B   C   D

Features Produced by Wind

WIND ACTION AND THE FEATURES IT PRODUCES

Wind action is very powerful in arid and semi-arid regions where rock waste is produced by weathering and is easily picked up by the wind. In humid regions rock particles are bound together by water droplets and plant roots, and hence there is little wind erosion.

Wind erosion consists of *abrasion* which breaks up rocks and produces rock pedestals, zeugens, yardangs and inselbergs, and *deflation* which blows away rock waste and thus lowers the desert surface producing depressions, some of which are very extensive.

Wind transport causes fine particles of rock waste, called desert dust, to be carried great distances. Coarser particles called sand are bounced over the surface for short distances.

Wind deposition gives rise to dunes which are made of sand and loess which is made of desert dust.

Types of tropical and temperate desert surface

Most of the world's deserts are located in latitudinal belts of 15° to 30° north and south of the equator.

a) West Coast Desert

These mainly occur in the trade wind belt on the western sides of continents where the winds are off-shore. On-shore local winds do blow across these coasts but they rarely, if ever, bring rain because they have to cross cool currents which parallel the coasts in these latitudes. The cool currents cause condensation to take place in the on-shore winds which produces mist, fog or light rains. By the time they reach the coasts the winds are dry.

b) Continental Deserts

These occur in the interior of continents where the winds have already travelled a considerable distance across the land and in doing so, have lost much of their moisture as happens when winds blow over a dry land or over high mountains. The day temperatures of these continental areas are very high because they are so far from the moderating influence of the sea. The night temperatures are low because the absence of clouds causes these areas to lose their heat rapidly. Deserts of this type occur mainly in west coast tropical latitudes but they do extend into the temperate zones. Sometimes these deserts occur in intermontane plateau regions. The deserts of Arizona and Nevada in the Rockies are of this type and they are sometimes known as Mountain Deserts. A desert often occurs because of a combination of these factors e.g. the Sahara and the Australian Desert are partly of the west coast type and partly of the continental type.

Although the dominant agent of erosion in deserts is wind, rain does occur which results in both water erosion and water deposition. The following types of desert surface can be recognised.

Sandy Desert – called *erg* in the Sahara and *koum* in Turkestan. This is an undulating plain of sand produced by wind deposition.

Stony Desert – called *reg* in Algeria and *serir* in Libya and Egypt. The surface is covered with boulders and angular pebbles and gravels which have been produced by diurnal temperature changes.

Rocky Desert – called *hamada* in the Sahara. The bare rock surface is formed by deflation which removes all the small loose rock particles. A part of the Sahara Desert in Libya has extensive areas of this type of desert.

Badlands – This is quite a different type of desert to the three just mentioned in that it develops in semi-desert regions mainly as a result of water erosion produced by violent rain storms. The land is broken by extensive gullies and ravines which are separated by steep-sided ridges. Excellent examples of this type of desert occur in the region extending from Alberta to Arizona (North America).

Features produced by Wind Erosion

Wind abrasion attacks rock masses and sculptures them into fantastic shapes. Some of these, because of their shape, are called rock pedestals.

I Rock Pedestals

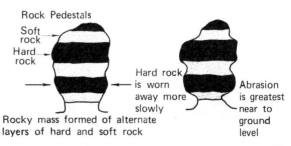

Rock Pedestals
Soft rock
Hard rock
Hard rock is worn away more slowly
Abrasion is greatest near to ground level
Rocky mass formed of alternate layers of hard and soft rock

Rock Pedestals in the Lut Desert in Iran

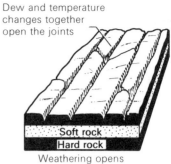

Dew and temperature changes together open the joints

Weathering opens up the joints

II Zeugens

Wind abrasion turns a desert area which has a surface layer of hard rock underlain by a layer of soft rock into a 'ridge and furrow' landscape. The ridges are called zeugens. A zeugen may be as high as 30 metres (100 feet). Ultimately they are undercut and gradually worn away.

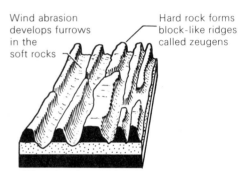

Wind abrasion develops furrows in the soft rocks

Hard rock forms block-like ridges called zeugens

Wind abrasion continues the work of weathering

III Yardangs

Bands of hard and soft rocks which lie parallel to the prevailing winds in a desert region are turned into another type of 'ridge and furrow' landscape by wind and abrasion. **The belts of hard rock stand up as rocky ribs up to 15 metres (50 feet) in height and they are of fantastic shapes. They are called yardangs. Yardangs are very common in the central Asian deserts and in the Atacama Desert.**

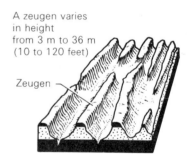

A zeugen varies in height from 3 m to 36 m (10 to 120 feet)

Zeugen

Wind abrasion slowly lowers the zeugens and widens the furrows

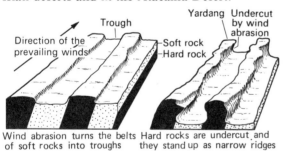

Direction of the prevailing winds

Trough

Yardang Undercut by wind abrasion

Soft rock
Hard rock

Wind abrasion turns the belts of soft rocks into troughs

Hard rocks are undercut and they stand up as narrow ridges called yardangs

IV Depressions

Some depressions produced by wind deflation reach down to water-bearing rocks. A swamp or an oasis then develops.

The Qattara Depression is 122 metres (400 feet) below sea level. It has salt marshes and the sand excavated from it forms a zone of dunes on the leeside.

Fault-produced depression

The formation of a depression may first be caused by faulting. The soft rocks thus exposed are excavated by wind action.

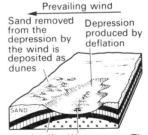

Prevailing wind

Sand removed from the depression by the wind is deposited as dunes

Depression produced by deflation

SAND

Aquifer Water seeps out of aquifer and forms swamp or an oasis

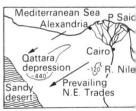

Mediterranean Sea
Alexandria P Said

Cairo

Qattara depression
-440

R. Nile

Sandy desert

Prevailing N.E. Trades

The Qattara Depression is 122 m (400 ft) below sea level. It has salt marshes and the sand excavated from it forms a zone of dunes on the leeside.

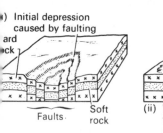

(i) Initial depression caused by faulting

Hard rock

Faults · Soft rock

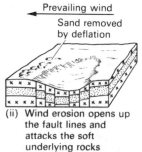

Prevailing wind

Sand removed by deflation

(ii) Wind erosion opens up the fault lines and attacks the soft underlying rocks

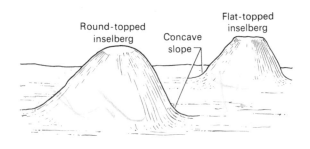

Round-topped inselberg · Concave slope · Flat-topped inselberg

Inselberg

In some desert regions erosion has removed all of the original surface except for isolated pieces which stand up as round-topped masses, called inselbergs. Some of them are probably the remains of plateau edges which have been cut back by weathering after which the weathered rock waste has been removed by sheet wash. Others may be the result of wind erosion or the combined action of wind and water erosion.

Inselbergs are common in the Kalahari Desert, parts of Algeria, north-west Nigeria and Western Australia.

Features produced by Wind Deposition

Strong winds occasionally blow across desert surfaces and when they occur they carry vast amounts of desert dust (very fine sand particles). This movement produces dust storms and in time it results in the transport of enormous quantities of fine material from one part to another part of a desert or from a desert to a neighbouring region. Slight movements of air, called *wind eddies*, bounce grains of sand forward, which, when deposited, may form gentle ripples or sandy ridges, called *dunes*.

Sand dunes

There are two types of dune, *barchan* (barkhan) and *seif*. A barchan is crescent-shaped, lies at right angles to the prevailing wind and is much more common than a seif which is long and usually straight, and which is parallel to the prevailing wind. A barchan usually develops from the accumulation of sand caused by a small obstruction such as a rock or a bit of vegetation. As the mound of sand grows bigger its two edges are slowly carried forward down-wind and the typical crescent shape develops. The windward face of the dune is gently-sloping but the leeward side is steep and is slightly concave. This is caused by wind eddies which are set up by the prevailing wind (see diagram). A barchan moves slowly forward as grains of sand are carried up the windward face and slip down the leeward side. Barchans range in height from a few metres to over thirty metres and they may occur singly or in groups. There are good examples

Ayers Rock, Northern Territory, Australia. This inselberg is 988 metres (1050 yards) long, 1610 metres (1 mile) wide and 340 metres (1100 feet) high.

in the deserts of the Sahara and Turkestan.

A *seif dune* forms when a cross wind develops to the prevailing wind and the corridors between the dunes are swept clear of sand by this wind. Eddies blow up against the sides of the dunes and it is these which drop the sand grains and thus build up the dunes. The dunes are lengthened by the prevailing wind. Seif dunes are often several hundred metres high and many kilometres long. Some even reach a length of 160 kilometres (100 miles). Good examples of seif dunes occur in the Thar Desert, the desert of Western Australia and south of the Qattara Depression.

Loess Deposits

The wind blows fine particles out of the deserts each year. Some are blown into the sea, the rest are deposited on the land where they accumulate to form loess. Loess is friable and easily eroded

69

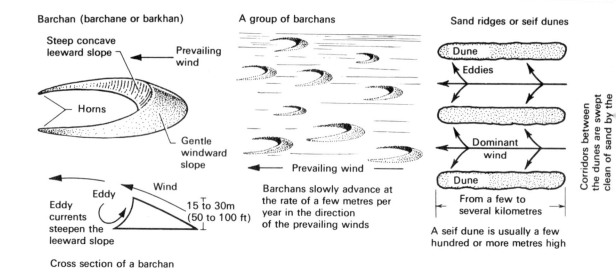

Barchan (barchane or barkhan)

Steep concave leeward slope — ← Prevailing wind

Horns

Gentle windward slope

Wind

Eddy

Eddy currents steepen the leeward slope

15 to 30m (50 to 100 ft)

Cross section of a barchan

A group of barchans

← Prevailing wind →

Barchans slowly advance at the rate of a few metres per year in the direction of the prevailing winds

Sand ridges or seif dunes

Dune

Eddies

Dominant wind

Dune

From a few to several kilometres

Corridors between the dunes are swept clean of sand by the

A seif dune is usually a few hundred or more metres high

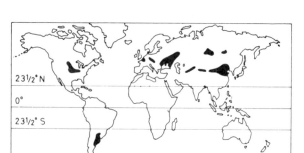

23½° N

0°

23½° S

Loess regions of the world

In northern China the loess has been intense[ly] eroded by rivers to give a 'badland' landscap[e]. The photograph on the right shows deep gorg[e]-like valleys which have been cut into the loes[s]. The centre of the photograph shows a loess pla[in] which is under cultivation.

Many of the loess hills are terraced for crop[s]. Most of the people in this loess region live in house[s] which have been carved out of the loess cliffs.

by rivers. There are extensive loess deposits in Northern China. These are formed of desert dust blown out of the Gobi Desert to the west. The loess deposits of central Europe were probably deposited in the last Ice Age when out-blowing winds carried fine glacial dust from the ice-sheets of northern Europe. The loess deposits of the Pampas have been derived from the deserts to the west.

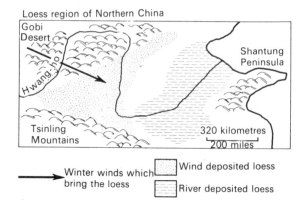

Loess region of Northern China

Gobi Desert

Hwang ho

Shantung Peninsula

Tsinling Mountains

320 kilometres
200 miles

→ Winter winds which bring the loess

Wind deposited loess

River deposited loess

Loess cave dwellers of Shansi Province in Northern Chin[a]. These cave dwellings are warm in winter and cool in summe[r], however just slight earthquakes can cause serious landslid[es] and the consequent loss of many lives

WATER ACTION IN DESERTS

A desert region may receive no rain for several yea[rs] and then a sudden downpour of from 100 to 25[0] millimetres (3·9 to 9·75 in) occurs. These rare b[ut] heavy rain storms give birth to rushing torrents o[n] steep slopes and sheet flood water on gentle slope[s]. The run-off on steep slopes is usually via *rills* (sha[llow] ow grooves) which lead into gullies which in turn connect with steep-sided, deep and often flat-floored valleys, called *wadis* or *chebka* (in Algeria). During sudden rain storms, flood waters rush down wadis as *flash floods*. These are short-lived and the large quantities of sediment which they carry are deposited in bulk giving rise to alluvial fans and delta-like deposits at the foot of steep slopes, e.g. where a tributary joins a wadi, and where the flood waters dry up, e.g. at the exit of a wadi.

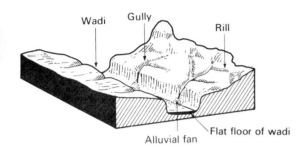

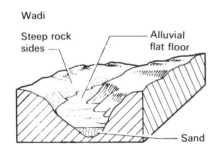

In intermontane desert basins, e.g. Tarim Basin, intermittent rivers drain into the centre of the basin. The alluvial fans which build up around the edge of such a basin may eventually join together to form a continuous depositional feature, sloping gently to the centre of the basin. This feature is called a *bahada* or *bajada*. Sometimes the centre of a basin is occupied by temporary salt lakes, caused by high evaporation, e.g. the shotts of North Africa and playas of North America. In some cases the lakes

Occasional rainfall gives rise to intermittent rivers

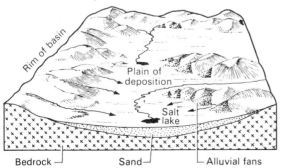

Crescent Dunes in North Africa

dry up and only the salt beds of these remain. These are called *salinas*.

Intermontane Desert Basin

As the edges of desert and semi-desert highlands get pushed back by erosion and weathering, a gently sloping platform develops. This is called a *pediment*. The slope of the land changes abruptly where a pediment joins the highland mass.

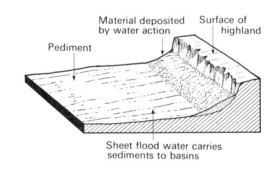

Wadis and Rocky Ridges in Jordan

71

Mount Sinai in Egypt

Reg of Sinai

Deserts and rivers

Most deserts are regions of inland drainage, i.e. their rivers and streams never reach the sea. Very few rivers persist throughout the year in deserts but there are some significant exceptions. The Nile, in North Africa, the Tigris-Euphrates, in Southwest Asia, and the Colorado, in the U.S.A., are three of the best examples of rivers which cross desert areas and which are permanent rivers. This is because these rivers rise in regions of rain which falls throughout the year and which is sufficient to sustain a permanent flow of water across the desert areas.

The Value of Wind-blown Deposits to Man

1 Loess deposits are unusually fertile. In northern China the loess region has been cultivated for 4000 years. In the Ukraine (north of the Black Sea), in the Pampas of Argentina, and in the High

This map shows those deserts where wind action is important.

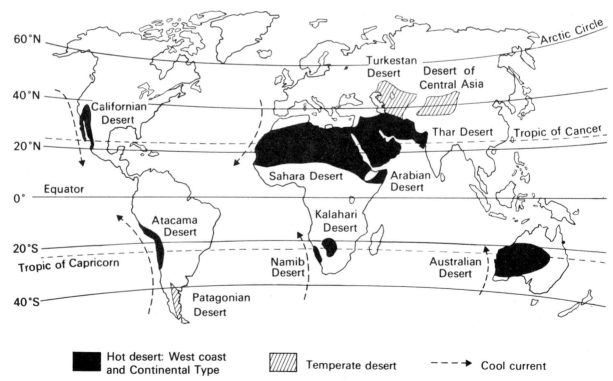

Plains of the U.S.A., the loess soils are very fertile and are used for grain growing.

Loess deposits of both China and Europe are used as a means for building dwellings. These are cut into the deposits. Their chief advantages are the ease with which they are built and their warmness in winter and their coolness in the summer. Their main disadvantage is their instability (they often collapse during even mild earthquakes).

EXERCISES

For each of any *three* of the following features:
(a) rock pedestal
(b) yardang
(c) zeugen
(d) inselberg
(e) deflation depression
  (i) draw a large diagram to show the main characteristics of the feature, and name these
  (ii) carefully explain how the feature was produced
  (iii) name *one* region where an example of the feature may be seen.

2 Name *two* types of depositional feature which develop under arid conditions and for *one* of these briefly explain how it forms.

Objective Exercises

1 Wind can effect erosion through the process of
A abrasion
B exfoliation
C extrusion
D solution
E attrition

A B C D E

2 Which one of the following features is **not** formed by wind erosion
A yardang
B rock pedestal
C zeugen
D barchan
E hamada

A B C D E

3 Which of the following features has been produced by wind deflation?
A Lake Toba (Sumatra)
B Lake Chad
C Lake Baikal
D Lake Victoria
E Qattara Depression

A B C D E

4 Loess consists of fine rock particles which
A are often deposited on the flood plains of large river valleys
B frequently occur in regions adjacent to deserts which lie under winds blowing from the deserts
C result from the weathering of limestone rocks
D are deposited by waves along arid coastlines
E are formed from the weathering of volcanic rocks

A B C D E

5 Which one of the following landforms is **not** the product of denudation in a hot arid region?
A seif
B reg
C grike
D erg

A B C D

6 Although hot deserts have a very low annual rainfall, occasional heavy rain storms do occur and these sometimes produce steep-sided, flat-floored valleys called
A gorges
B wadis
C canyons
D dry valleys

A B C D

7 Two types of plains frequently occur between the borders of desert basins and the surrounding highlands. They are called bahadas (bajadas) and pediments. Bahadas are
A alluvial fans which have joined together to form a sloping plain around the desert basin
B located at the entrance to a wadi
C sloping rock platforms produced by wind erosion
D the steep slopes of the surrounding highlands

A B C D

8 A pediment is
A a rocky surface
B made of loose gravel
C the remnant of receding mountain slopes
D composed of deposits of sediments

A B C D

9 Which of the following rivers crosses an extensive desert before it reaches the sea?
A Mississippi
B Amazon
C Hwang-ho
D Colorado

A B C D

7   Features Produced by Waves

SEA ACTION AND THE FEATURES IT PRODUCES

Coasts are forever changing—some are retreating under wave erosion and others are advancing under wave deposition. There are many different types of coasts; some are steep, some are gentle, some are sandy, some are rocky. The character of a coast results from two or more of the following factors:

1 Wave action
2 Nature of the rocks forming the coast
3 Slope of the coast
4 Changes in the level of sea or land
5 Volcanic activity
6 Coral formations
7 The effects of glaciers
8 The action of man.

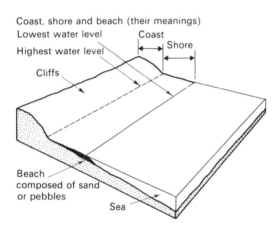

Coast, shore and beach (their meanings)
Lowest water level
Coast
Highest water level
Shore
Cliffs
Beach composed of sand or pebbles
Sea

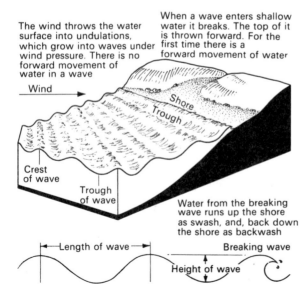

The wind throws the water surface into undulations, which grow into waves under wind pressure. There is no forward movement of water in a wave

When a wave enters shallow water it breaks. The top of it is thrown forward. For the first time there is a forward movement of water

Wind
Shore
Trough
Crest of wave
Trough of wave

Water from the breaking wave runs up the shore as swash, and, back down the shore as backwash

Length of wave
Height of wave
Breaking wave

Definition of Terms

Find out the meanings, of *coast*, *shore* and *beach* by studying the diagrams on the left. The highest water level refers to the level reached by the most powerful storm waves. The *coastline* is the margin of the land. This is also the *cliff line* on rocky coasts. The height and power of a wave depend upon the strength of the wind and the *fetch* (distance of open water over which it blows). The stronger the wind and the greater the fetch, the more powerful the wave. Storm waves are particularly powerful.

How waves are caused

Waves are caused by winds and the following diagrams show how this takes place, how a wave grows and the movement inside a wave.

1 In this diagram the wind, which is shown as four layers, blows over a sea surface. The surface exerts a frictional drag on the bottom layer, and this layer exerts a frictional drag on the layer above it, and so on. The top layer has the least drag exerted on it which means that the layers of air move forward at different speeds. The air tumbles forward and finally develops a circular motion. This motion exerts downward pressure (D.P.) on the surface at its front and an upward pressure (U.P.) at its rear. The surface begins to take on the form of a wave.

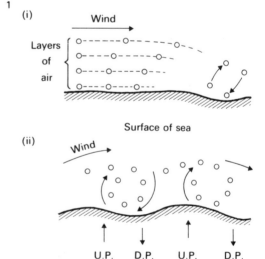

1

(i)

Wind

Layers of air

Surface of sea

(ii)

Wind

U.P.   D.P.   U.P.   D.P.

2 This diagram shows the wind pressing on the back of a developing wave, thus causing it to steepen. The back of the wave tumbles forward but it moves back later and slows the forward movement of the front of the wave. This causes the wave to grow bigger.

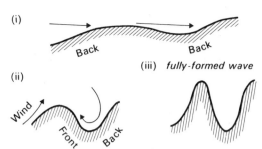

(i) Back Back

(ii) Wind Front Back

(iii) *fully-formed wave*

3 This diagram shows the four component movements in a wave. Any particle of water at A moves to A1, to A2, to A3 and then back to A. It is therefore the wave form which moves and not the water.

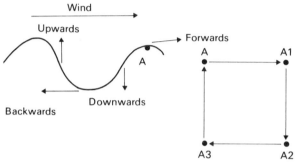

Wind

Upwards

Forwards

A

Downwards

Backwards

A   A1

A3   A2

4 When a wave enters shallow water its top falls forward and its water is thrown forward. The water thrown up the beach by breaking waves is called the *swash*. When the swash drains back down the beach it is called the *backwash*.

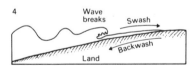

4   Wave breaks   Swash   Backwash   Land

Types of waves

The swash of a wave is always more powerful than the backwash because it has the force of the breaking wave behind it. When waves break at a rate of ten or less a minute, each breaking wave is able to run its course without interference from the wave behind it. These waves are called *constructive waves*. When waves break more frequently, especially over fifteen a minute, then the backwash of a wave runs into the swash of the wave behind. The force of the swash is therefore reduced in comparison with the force of the backwash. These waves remove pebbles and sand from a coast. They are *destructive waves*.

Waves as Agents of Erosion

Wave erosion consists of three parts:

1 *Corrasive action:* boulders, pebbles and sand are hurled against the base of a cliff by breaking waves and this causes undercutting and rock break-up.

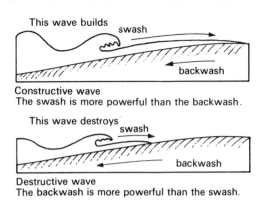

This wave builds   swash   backwash

Constructive wave
The swash is more powerful than the backwash.

This wave destroys   swash   backwash

Destructive wave
The backwash is more powerful than the swash.

2 *Hydraulic action:* Water thrown against a cliff face by breaking waves causes air in cracks and crevices to become suddenly compressed. When the wave retreats the air expands; often explosively. This action causes the rocks to shatter as the cracks become enlarged and extended.

3 *Attrition:* Boulders and pebbles dashed against the shore are themselves broken into finer and finer particles.

Features Produced by Wave Erosion
Cliffs and Wave-cut Platforms
The Strandflat off the west coast of Norway is a good example of a wave-cut platform. This platform is over 50 kilometres (30 miles) wide.

Stages in the development of a cliff and wave-cut platform

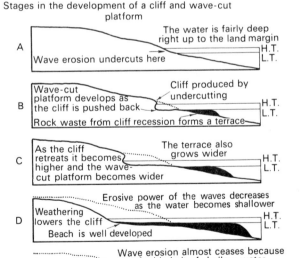

A   The water is fairly deep right up to the land margin   H.T. L.T.
Wave erosion undercuts here

B   Wave-cut platform develops as the cliff is pushed back   Cliff produced by undercutting   H.T. L.T.
Rock waste from cliff recession forms a terrace

C   As the cliff retreats it becomes higher and the wave-cut platform becomes wider   The terrace also grows wider   H.T. L.T.

D   Weathering lowers the cliff   Erosive power of the waves decreases as the water becomes shallower   H.T. L.T.
Beach is well developed

E   Cliff has almost disappeared   Wave erosion almost ceases because of wide belt of shallow water   Wave-cut platform is buried under beach deposits   H.T. L.T.

H.T.   High tide   L.T.   Low tide

Types of Cliffs

The rocks of some cliffs are in layers which slope landwards (fig. A). In other cliffs the rock layers slope seawards and blocks of rock loosened by erosion easily fall into the sea. These cliffs are often very steep and overhanging (fig. B). Landslides are quite common on cliffs more especially on those formed of alternate layers of pervious and impervious rocks.

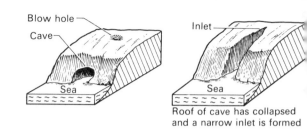

Roof of cave has collapsed and a narrow inlet is formed

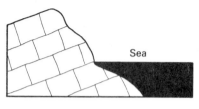

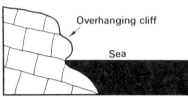

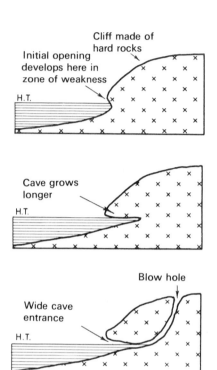

A Blow Hole in Scotland

Part of the Devon Coast of S.W. England

Caves, Arches and Stacks

These are minor features produced by wave action during the process of cliff formation. A cave develops along a line of weakness at the base of a cliff which has been subjected to prolonged wave action. It is a cylindrical tunnel which extends into the cliff, following the line of weakness, and whose diameter decreases from the entrance. If a joint extends from the end of the tunnel to the top of the cliff, this becomes enlarged in time and finally opens out on the cliff top to form a *blow hole*. The roof of the cave ultimately collapses and a long narrow sea inlet forms.

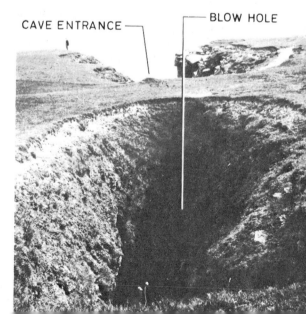

76

Caves which develop on either side of a headland such that they ultimately join together, give rise to a natural *arch*. When the arch collapses, the end of the headland stands up as a *stack*. In time this is completely removed by wave erosion.

The diagrams below show the stages in the development of an arch and a stack.

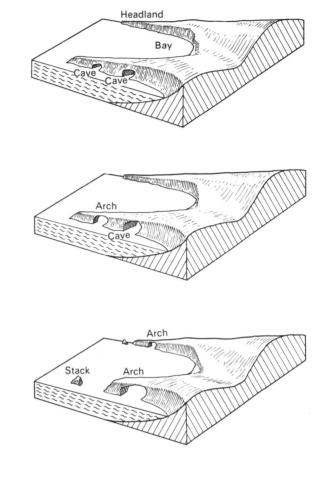

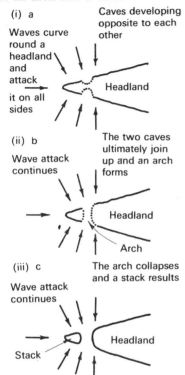

(i) a

Waves curve round a headland and attack it on all sides

Caves developing opposite to each other

Headland

(ii) b

Wave attack continues

The two caves ultimately join up and an arch forms

Headland

Arch

(iii) c

Wave attack continues

The arch collapses and a stack results

Stack

Headland

Natural stack and bay on the Gaspé Coast of Quebec (Canada)

Wave Transport

All the material carried by breaking waves is called the *load*. Some of this comes from rivers entering the sea, some from landslides on cliffs, and the rest comes from wave erosion. The load consists of mud, sand and shingle.

The swash and backwash push and drag material up and down the beach. When waves break obliquely to the shore the swash moves obliquely up the beach but the backwash runs back at right angles to the shore as shown in *fig. c*. You will see from this diagram that material is gradually carried along the beach by these two actions which together constitute *longshore drift*. The removal of material by longshore drift can be stopped by building groynes or walls out to sea (*fig. b*).

Material is also carried off-shore into deeper water by the *undertow*. This is an under current which balances the piling up of water along the coast by breaking waves and high tides.

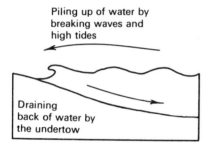

Fig. a

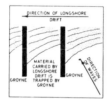

Fig. b

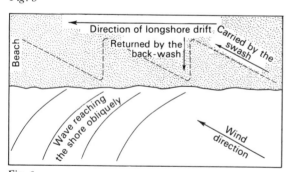

Fig. c

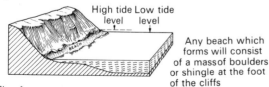

Fig. d

Wave-deposited material on the Devon Coast, England

Erosion is Dominant on a Highland Coast. Examine *fig. d*. The only depositional feature is a narrow beach.

Deposition is Dominant on a Lowland Coast. Examine *fig. e* and compare it with *fig. d*. Along the highland coast there is deep water close to the land.

The photograph above shows that the sea, like the wind and rivers, sorts its load on deposition. Moving down the beach the sequence of deposits is boulders, pebbles, sand and mud. The coast is gently sloping. This is shown by the wide expanse of beach.

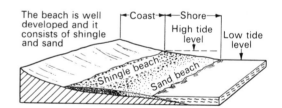

Fig. e

Wave Depositional Features

The chief of these are:
 (i) Beaches
 (ii) Spits and bars
 (iii) Mud flats.

Beach: The main action of constructive waves is to deposit pebbles, sand and mud, which, when deposited along a coast, form a gently sloping platform, called a beach. The material of which a beach is composed is transported along a coast by longshore drift. Beaches usually lie between high and low water levels but storm waves along some coasts throw pebbles and stones well beyond the normal level reached by waves at high tide. The material deposited in this way produces a ridge called a *storm beach*.

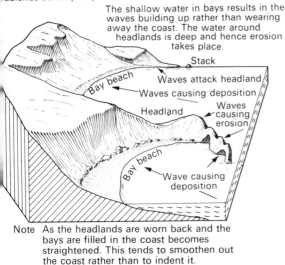

ys and coves between
adlands develop bay beaches

The shallow water in bays results in the waves building up rather than wearing away the coast. The water around headlands is deep and hence erosion takes place.

Stack
Waves attack headland
Waves causing deposition
Bay beach
Headland
Waves causing erosion
Bay beach
Wave causing deposition

Note As the headlands are worn back and the bays are filled in the coast becomes straightened. This tends to smoothen out the coast rather than to indent it.

Fig. f

Wave action in bays is usually not strong and deposition is the dominant action. Beaches called *bay-head beaches* develop at the heads of bays (*fig. f*). These beaches do not extend to the headlands where wave erosion is dominant. Good examples of beaches occur along the east coast and parts of the west coast of Peninsular Malaysia. A particularly fine example of bay-head beaches occurs between Port Dickson and Cape Rachado on the west coast of Peninsular Malaysia.

Spit: Material which is eroded from a coast may be carried along the coast by longshore drift and deposited further along the coast as a spit or bar. This is likely to happen along indented coasts and coasts broken by river mouths.

A spit is a low, narrow ridge of pebbles or sand joined to the land at one end with the other end terminating in the sea. A spit sometimes develops at a headland and projects across a bay. As waves swing into the bay obliquely, the end of the spit becomes curved or hooked (*fig. g*).

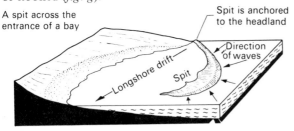

A spit across the entrance of a bay
Spit is anchored to the headland
Direction of waves
Longshore drift
Spit

Fig. g

When longshore drift operates across a river's mouth a zone of slack water develops between longshore drift and the river and any material carried by longshore drift is deposited. The deposited material forms a spit which may, in time, extend across the mouth of the river. When this happens the river's

outlet may be diverted (*fig. h*) or the river's mouth may be converted into a *lagoon* (*fig. i*). A good example of a spit across a river's mouth occurs at the mouth of Sungai Kelantan (Peninsular Malaysia). A spit may also develop across a bay.

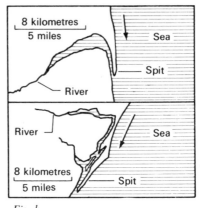

8 kilometres
5 miles
Sea
Spit
River

River
Sea
8 kilometres
5 miles
Spit

Fig. h

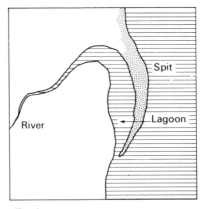

Spit
River
Lagoon

Fig. i

Bar: This is very similar to a spit. A common type of bar is that which extends right across a bay. This starts as a spit growing out from a headland but ultimately it stretches across the bay to the next headland. This type of bar is called a *bay-bar*. Many bay-bars do have breaks in them where tidal

Chesil Beach in S.W. England

Examples of some Bars

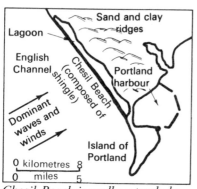

Chesil Beach is really a tombolo because it joins an island to the mainland.

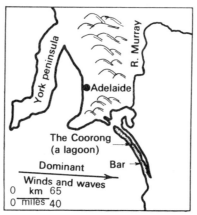

Coast of South Australia

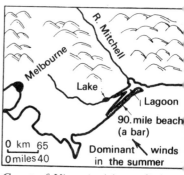

Coast of Victoria (Australia)

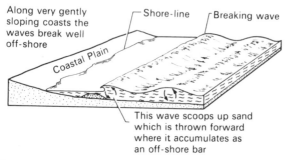

action prevents the bar from being continuous. Along the coast of Poland, bay-bars are called *nehrungs*. When a bar links an island to the mainland it is called a *tombolo*. Chesil Beach, in southern England, is a very good example.

Off-shore Bar: This develops only along very gently sloping coasts such as the southern part of the Atlantic Coast of North America. The two diagrams at the bottom of the page will help you to see how an off-shore bar can be built by wave action.

Mud flats: Tides tend to deposit fine silts along gently shelving coasts, especially in bays and estuaries. The deposition of these silts together, perhaps, with river alluvium, results in the building up of a platform of muds called a mud flat. Salt-tolerant plants soon begin to colonise the mud flat which in time becomes a swamp or marshland. In

tropical regions, mud flats often become mangrove swamps. Mud flats are usually crossed by winding channels kept clear of vegetation by tidal action. At low tide these channels often contain little, if any, water.

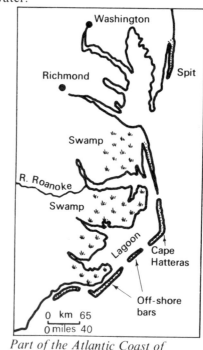

Part of the Atlantic Coast of North America

Along very gently sloping coasts the waves break well off-shore

This wave scoops up sand which is thrown forward where it accumulates as an off-shore bar

The off-shore bar has become wider and higher. The water between the bar and the shore-line is called a lagoon

In time the lagoon becomes filled in with sediments forming swamp or marsh and finally dry land

80

TYPES OF COASTS

Neither the level of the land nor the level of the sea remains unchanged for long periods of time. During the last Ice Age the level of the sea was lower than it is today because large quantities of water were locked up in the ice masses which covered extensive parts of Europe and North America. The Ice Age was followed by a gradual return to warmer conditions and the ice sheets melted and their waters returned to the sea. The level of the sea rose and some coastal regions were submerged. Land masses too change their level. Sometimes coastal regions were depressed; sometimes they were uplifted. After the disappearance of the ice sheets parts of the land masses were slowly uplifted through the removal of the great weight of ice.

It will be seen that coastal regions may be either submerged or uplifted by changes in land or sea levels. There are therefore two basic types of coast: *submerged* and *emerged*. Each can be sub-divided into highland and lowland types to give:

Submerged Coasts
1 Highland type
2 Lowland type

Emerged Coasts
1 Highland type
2 Lowland type

Submerged Highland Coasts
There are three main types:
1 Ria Coast
2 Longitudinal Coast
3 Fiord Coast

Ria Coast: When a highland coast is submerged the lower parts of its river valleys become flooded. These submerged parts of the valley are called rias. Such rias are common in S.W. Ireland, S.W. England, N.W. Spain, and Brittany.
Due to submergence the coast becomes indented and the tips of headlands may be turned into islands.

Longitudinal Coast: When a highland coast whose valleys are parallel to the coast is submerged, some of the valleys are flooded and the separating mountain ranges become chains of islands. These valleys are sometimes called *sounds*, e.g. Puget Sound in Washington (U.S.A.). This type of coast occurs in Yugoslavia and along parts of the Pacific coasts of North and South America.

Fiord Coast: When glaciated highland coasts become submerged the flooded lower parts of the valleys are called *fiords*. The three diagrams on page 82 show how the fiords have developed. During

Ria Coast

Before submergence

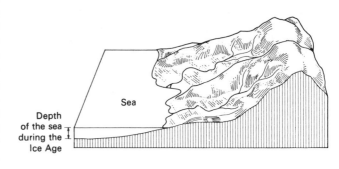

After submergence

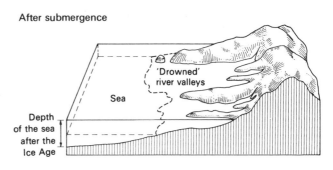

Longitudinal Coast

Before submergence

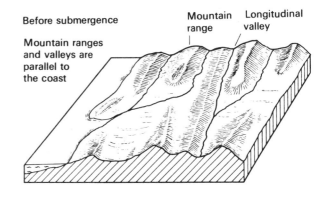

After submergence

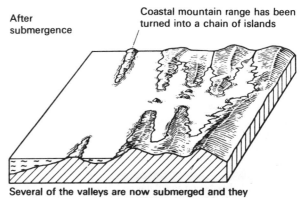

Several of the valleys are now submerged and they form long narrow inlets parallel to the coast

81

glaciation the river valleys become widened and deepened. After the glaciers have disappeared and the sea has risen the steep-sided valleys are 'drowned'. Notice that the water inside the fiord is much deeper than it is at the entrance of the fiord. Fiords have steeper sides and deeper water than rias. All the fiord coasts lie in the belt of prevailing westerly winds and are on the western sides of land masses. It was in these regions that vast amounts of snow and ice accumulated in the Ice Age. Some of the best examples of fiord coasts occur in Chile, South Island of New Zealand, Greenland, Norway and British Columbia.

Ria of the River Yealm on the South Devon Coast

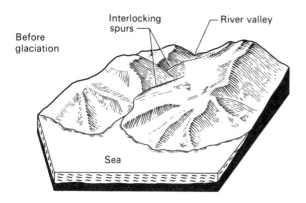

Before glaciation

Interlocking spurs

River valley

Sea

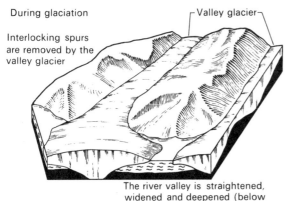

During glaciation

Interlocking spurs are removed by the valley glacier

Valley glacier

The river valley is straightened, widened and deepened (below sea level) by the glacier

Trollfiord in Northern Norway. Compare and contrast this with the photograph above

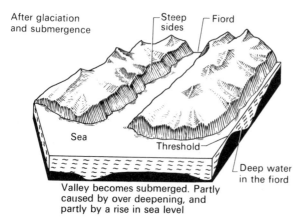

After glaciation and submergence

Steep sides

Fiord

Sea

Threshold

Deep water in the fiord

Valley becomes submerged. Partly caused by over deepening, and partly by a rise in sea level

The Value of Rias and Fiords to Man

1 Both rias and fiords often provide natural 'harbours'.

2 It is often extremely difficult to get inland from the head of a fiord because of the mountainous country. A fiord, therefore, is not very useful as a site for a port.

3 It is usually easy to get inland from the head of a ria and because of this a ria is sometimes the site of a port.

4 Settlement is difficult along the sides of a fiord because there is little or no level land. Fiord settlements occur at the head of a fiord where there is level land.

Wave Action Alters a Submerged Highland Coast

I. In the beginning

Ria or 'drowned' valley

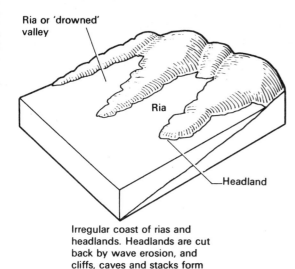

Ria

Headland

Irregular coast of rias and headlands. Headlands are cut back by wave erosion, and cliffs, caves and stacks form

II. Stage of youth

Spit

Bay beach

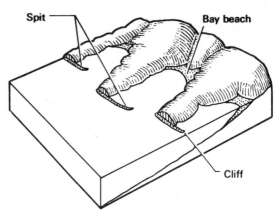

Cliff

Wave deposition is more important than erosion. Spits and bay beaches are formed. The coast is becoming straighter for erosion of the headlands still goes on

III. Stage of early maturity

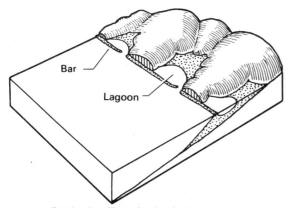

Bar

Lagoon

Erosion is still cutting back the headlands. Bars extend across the bays which are now turned into lagoons. These are being filled in with sediments and marshes form

IV. Stage of late maturity

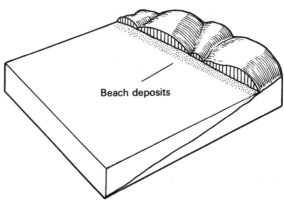

Beach deposits

The coast is now cut back beyond the heads of the bays, and it is now almost straight

Submerged Lowland Coast

A rise in sea level along a lowland coast causes the sea to penetrate inland along the river valleys, often to considerable distances. The flooded parts of the valleys are called *estuaries*. Marshes, swamps and mud flats can often be seen in estuaries at low tide. The Baltic coasts of Poland and Germany and the Dutch coast are good examples of estuarine coasts.

Emerged Highland Coast

An old sea beach backed by a sea cliff lying from 7·5 metres to 30 metres (25 to 100 feet) above sea level often characterises this type of coast. These two features could only have been produced by sea action, but since the sea no longer reaches them, it is evident that there has been a change in either sea level or the level of the land. Raised beaches are common in western Scotland.

OLD SEA CLIFFS

OLD SEA BEACH

PRESENT DAY SEA CLIFFS

PRESENT DAY BEACH

Raised Beach in Scotland

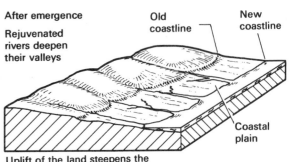

After emergence

Rejuvenated rivers deepen their valleys

Old coastline

New coastline

Coastal plain

Uplift of the land steepens the gradients of the rivers and they deepen their valleys. The rivers are said to be rejuvenated (made young again)

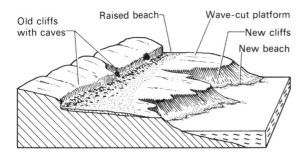

Old cliffs with caves

Raised beach

Wave-cut platform

New cliffs

New beach

Emerged Lowland Coast

This forms when a part of the continental shelf emerges from the sea and forms a coastal plain. The coast has no bays or headlands and deposition takes place in the shallow water off-shore, producing off-shore bars, lagoons, spits and beaches. Examples of this type occur along the south-east coast of the U.S.A. and the north coast of the Gulf of Mexico. The development of ports is difficult.

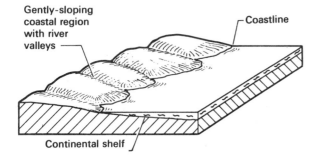

Before emergence

Gently-sloping coastal region with river valleys

Coastline

Continental shelf

EXERCISES

1 Briefly explain the meanings of the following terms:
   (a) swash
   (b) backwash
   (c) longshore drift
   (d) constructive wave
   (e) beach
   Choose any *three* features, and for *each*:
   (i) draw a well-labelled diagram to illustrate its main characteristics
   (ii) say how it develops
   (iii) state the type of coast where it occurs.

2 Choose *three* of the following physical features:
   (a) cliff
   (b) headland
   (c) stack
   (d) cave
   (e) blow hole
   For *each* feature
   (i) describe its appearance and mode of formation
   (ii) name and locate an area where an example may be seen.
   Illustrate your answer with relevant diagrams.

3 Describe the main destructive and constructive processes at work along coasts and make a list of some of the features which these processes produce. You should give specific examples where possible and your answer should be illustrated with well-labelled diagrams.

4 (a) By using well-annotated diagrams, state the main differences between the following coastlines:
   (i) ria coast
   (ii) fiord coast
   (iii) longitudinal coast
   (iv) estuarine coast
   (b) For *each* type of coastline, name *one* region where an example of it may be seen.

84

Objective Exercises

1 Which one of the following features is **not** produced by wave erosion?
A headland
B stack
C beach
D cliff
E blow hole

A B C D E

2 The base of a cliff is undercut by rocks and sand being hurled at it by breaking waves. This process is called
A attrition
B corrasion
C solution
D hydraulic action
E abrasion

A B C D E

3 Various coastal features are formed before a stack is finally produced. In what order do the features produced take place?
A cave, headland, arch, stack
B headland, cave, arch, stack
C arch, cave, headland, stack
D headland, arch, cave, stack
E cave, arch, headland, stack

A B C D E

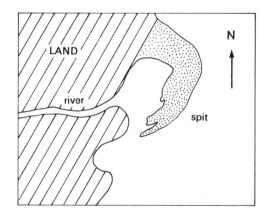

This is a diagram of a coastal area.

4 This direction of longshore drift as suggested by the spit is mainly
A north to south
B south to north
C west to east
D east to west
E south-west to north-east

A B C D E

5 All of the following features are produced by wave deposition **except**
A sand bar
B beach
C stack
D spit
E mud flat

A B C D E

6 Which one of the following features is **not** predominantly due to wave deposition?
A off-shore bar
B beach
C mud flat
D spit

A B C D

7 Which of the coasts listed below is predominantly a fiord coast
A coast of Norway
B western coast of South America
C western coast of England
D western coast of Australia

A B C D

8 A coastline which exhibits drowned river valleys is called a
A fiord coast
B submerged coast
C ria coast
D lowland coast

A B C D

9 Raised beaches usually occur along
A emerged highland coasts
B emerged lowland coasts
C submerged highland coasts
D submerged lowland coasts

A B C D

10 The two features marked '1' and '2' in the diagram below represent respectively
A stack and cliff
B headland and beach
C arch and stack
D headland and stack

A B C D

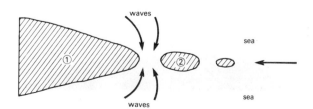

85

8   Coral Reefs and Islands

CORAL COASTS

Nature of coral

It is a limestone rock made up of the skeletons of tiny marine organisms called *coral polyps*. The tube-like skeletons in which the organisms live extend upwards and outwards as the old polyps die and new ones are born. Coral polyps cannot grow out of water and they are therefore formed below the level of low tide. Coral polyps thrive under these conditions:

1 sea temperature of about 21°C (70°F).
2 sunlit, clear salt water down to a depth of about 55 metres (180 feet).

Extensive coral formations develop between 30°N. and 30°S. especially on the eastern sides of land masses where warm currents flow near to the coasts. They do not develop on the western coasts in these latitudes because of the cool currents which flow along these coasts.

Types of Coral Formation

Coral masses are called *reefs* and there are three types:

1 *Fringing Reef*:   a narrow coral platform separated from the coast by a lagoon which may disappear at low water.
2 *Barrier Reef*:   a wide coral platform separated from the coast by a wide, deep lagoon.
3 *Atoll*:   a circular coral reef which encloses a lagoon.

Structure of a Coral Reef

Most reefs are fairly narrow and the coral platform lies near to low water level. The seaward edge is steep and pieces of coral broken off by wave action are thrown up onto the platform where they form a low mound. On the landward side of this the breaking waves deposit sand in which the seeds of plants, such as coconut, readily germinate. Coral atolls in the Pacific are of this type.

Section across a Coral Reef

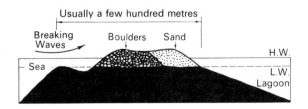

Fringing Reef

This type of reef consists of a platform of coral which is connected to, and which is built out from, a coast. The surface of the platform is usually flat or slightly concave and its outer edge drops away steeply to the surrounding sea floor. A shallow lagoon usually occurs between the coast and the outer edge of the reef.

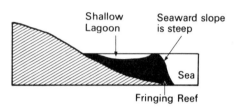

Barrier Reef

This reef is similar to a fringing reef except that it is situated several miles off the coast and is separated from it by a deep water lagoon. The coral of a barrier reef is often joined to the coast although the lagoon may be too deep for coral to grow on its bed. The origin of reefs is discussed later and this offers a possible explanation to this phenomenon. Some barrier reefs lie off the coasts of continents, e.g. Great Barrier Reef along the east coast of Australia. Others occur around islands forming a continuous reef except for openings on the leeward side. In some cases, fringing reefs develop on the inner side of lagoons which lie between a barrier reef and the coast of the island that it encircles.

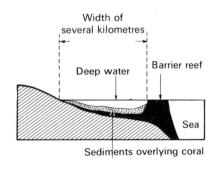

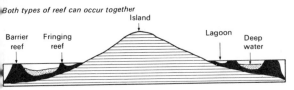
Both types of reef can occur together

Atoll

The diagrams below and on the right show drawings of a typical atoll. Atolls are particularly common in the Pacific and Indian Oceans. Some atolls are very large, e.g. that of Suvadiva in the Maldives. The lagoon of this atoll is 64 kilometres (40 miles) across and its reef extends for 190 kilometres (120 miles). The Gilbert and Ellice Islands of the Pacific are all atolls.

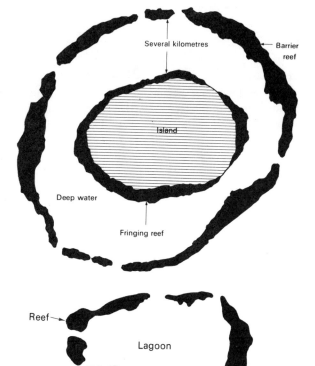

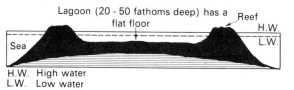
H.W. High water
L.W. Low water

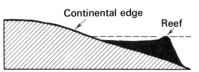

An Atoll in the Pacific Ocean

The Origin of Coral Reefs

Many theories have been put forward to explain the origin of reefs, and in common with so many theories explaining the possible origin of other landforms, no one can say which is the true one.

All that we can do is to examine these theories and accept that which sounds the most reasonable.

Theory I (Daly's Theory). This is based on the changing level of the sea during and after the last Ice Age.

Fringing Reef   *Barrier Reef*   *Atoll*

Before the Ice Age

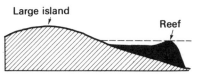

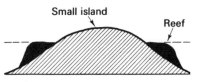

During the Ice Age

As sea level fell wave erosion removed most of the reef.

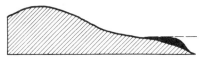

As sea level fell wave erosion cut a terrace into the reef and the edge of the island.

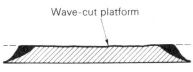

Wave-cut platform

As sea level fell wave erosion removed the top of the island and turned it into a wave-cut platform.

After the Ice Age

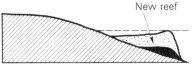

New reef

A rising sea level plus a return to warmer conditions caused the reef to grow up and form a barrier reef.

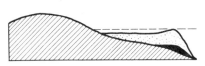

A rising sea level plus a return to warmer conditions caused the reef to grow up again.

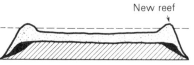

New reef

A rising sea level plus a return to warmer conditions caused the reef to grow up again and form an atoll.

Theory II (Darwin's Theory). This depends on the subsidence of land masses.

As the island subsides the coral reef grows upwards and outwards keeping pace with the subsidence.

According to this theory fringing reefs pass into barrier reefs which in turn pass into atolls. This theory was first suggested by Charles Darwin. The two theories mentioned above can only be used with reference to coral island reefs.

Fringing Reef

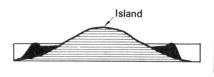

Island

Barrier Reef

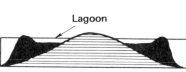

Lagoon

Atoll

Coral broken from the reef by the waves is deposited inside the reef where it forms the floor of the lagoon

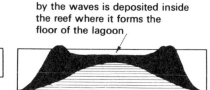

EXERCISES

1 Select *two* of the following coral features: atoll, fringing reef and barrier reef, and for *each* feature selected:
   (i) briefly explain how it may have originated and describe its characteristics
   (ii) name *one* example, or *one* region where an example may be seen
   (iii) draw a well-labelled diagram to illustrate its appearance.
2 Choose *two* of the following physical features: a volcano, a barrier reef, a delta and a raised beach, and for *each* feature chosen:
   (i) describe its appearance
   (ii) draw an annotated diagram to show its main characteristics
   (iii) name *one* example, or *one* region where an example may be seen
   (iv) suggest how it was formed.

Objective Exercises

1 What is the name given to an almost circular coral reef inside which is a lagoon?
   A fringing reef
   B barrier reef
   C atoll
   D coral island

   A B C D
   ·· ·· ·· ··

2 Which one of the following conditions is **not** important for the growth of coral?
   A wave-free salt water
   B clean salt water
   C warm seas (about 21°C)
   D plenty of sunlight

   A B C D
   ·· ·· ·· ··

9   Features Produced by Glaciers

ICE ACTION AND THE FEATURES IT PRODUCES

When the temperature of the air falls below 0°C (32°F) some of the water vapour condenses and freezes into ice crystals which fall to earth as *snow*. Many regions in the high latitudes receive snow in the winter season but in most of them the snow melts in the following summer. If some of it fails to melt then a perpetual cover of snow results. This happens in Greenland, Antarctica and on the tops of some high mountains. The level above which there is a perpetual snow cover is called the *snowline*. The height of this ranges from sea level around the Poles to 4800 metres (15 700 feet) in the mountains of East Africa which are on the equator. When the accumulation of snow in a region increases from year to year it gradually turns into ice by its own weight. About 1 000 000 years ago the climates of regions in the high latitudes began to get colder and colder and not all of the winter snowfall melted in the following summers. The accumulations of snow increased in area and in depth in the polar regions, in the northern part of North America and the north-western part of Europe. The snow of these vast *snow fields* gradually turned into ice which extended over most of the lowlands and some of the mountains. Masses of ice which cover large areas of a continent are called *ice sheets*, and those which occupy mountain valleys are called *valley glaciers*. Today ice sheets occur in Antarctica and Greenland, and valley glaciers in the Himalayas, Andes, Alps and Rockies. The period when the high latitudes were buried beneath ice sheets is known as the *Ice Age*. With the return to warmer conditions most of the ice melted. However, there are still extensive regions around the Poles and smaller areas in the mountain systems named above which still have glaciers, and these regions are therefore still in the Ice Age.

Ice action greatly changes the appearance of a region. Highlands are subjected to erosion and lowlands to deposition. In many parts of the northern continents which are now free from ice, striking features of both glacial erosion and deposition can be clearly seen. With the melting of the ice at the end of the Ice Age enormous quantities of water were set free. Some of this collected in hollows or was held back by glacial deposits and formed lakes. The Great Lakes of North America and the lakes of Finland were formed in this way. Most of the melt waters, however, flowed as rivers into the sea.

These rivers carried large quantities of morainic deposits which were later spread over the land outside the regions which lay under the ice. Here the deposits formed extensive plains called *outwash plains*. These are usually very sandy.

GLACIAL EROSION AND THE FEATURES IT PRODUCES

Glacial Erosion. Consists of two processes: (i) *plucking* (the tearing away of blocks of rock which have become frozen into the base and sides of a glacier), and (ii) *abrasion* (the wearing away of rocks beneath a glacier by the scouring action of the rocks embedded in the glacier).

Erosional Features. The most important of these are: U-shaped valley; hanging valley; cirque (corrie); arête; pyramidal peak. These features are chiefly produced by valley glaciers. Study diagram on page 91.

VALLEY GLACIERS

Snow falling in mountainous regions accumulates in depressions on mountain slopes which are not facing the sun. As the snow accumulation increases, a large part of it is turned to ice and there comes a time when there is more ice and snow than the depression can hold and some of it moves downslope to lower levels.

The movement of ice in the depression causes considerable erosion on the floor and on the sides of the depression. Eventually the depression is turned into a deep hollow. Erosion on the depression's floor is achieved by abrasion of the heavy accumulation of boulders embedded in the base of the ice mass. The floor becomes concave and the edge of the depression becomes ridge-like. This is called the *lip*. On the sides and back of the depression a different type of erosion takes place. The ice becomes frozen to the sides, especially to the back wall and when the ice moves forward it pulls pieces of rock out of the wall. This action is called plucking and it results in the back wall becoming very steep. When fully formed the hollow has the appearance of an arm chair and it is called a *corrie* or *cirque*. Sometimes corries develop on adjacent mountain slopes and when fully formed only a knife-edge ridge, called an *arête* separates them. If corries develop on all sides of a mountain then the top of the mountain is reduced to a jagged peak, called

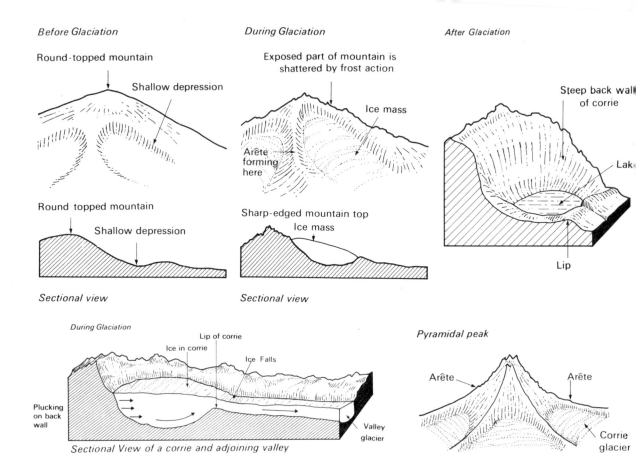

Before Glaciation

Round-topped mountain

Shallow depression

During Glaciation

Exposed part of mountain is shattered by frost action

Ice mass

After Glaciation

Steep back wall of corrie

Arête forming here

Lake

Round topped mountain

Shallow depression

Sharp-edged mountain top

Ice mass

Lip

Sectional view

Sectional view

During Glaciation

Lip of corrie

Ice in corrie

Ice Falls

Plucking on back wall

Pyramidal peak

Arête

Arête

Valley glacier

Corrie glacier

Sectional View of a corrie and adjoining valley

a *pyramidal peak*, by the steepening of the back walls of the corries. The arêtes between the corries are pronounced and these, together with the pyramidal peak, are sharpened by frost action.

Ultimately, climatic changes may cause the ice to melt and disappear. When this happens a deep lake or *tarn* usually occupies a corrie. This overflows the lip, into the valley below, giving rise to a waterfall. A corrie glacier often extends down the valley as a valley glacier. As the glacier passes over the lip of the corrie, it becomes broken by vertical cracks. called *crevasses* which produce a series of ice falls on the surface. As it moves down the valley large quantities of boulders and rocks, produced by weathering, especially frost action, fall onto its surface and work their way between the valley sides and the glacier, and the valley bottom and the glacier. It is this material which gives a glacier its powers of erosion. Any material carried along by a glacier is called *moraine*.

How a glacier moves

If the accumulation of snow above the snowline increases from year to year, then a glacier will form. This will expand, and in time, extend below the snowline as a valley glacier in mountain regions. The glacier will continue to move down the valley as more ice accumulates above the snowline. Eventually a glacier will reach a point in the valley where the temperature is high enough to cause ice to melt and the glacier usually does not extend down-valley beyond this point. If the melting of the ice at the glacier *snout* (front) is balanced by the addition of new ice at the head of the glacier, then the glacier remains stationary. If ice is added faster than it melts, then the glacier will gradually extend farther down the valley. But when more ice melts than is added, the glacier will slowly retreat up the valley. These changes in the length of a glacier suggest a movement of a glacier in its valley. But the movement is not of the glacier as a whole. The pressure at the bottom, at the sides, and in the middle of a glacier is very great and this causes some of the ice to melt. The water so formed is turned back to ice by the intense cold very quickly but there is just sufficient time for the water to trickle a short distance. Throughout a glacier bits of ice are melting, trickling down-valley and then turning back to ice the whole time. This means that within the glacier there is a gradual down-valley movement.

The Rhône Glacier in Switzerland

The glacier is fed by the snow field which occupies the upper basin. Try to locate an ice fall, an arête, a cirque and a pyramidal peak.

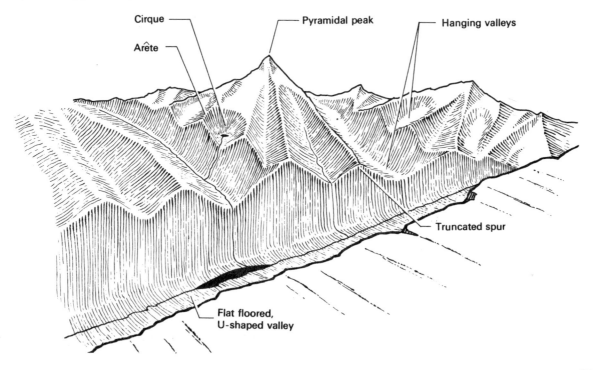

If a glacier extending down a valley enters a part of the valley which is wider than the rest, the ice of the glacier spreads out to fill the valley. This causes the upper layers of ice of the glacier to crack along lines parallel to the valley sides. These cracks are often very deep. Similar cracks develop in a glacier's surface when it passes over a part of the valley floor which is steeper than the rest. These cracks develop at right angles to the valley sides. All of these cracks are called *crevasses*.

How a glacier shapes a valley

As the amount of ice in a valley increases by addition of ice at the rear and by tributary glaciers entering the main valley, the power to erode by a valley glacier also increases. This results in a glacier deepening, straightening and widening a river valley. It cuts back the ends of spurs turning them into *truncated spurs*. One of the most noticeable effects of valley glacier erosion is the over-deepening of the valley which gives it a characteristic U shape.

U-shaped Valley

These diagrams show how a glaciated valley may have formed. Inter-valley divides are sharpened to give arêtes and pyramidal peaks. Examine the two contour maps and compare these with their respective block diagrams.

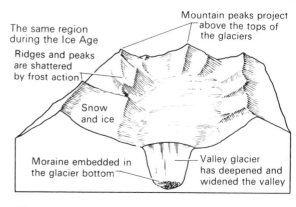

The same region during the Ice Age

Mountain peaks project above the tops of the glaciers

Ridges and peaks are shattered by frost action

Snow and ice

Moraine embedded in the glacier bottom

Valley glacier has deepened and widened the valley

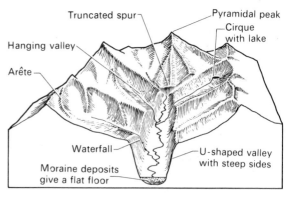

The same region after the glaciers have melted

Truncated spur

Pyramidal peak

Cirque with lake

Hanging valley

Arête

Waterfall

U-shaped valley with steep sides

Moraine deposits give a flat floor

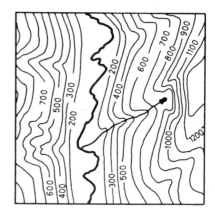

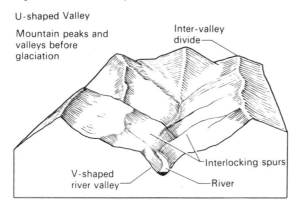

U-shaped Valley

Mountain peaks and valleys before glaciation

Inter-valley divide

V-shaped river valley

Interlocking spurs

River

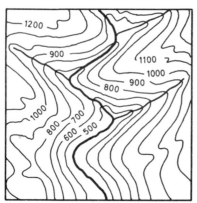

Hanging valleys

Vertical erosion of the main valley is much greater than that of the tributary valleys which contain many small glaciers or no glaciers at all. After the glaciers have retreated the floor of the main valley lies far below the floors of the tributary valleys which, as a result are called *hanging valleys*. Rivers and streams which usually take over the valleys after the glaciers have disappeared, often appear to be too small for the valleys. The streams of hanging valleys join the main river via waterfalls, which may be several hundred metres high. Waterfalls from hanging valleys sometimes build up alluvial fans of coarse materials.

The Lauterbrunnen Valley in Switzerland

Identify four important features in this valley which have been produced by glacial action.

Alluvial fan in a lake in Switzerland

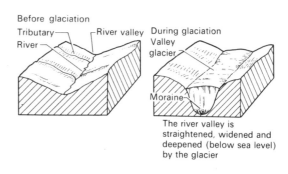

Before glaciation

Tributary — River valley

River

During glaciation

Valley glacier

Moraine

The river valley is straightened, widened and deepened (below sea level) by the glacier

After glaciation

Hanging valley

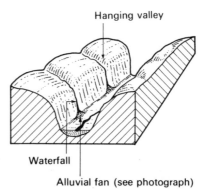

Waterfall

Alluvial fan (see photograph)

Effects of Glaciers on Mountain Peaks and Ridges

We have seen that mountain peaks and ridges which project above the surface of glaciers and snow fields are shattered by frost action to give sharp-edged features. Sometimes, however, whole mountain regions are completely buried by glaciers. When this happens, there is virtually no frost action and the rugged peaks and jagged slopes are smoothed off as shown in the following diagrams. Compare these mountains with those of the diagram at the bottom left of page 91 which projected above the level of the glaciers and which were shattered by frost.

The Dent Blanche in Switzerland. Mt. Bernina and Mt. Roseg

This photograph shows a pyramidal peak, cirques and arêtes. The cirques contain snow fields. The glacier of the cirque in the foreground forms ice falls as it enters the valley below the cirque. Notice the sharpness of the arête on the right.

The Mourne Mountains in Ireland – compare with the photograph above

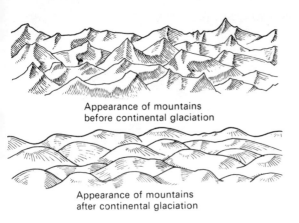

Appearance of mountains
before continental glaciation

Appearance of mountains
after continental glaciation

GLACIAL DEPOSITION AND THE FEATURES IT PRODUCES

A valley glacier carries large quantities of rock waste called *moraine*. Some of this is torn by the glacier from the bottom and sides of the valley and becomes embedded in the glacier. The rest falls onto the glacier surface from the mountain slopes where it has accumulated under frost action. The moraine which forms along the sides of a glacier is called *lateral moraine*; that along the front of a glacier is called *terminal moraine* and that at the bottom of a glacier is called *ground moraine*. When two glaciers join together, their inner lateral moraines coalesce (or join together) to give a *medial moraine*.

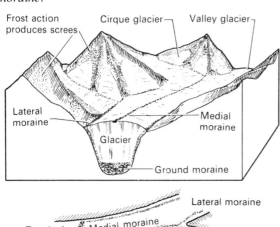

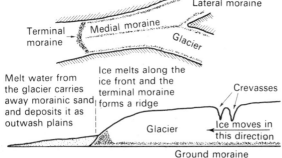

A terminal moraine only builds up when a glacier is stationary, that is, when the amount of ice being added at a glacier's snout is balanced by the amount

LATERAL MORAINE   MEDIAL MORAINE

Nabesna Glacier in Alaska

of ice which is melting there. Some terminal moraines are very impressive. Material from a terminal moraine is always carried down-valley by the melt waters issuing from a glacier's snout, and is deposited as a layer called an *outwash plain*.

These types of moraine can often be clearly seen in glaciated valleys long after the glaciers have retreated.

Examine the photograph of the Nabesna Glacier in Alaska and see how many of these features you can recognise:

   (i) medial moraine
  (ii) truncated spur
 (iii) hanging valley
  (iv) cirque
   (v) pyramidal peak
  (vi) arête
 (vii) snow field
(viii) lateral moraine.

ICE SHEETS

Ice sheets tend to move more slowly than valley glaciers but the erosion they cause can give rise to spectacular features depending on the structure and topography of the region.

Features produced by erosion

The ancient shields of Finland and northern Canada are two regions which were glaciated by ice sheets. In both regions hills were rounded and worn smooth and hollows were excavated in outcrops of soft rocks. These hollows are now the sites of lakes. Outcrops of harder rocks were smoothed on the side facing the ice to give a gentle slope, and plucked on the side facing away from the ice to give a steep

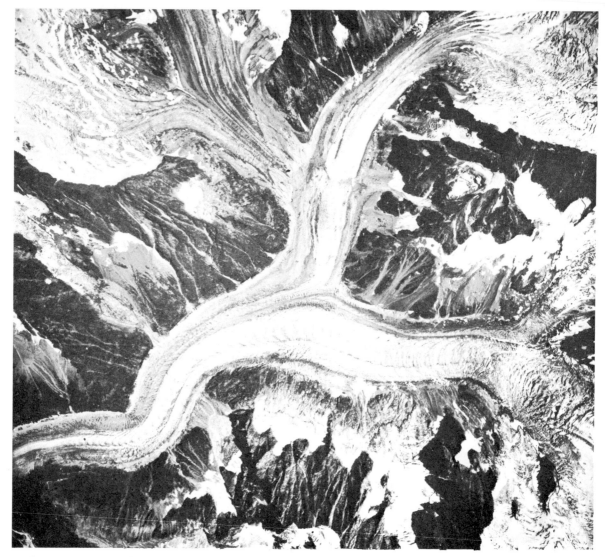

Mer de Glace in the French Alps

This glacier is $5\frac{1}{2}$ kilometres ($3\frac{1}{2}$ miles) long and at its end the melt waters give rise to a river. How many glacier features can you recognise on this photograph?

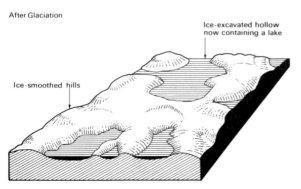

After Glaciation

Ice-excavated hollow now containing a lake

Ice-smoothed hills

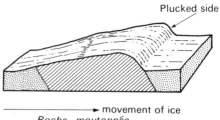

Plucked side

movement of ice

Roche moutonnée

slope. Because of their characteristic shape they are called *roches moutonnées*. Outcrops of harder rocks, especially volcanic rocks were also shaped by ice erosion to give features similar to roches moutonnées, except that the sides facing the ice was steepened. The outcrop protected the softer rocks on the down-stream side from being eroded. These features, which are called *crag and tail*, have a gently sloping down-stream side.

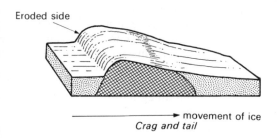

Eroded side

movement of ice
Crag and tail

Features produced by deposition

One of the most conspicuous features of lowlands which have been glaciated by ice sheets is the widespread morainic deposits. Thick deposits of clay containing angular rock particles, cover large parts of glaciated lowlands. Because of the numerous boulders in the clay, these deposits are called *boulder clay* deposits. The deposits are sometimes several hundred metres thick and their surface is marked by long rounded hills, called *drumlins*. Large blocks of rock, which are of a material quite different to that of the rocks of the region, often occur in regions which lay under ice sheets. These blocks are known as *erratics*, and they were obviously uprooted in one region and deposited in another.

This is what an ancient shield looks like after it has been glaciated. Rock hollows containing lakes, eskers and terminal moraines, often covered with pine trees, characterise the landscape.

The Lake Plateau of Finland

The edge of a boulder clay deposit is sometimes marked by a terminal moraine which will be well developed if the ice remained stationary for a long time. Glacial deposits quite different to boulder clay often occur on the outside of the terminal moraine. They are well-sorted, that is, coarse materials such as gravels occur immediately outside the moraine while finer deposits, such as sand, occur further away. Such deposits have been carried away from the ice sheets by melt waters emerging from the ice front. The deposits are water-sorted and build up outwash plains.

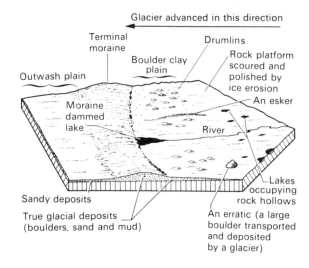

97

Rivers and streams occur inside most glaciers and these are heavily loaded with rock debris. As an ice front retreats the rivers build up ridge-like deposits called *eskers*. They develop on top of the boulder clay deposits.

Boulder clay plains, drumlins, terminal moraines, outwash plains and eskers rarely occur in such an ordered manner as has perhaps been suggested. In many regions ice sheets advanced, later retreated and then advanced again thus mixing up and sometimes obliterating the originally deposited materials. Also, since the final retreat of the ice, river erosion and weathering have modified and sometimes partially covered up or removed many of the glacial depositional features.

Note Valley glaciers also produce roche moutonnées, drumlins, eskers, etc.

The Value of Glaciated Regions to Man

I *Glacial Features of Value to Man*

(i) Boulder clay plains are sometimes very fertile, e.g. East Anglia in Great Britain and parts of the Dairy Belt of North America.

(ii) Old glacial lake beds are invariably fertile. Extensive areas of the Canadian Prairies producing vast amounts of wheat each year owe their prosperity to the rich alluviums which once collected on the floors of glacial lakes.

(iii) Some glacial lakes, e.g. the Great Lakes of North America, are of real value as natural routeways.

(iv) Waterfalls issuing from hanging valleys are sometimes suitable for the development of hydro-electric power. Both Norway and Switzerland develop large amounts of H.E.P. from such waterfalls.

(v) Glaciated mountain regions attract tourists, especially during the winter season when heavy snowfalls make skiing and other sports possible.

(vi) Some glacial lakes have cut deep overflow channels where they have drained away. Some of these channels today form excellent routeways across difficult country, e.g. the Hudson – Mohawk Gap which leads down to New York.

(vii) Many glaciated valleys have benches or 'alps' high up on their sides. During the summer these 'alps' have good pasture, but during the winter they are covered with snow. Cattle are grazed on the alpine pastures during the summer and are brought down to the sheltered valley bottom pastures during the winter. This movement of animals is called *transhumance*, and it goes on in Switzerland, Norway and other mountainous countries.

II *Glacial Features of Little Value to Man*

Now let us look at the disadvantageous aspects of glaciation.

(i) Boulder clay deposits in some regions, e.g. Central Ireland, have produced a marshy landscape which is of little or no value to agriculture.

(ii) Many outwash plains contain infertile sands which give rise to extensive areas of waste land. It is true that this is sometimes of recreational value but from an agricultural standpoint such regions are negative.

(iii) Extensive areas of land are sometimes turned into myriads of lakes by morainic deposits. Such lake landscapes offer little scope for development by Man.

EXERCISES

1 The following features often occur in glaciated regions: cirque (corrie), moraine, hanging valley, pyramidal peak, arête. Choose *three* of these features and for *each*,
  (i) briefly explain how it may have originated
  (ii) show its appearance by means of a well-labelled diagram
  (iii) name *one* region where an example may be found.

2 With the aid of diagrams, describe *three* of the following and explain how they may have been formed: esker, drumlin, terminal moraine, crag and tail, roche moutonnée.

3 (a) Outline the main differences between continental ice sheets and valley glaciers.
  (b) Briefly explain the main differences between ice action in mountain regions and ice action in lowland regions and name the characteristic physical features produced in each region.

4 By using well-labelled diagrams, explain the main differences between *three* of the following pairs of features:
  (i) truncated spur and interlocking spur
  (ii) crevasse and ice fall
  (iii) terminal moraine and boulder clay
  (iv) ice sheet and valley glacier
  (v) glaciated valley and river valley.

5 Briefly explain *three* of the following:
  (i) the sides of a valley glacier move more slowly than its middle
  (ii) boulder clay differs from an outwash plain in that its material is unsorted
  (iii) glaciated valleys are distinctly U-shaped
  (iv) the floor of a corrie is usually concave
  (v) tributary valleys usually join a glaciated valley high above the floor of the glaciated valley.

6 Choose *two* of the following landforms: a rift valley, an atoll, a glaciated valley, a fiord, and for *each* of the two chosen:
  (i) describe its characteristic features
  (ii) briefly explain how it has originated
  (iii) show the landform by means of an annotated diagram
  (iv) name *one* region where the landform may be seen.

Objective Exercises

1 The erosive action of a valley glacier depends most upon
A the gradient and width of its valley
B the width and thickness of the ice
C the height of the glacier above sea level
D the length of the glacier
E the rate of movement of the glacier

A B C D E

2 Crevasses, which are cracks in the surfaces of glaciers, are produced by
A the glacier moving over level land
B the melting of the ice
C differential movement in the ice
D the thickness of the ice
E rain action on the surface of the glacier

A B C D E

3 Which one of the following erosional features proves that ice can move uphill?
A cirque
B arête
C truncated spur
D alp
E ice fall

A B C D E

4 A region which has not been glaciated may show
A U-shaped valleys
B corries in the mountains
C waterfalls rushing down the steep sides of valleys
D ox-bow lakes
E hanging valleys

A B C D E

5 Which one of the following is most associated with ice action?
A deltaic plain
B loess deposits
C outwash plains
D tombolo
E laterite plain

A B C D E

6 The first requirement for the formation of a glaciated U-shaped valley is
A a glacier
B heavy and continuous falls of snow
C a mountainous terrain
D a river valley

A B C D

7 All of the following features are produced by glacial deposition **except**
A drumlin
B moraine
C esker
D roche moutonnée

A B C D

8 Of the several features produced by the action of ice, some are the **product** of both erosion and deposition. Which of the following features is of this type?
A arête
B hanging valley
C erratic
D roche moutonnée

A B C D

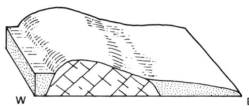

W   E

9 This is a diagram of a crag and tail. In which direction did the ice predominantly move?
A east to west
B west to east
C south to north
D north to south

A B C D

10 The diagram below represents a glacier emerging from a valley near the foot of the mountains. The end of the glacier has melted without it retreating, and has deposited a feature at X which is called a
A medial moraine
B boulder clay
C terminal moraine
D drumlin

A B C D

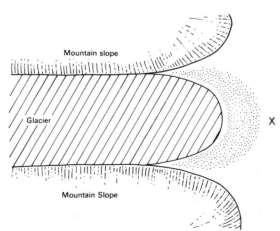

Mountain slope

Glacier   X

Mountain Slope

10   The Oceans

The oceans and seas together cover about 70 per cent of the world's surface. There are five oceans and all are joined with one another. These are: Southern Ocean; Indian Ocean; Pacific Ocean; Arctic Ocean; Atlantic Ocean.

NATURE OF THE OCEAN FLOOR

Ocean floors, like continental surfaces, have relief features the chief of which are shown in the diagram. The edges of continents slope gently downwards under the surrounding oceanic waters. This part of the ocean floor is called the *continental shelf*. Along some coasts it is so narrow as to be almost absent. The best developed continental shelves are shown in the diagrams on page 101. Seawards of the shelf the ocean floor slopes more steeply, and this part of the floor forms the *continental slope*. The bed of the ocean sometimes rises up to give *ridges* some of which may appear above the surface of the ocean as *oceanic islands*. Below the ocean bed there are troughs and basins which are known as *deeps* or *trenches*. All the slopes shown in the diagram below are greatly exaggerated but the depths given are true.

Note Do not confuse *oceanic islands* with *continental islands*. The latter rise from the continental shelf.

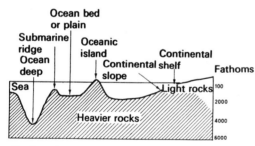

Generalised section across an ocean floor

Continental Shelves of the World
Northern Hemisphere
Eurasia
1 Along the coast of N.W. Europe
2 Along the coast of Siberia
3 The floor of the Yellow Sea
4 The floor of the Java Sea and the southern part of the S. China Sea

North America
1 Along the north-east coast of North America
2 The floor of Hudson Bay
3 Along the Gulf Coast of North America

Southern Hemisphere

Australasia
1 The floor of the Gulf of Carpentaria
2 The floor of the Australian Bight

South America
Along the coast of Patagonia

Africa
Very poorly developed

Value to Man
1 Sunlight easily penetrates the seas on continent shelves, and therefore there is an abundance o *plankton* or small green marine plants whic usually results in an abundance of fish.
2 They increase the height of tides thus improvir shipping facilities.

Ocean Deeps and Ridges (they all lie below 4000 metres)

Notice how the deeps flank the east coast of Asia and compare the location of the deeps with th location of Young Fold Mountains (diagram, pag 26) and of volcanoes (diagram top of page 37).

Deeps

Pacific Ocean
1 Mariana Trench (12 000 metres approx.)
            (39 240 feet approx.)
2 Philippine Trench (11 980 metres approx.)
            (39 175 feet approx.)
3 Tonga Trench (10 300 metres approx.)
            (33 680 feet approx.)
4 Japanese Trench (9300 metres approx.)
            (30 400 feet approx.)
5 Aleutian Trench (8380 metres approx.)
            (27 400 feet approx.)

Atlantic Ocean
1 Puerto Rico Deep (9625 metres approx.)
            (31 475 feet approx.)
2 Romanche Deep (8060 metres approx.)
            (26 355 feet approx.)
3 South Sandwich Trench (9090 metres approx.)
            (29 725 feet approx.)

Indian Ocean
Sunda Trench (8140 metres approx.)
            (26 620 feet approx.)

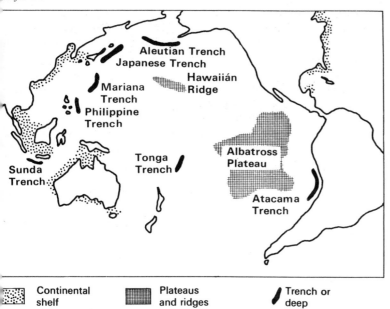

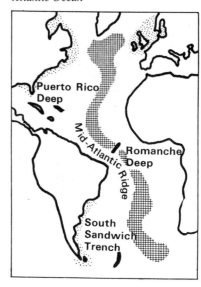

Continental shelf	Plateaus and ridges	Trench or deep	

Ridges

Pacific Ocean
Hawaiian Ridge
Albatross Plateau

Atlantic Ocean
Mid-Atlantic Ridge (arising from this are the oceanic islands of the Azores, Tristan da Cunha and Ascension).

Indian Ocean
A ridge extending from S. India to Antarctica.

THE NATURE OF SALT WATER

Salinity

Sea water contains mineral salts. Two important salts are sodium chloride (NaCl) and calcium bicarbonate (CA(HCO$_3$)$_2$). The latter provides marine organisms with calcium carbonate (CaCO$_3$) which is necessary to the formation of shells and bones. When water evaporates the salts are left behind. The saltiness of a sea depends upon the amount of evaporation taking place from its surface, and the amount of fresh water brought into it by rivers. Semi-enclosed seas do not mix freely with the oceans, so it is there that most variation is found. The Mediterranean and Red Seas are more salty than the oceans because they are located in regions of high temperatures and few rivers discharge into them. The Black and Baltic Seas are much fresher than the oceans. These seas are located in regions of low temperatures and fresh water is brought into them by rivers and melting ice. Inland seas like the Great Salt Lake (N. America) and the Dead Sea (Jordan) are very salty indeed.

Temperature

Water is heated by the sun's rays much more slowly than is land. It also loses heat to the air more slowly than does land. This causes the temperature of sea water to vary only slightly from season to season. The temperature of surface sea water ranges from about $-2°C$ ($28°F$) in polar regions to $26°C$ (about $79°F$) in equatorial regions. The bottom water of the oceans is always cold, the temperature being about $1°C$ (about $34°F$).

WATER MOVEMENTS IN THE OCEANS

There are two types of movement:

1 *Horizontal*, i.e. ocean currents
2 *Vertical*, i.e. the rising of bottom water and the sinking of surface water.

These movements result from the combined action of:

1 *Density* (particularly important in vertical movements)
2 *Winds* (particularly important in horizontal movements).

The density of sea water depends upon the temperature and the amount of salt in the water. It falls when water is heated, as in the tropics; when there is a large inflow of fresh water, as in the Baltic (brought by rivers), and in the polar seas (from melting ice). It rises when evaporation is high and rainfall is low, as in the Red Sea; when water is cooled, as in the polar seas, and when water freezes (salts remain in the sub-surface water which does not freeze) as in the polar seas.

101

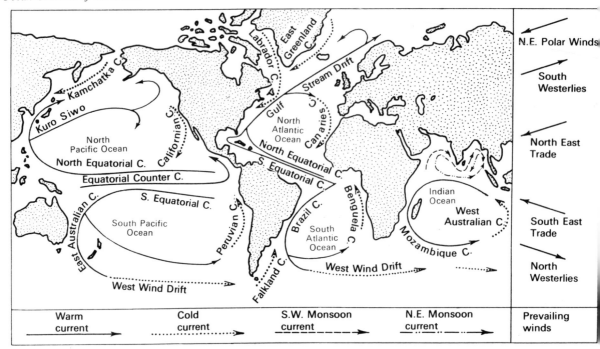

Warm current	Cold current	S.W. Monsoon current	N.E. Monsoon current	Prevailing winds

MOVEMENTS OF OCEAN WATERS

1 Tides

Both the sun and the moon, (the latter to a greater extent), exert a gravitational attraction on the earth's surface which causes a rising and falling motion to develop in the waters of the larger oceans. This rising and falling of the water surface produces a *tide*.

The Influence of the Moon

1 Water at H_2 is 'pulled' towards the moon more than the earth – therefore water piles up at H_2 forming a high tide.
2 The earth is 'pulled' towards the moon more than the water at H_1 – therefore water lags behind and piles up at H_1 forming a high tide.
3 The moon's 'pull' causes water to be drawn from L_1 and L_2 – therefore there are low tides there.

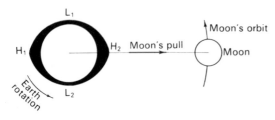

4 The rotation of the earth results in every meridian coming into the positions of two high tides and two low tides very nearly every 24 hours. The moon travels in its orbit in the same direction

as the earth is rotating and in consequence it takes about 24 hours 52 minutes, or one lunar day, for the sequence of two high and two low tides to be completed.

When the sun, earth and moon are in a straight line as they are at Full Moon and New Moon, the gravitational force is at its greatest because the sun and the moon are "pulling" together. At these times, high tides are very high and low tides are very low, and they are called *spring tides*.

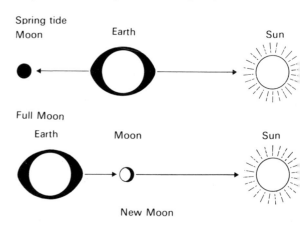

When the sun, earth and moon are not in a straight line, the sun and moon are not "pulling" together and the gravitational force is less. At Half Moon that is, when the sun and the moon are "pulling"

right angles, the force is at its least and the difference between high and low tides is not large. These tides are called *neap tides*.

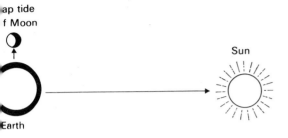

neap tide
of Moon

Sun

Earth

The difference between high tide and low tide is called the *tidal range* or *amplitude*. At spring tides this is high; at neap tides it is low.

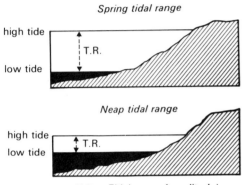

Spring tidal range

high tide

T.R.

low tide

Neap tidal range

high tide

T.R.

low tide

T.R. – *Tidal range (amplitude)*

Tides and Waves

At one time it was believed that the tide-producing forces resulted in the formation of two large tidal waves (H_1 and H_2 in the diagram on the opposite page) which progressively moved westward across the Southern Ocean. From these, minor tidal waves were sent out into the Pacific, Indian and Atlantic Oceans and their neighbouring seas. The two tidal waves were separated by low water (L_1 and L_2). Observations on tidal waves in different parts of the world show that this explanation can no longer be accepted. The present theory maintains that the oceans can be divided into zones in each of which the tide-producing forces cause the surface of the water to oscillate. This means that the water will rock bodily, rising and falling around the edges of the water body whilst near the centre there will be practically no rise or fall in water level (diagram below).

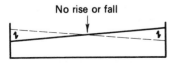

No rise or fall

Bores

When a tidal wave enters an estuary the wave increases in height as the estuary becomes increasingly shallow and narrow. Ultimately, the wave breaks and forms a wall of foaming water which often surges forward at several kilometres per hour. This usually happens when the tidal wave meets a river. Bores occur on these rivers: Hooghly, Tsien-tang-kiang (N. China) and Amazon. Bores occur in rivers which have large funnel-shaped estuaries where there is a large tidal range, and which face the direction of tidal surge.

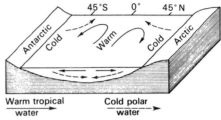

45°S 0° 45°N

Antarctic

Cold

Warm

Cold

Arctic

Warm tropical water

Cold polar water

2. Ocean Currents

The diagram above shows a general drift of tropical water towards the poles and a return flow of polar water towards the equator. The polar flow begins as a surface current but slowly it sinks to the ocean bottom. The winds greatly modify this simple pattern: indeed, ocean currents closely resemble prevailing winds in their direction and position. Earth rotation and the shapes of the continents also influence the direction of currents. Study the diagram below which shows the general pattern of currents for the Atlantic Ocean. Notice how water is piled up along the **Brazilian Coast**, and how off-shore

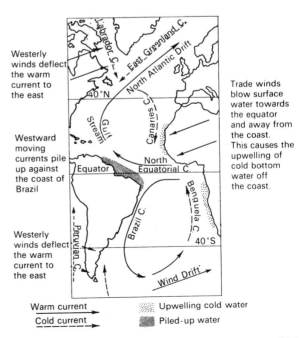

Westerly winds deflect the warm current to the east

Trade winds blow surface water towards the equator and away from the coast. This causes the upwelling of cold bottom water off the coast.

Westward moving currents pile up against the coast of Brazil

Westerly winds deflect the warm current to the east

Labrador C.

East Greenland C.

North Atlantic Drift

40°N

Gulf Stream

Canaries C.

Equator

North Equatorial C.

Benguela C.

Brazil C.

Peruvian C.

40°S

Wind Drift

Warm current

Cold current

Upwelling cold water

Piled-up water

trade winds cause cold water to upwell along the west coast of Africa. This water forms the Canaries and Benguela Currents. The Californian and Humboldt Currents along the west coast of the Americas are formed in a similar way.

The diagram on the right indicates in greater detail how this type of cool current develops. These four currents have a cooling effect on the climates of coastal regions. Warm, moist air moving towards the coast is cooled as it passes over the cool currents and some of its moisture is condensed. Banks of mist develop along the coast. The air enters the coastal regions but there is no rain because the air is now drier. Also it is hot over the land and the air heats up and absorbs, rather than gives out moisture. Coastal mists are very common along the coast of Chile.

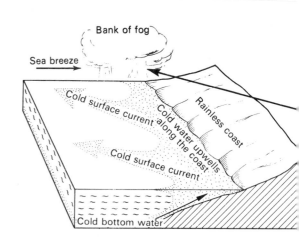

The Importance of the Oceans to Man
1 Oceans permit countries and regions to trade with one another. Goods can be moved in very large quantities by ships more cheaply than by any other means.
2 Some land margins would have colder winters if there were no warm currents in the nearby oceans. This would have an adverse effect upon such activities as agriculture.
3 The oceans contain a valuable source of food.

The map on page 268 shows the chief fishing grounds of the world. Notice that these occur in the continental shelf regions where cold and warm currents meet. Many fish feed on plankton. These plants require nitrates and phosphates, sunlight and well aerated water. Cold currents are rich in nitrates and phosphates, and epi-continental seas permit sunlight to reach almost to the bottom. The meeting of cold and warm currents causes the water to become well aerated. It is understandable therefore, why the areas shaded black form the richest fishing grounds in the world.

Ocean Currents can also Influence Climate

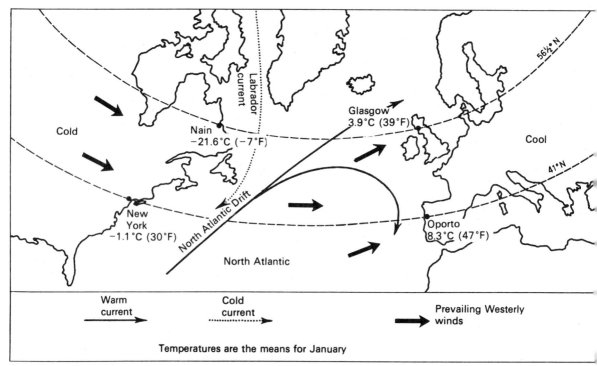

On an outline map of the world:
  (i) mark, name and show the direction of *two* warm and *two* cold ocean currents in each hemisphere
  (ii) mark and name *two* ocean currents whose directions are seasonally reversed
  (iii) shade *two* areas which are usually foggy for most of the year
  (iv) shade *two* coasts whose winter temperatures are raised by the combined influence of ocean currents and prevailing winds.

2 Briefly describe the relief of the ocean floor and try and account for its more important features.

3 Choose any *one* ocean current and for it:
  (i) draw a sketch-map to show its location
  (ii) briefly explain how it originates
  (iii) name any *one* other ocean current which is similar in location and origin.

4 Explain, with the aid of well-labelled diagrams, any *three* of the following features: continental shelf, continental slope, ocean deep, tidal bore.

5 Write a concise explanation, illustrated with diagrams or sketch-maps where appropriate, for *three* of the following:
  (i) important fishing grounds are usually located on continental shelves
  (ii) there are both horizontal and vertical movements of water in the oceans
  (iii) winds are the principal cause of ocean surface currents
  (iv) the salinity of ocean (or sea) water varies greatly.

6 With the aid of sketch-maps and diagrams write a brief account on *three* of the following:
  (i) the Benguela Current
  (ii) the high degree of salinity in the Dead Sea
  (iii) the seasonal change in direction of some ocean currents
  (iv) the continental shelf.

7 Write a brief description on the nature and occurence of tides making specific reference to spring and neap tides. Illustrate your answer with clear well-labelled diagrams.

Objective Exercises

1 Which one of the following statements best describes the causes of the movements of sea water?
  A The movement in the sea water is both vertical and horizontal.
  B Differences in surface water temperature result in the formation of currents.
  C Differences in temperature and density in the oceans and seas, and the effect of winds on surface water together result in the movement in sea water.
  D Ocean currents are deflected by the rotation of the Earth.
  E Differences in the density of sea water and the shape of land masses cause movements in sea water.

        A   B   C   D   E
        ·· ·· ·· ·· ··

2 Only one of the following currents has a warming influence upon the coast along which it flows. Which is it?
  A California Current
  B Benguela Current
  C Kuro Siwo Current
  D Humboldt Current
  E Labrador Current

        A   B   C   D   E
        ·· ·· ·· ·· ··

3 A 'wall' of water surging up an estuary is called a
  A tidal wave
  B bore
  C neap tide
  D current
  E swell

        A (B) C   D   E
        ·· ·· ·· ·· ··

4 There are several differences between the circulation pattern of the North Pacific Ocean and the South Pacific Ocean. Which one of the following is common to both?
  A clockwise circulation
  B cold current in the eastern sector
  C off-shore current caused by north-east trade winds
  D the westward movement of water in high latitudes connects with similar movements in neighbouring oceans

        A   B   C   D
        ·· ·· ·· ··

5 The winter temperatures of insular north-western Europe are higher than the winter temperatures of eastern Europe in the same latitudinal zone because
  A it is on the western side of a continent
  B it is near the sea
  C it lies under westerly winds which blow over the Gulf Stream Drift
  D it receives only light falls of snow

        A   B   C   D
        ·· ·· ·· ··

6 A fog is most likely to develop when
  A a warm moist wind blows over a warm current
  B a cold dry wind blows over a cold current
  C a warm moist wind blows over a cold current
  D a cold dry wind blows over a warm current

        A   B   C   D
        ·· ·· ·· ··

7 Which one of the following statements is **not** true?
  A An ocean current develops between two bodies of water which are of contrasting densities.
  B Most ocean currents tend to flow from cold water to warmer water.
  C Ocean currents along the west coasts of continents are usually cold currents.
  D Most of the important fishing grounds are where contrasting ocean currents meet.

        A   B   C   D
        ·· ·· ·· ··

11   Lakes

A lake can be defined as a hollow in the earth's surface in which water collects. Some lakes are of great size and are called seas, e.g. Caspian, Dead and Aral Seas. Although most lakes are permanent, some contain water in the wet season only. Lakes in basins of inland drainage (which are usually semi-arid) may contain water for a few months only out of a period of several years.

CLASSIFICATION OF LAKES ACCORDING TO ORIGIN

It is probably true to say that the majority of lakes have been formed by the action of glaciers and ice sheets. All the others have been formed by river, marine and wind action, by earth movements and vulcanicity, and by man.

Many lakes contain fresh water but some contain saline water, e.g. the Dead Sea. Some lakes are temporary, e.g. playas and other hot desert lakes, while others are more permanent. However, all lakes, with the exception of deep rift valley lakes, are eliminated in a comparatively short period of geological time, by the forces of nature.

There are several ways in which lakes can be classified and although most lakes fit neatly into specific categories in a specific classification, some lakes, because of their origins, may fit into two such categories.

1 Lakes Produced by Earth Movements

Tectonic forces cause sagging and faulting in the earth's crust thus producing depressions, some of which may become the site of lakes. Tectonic lakes can be put into definite groups.

(i) **Lakes formed by crustal warping** These lakes occupy basin-like depressions and good examples are: the Caspian Sea (U.S.S.R.), Lake Victoria (Africa), Lake Eyre (Australia) and Lake Chad (Africa). Lake Titicaca (Peru/Bolivia) is the highest tectonic lake in the world.

(ii) **Lakes formed by faulting** Most of these lakes occur in rift valleys, e.g. Lake Nyasa and Tanganyika (East Africa), the Dead Sea and Lake Baikal (U.S.S.R.) and Loch Ness (Scotland). These lakes are usually long and narrow and very deep. The levels of some lakes are below sea level, e.g. the Dead Sea which is 393 metres (1285 feet) below sea level (its floor is 817 metres (2672 feet) below sea level).

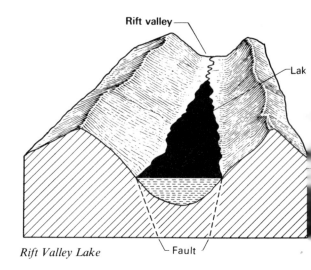

Rift Valley Lake

2 Lakes formed by Erosion

(i) **Glacially-eroded lakes**. Both valley glaciers and ice sheets can gouge out hollows and troughs on the earth's surface. These may later fill with water to form lakes. The main types of such lakes are as follows:

(a) *Cirque (corrie) lakes* Lakes often form in the armchair-like depressions, known as cirques. These lakes usually, though not always, feed mountain rivers. They are sometimes called *tarns*, and they are found in all **glaciated mountain regions**.

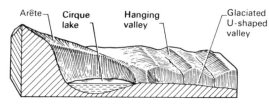

Cirque Lake   *Longitudinal section of a cirque*

(b) *Trough lakes* These occupy elongated hollows excavated in a valley bottom. They are sometimes called *ribbon lakes* because of their shape.

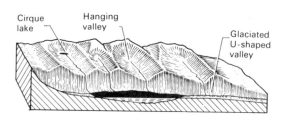

Glaciated Valley Trough Lake   *Ice-eroded trough contains a lake*

(c) *Rock basin lakes* These have been formed by ice-scouring action of ice sheets and valley glaciers which resulted in the formation of shallow hollows. They are numerous on the Baltic and Canadian Shields, and particularly good examples occur in Finland and parts of northern Canada.

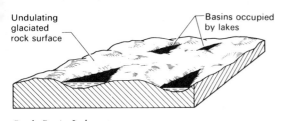

Rock Basin Lake

(ii) **Wind-eroded lakes**. Wind deflation sometimes produces extensive depressions in arid regions such as the deserts of the Southern Continents and central Asia. If the depression is excavated below the water-table then a lake will develop. The diagram shows that it is the action of eddy currents which scoop out the loose sand which is then blown away and deposited as dunes. The lakes of these depressions are not always true lakes and may be nothing more than muddy swamps. The Qattara Depression in Egypt is a good example. More permanent types of desert lakes develop when an aquifer is exposed. These are called *oases*. Some desert lakes dry up because of excessive evaporation and all that remains is a lake bed of salt. This is called a *playa* or *salt lake*.

Desert Depression Lake

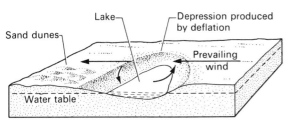

Oasis

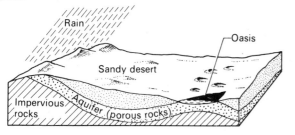

3 Lakes Produced by Deposition
River Deposits

(i) *Ox-bow lakes* Mature rivers meandering across a flood plain often produce cut-offs which in time get separated from the river to become ox-bow lakes. Such lakes are common in the lower valleys of the Mississippi, and similar rivers.

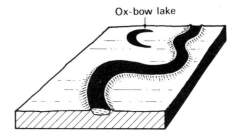

(ii) *Delta lakes* In the formation of deltas the deposition of alluvium by the rivers may cause a part of the sea to be surrounded thus turning it into a lagoon, or it may isolate a part of a distributary thus producing a lake. Lakes are common in large deltas and a good example is Etang de Vaccares in the Rhône Delta.

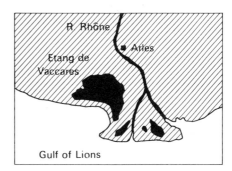

Glacial Deposits

(i) *Moraine-dammed lakes* Terminal moraines are sometimes deposited across a valley where they form a ridge which dams back the flow of water. Lakes result and these are called moraine-dammed lakes. Such lakes are common in the Lake District (England). Lake Garda (Italy) has been formed in this way.

Moraine-dammed Lake

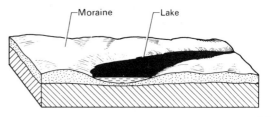

(ii) *Boulder clay lakes* Boulder clay is sometimes poorly drained and contains many depressions; lakes will often form in the latter.
Examples of such lakes occur in Northern Ireland.

Marine Deposits

Haffs These lakes have been formed by sand bars extending along a coast and cutting off indentations in the coast thus producing lagoons. Haffs are well developed on the southern coast of the Baltic Sea (see page 80). Similar lakes develop along Les Landes coast of south-west France.

4 Lakes Produced by Vulcanicity

(i) *Crater and Caldera lakes* In some violent volcanic eruptions the top of a volcano may be blown off leaving a large crater, or if this is enlarged by subsidence, a caldera. Some examples are: Crater Lake in Oregon (U.S.A.), Lake Toba in northern Sumatra and Lake Knebel (Iceland).

Crater Lake

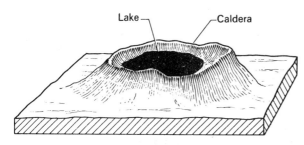

(ii) *Lava-blocked lakes* Lava flows from some volcanoes block river valleys and cause lakes to form. The Sea of Galilee was formed by a lava flow blocking the Jordon Valley.

Lava-dammed Lake

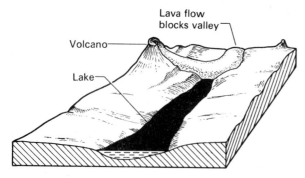

(iii) *Lava-subsidence lakes* Some lava flows are hollow and if the lava crust collapses, a depression is left which may become the site of a lake. Myvatn, in Iceland, is an example of such a lake.

5 Other Types of Lake

(i) *Solution lakes* Some rocks, especially chalk and limestone, are dissolved and removed by rain water percolating down from the surface. This action produces underground caverns in limestone regions and some of these caverns, whose floors are near to the base of the limestone layer, often contain lakes. If the cavern roof collapses the lakes become exposed. Such lakes are usually long and narrow, e.g. Lac de Chailexon in the Jura Mountains. Sometimes solution, perhaps aided by subsidence, produce large depressions called *poljes*. These often contain lakes, e.g. Lake Scutari in Yugoslavia. In Cheshire (England), rock salt has been removed by solution and shallow depressions have formed which now contain lakes, called '*flashes*'.

(ii) *Barrier lakes* These can be formed by ice, lava or moraine damming a valley, as we have already seen, but damming can also be effected by landslides, avalanches or screes. Lakes formed in this way are rarely as permanent as those formed by moraine-damming.

(iii) *Beaver lakes* Some animals, especially beavers, build dams across streams and so create small lakes. Beaver Lake in Yellowstone National Park (U.S.A.) is an example.

(iv) *Mining ponds* Opencast mining results in the development of large depressions which sometimes collect water and form lakes. Tin mining in West Malaysia has created many such lakes.

(v) *Man-made lakes* By building dams across rivers man has deliberately constructed artificial lakes called reservoirs, e.g. Lake Mead on the Colorado River (U.S.A.)

The Value of Lakes to Man

Lakes have played an important role in the cultural and social life of peoples in many parts of the world for a very long time, and in regions where lakes are abundant, e.g. Finland and Sweden, this is especially true. The most common functions of lakes are as follows:

1 *For communication* Lakes and connecting river systems often form important natural routes for the movement of people and goods. In recent years lakes systems have been improved in some countries to enable easy movement of large ships. The Great Lakes and the St. Lawrence Waterway of North America form the longest inland water transport system in the world.

2 *For hydro-electric power development* Lakes, both natural and man-made, can, and often are used for generating hydro-electric power. Good examples are the H.E.P. plant at Jinja (Uganda) which uses water from Lake Victoria, and the

one at Grand Coulee on the Columbia River (U.S.A.) where man has created a lake which is over 80 kilometres (50 miles) long.

For regulating the flow of rivers Rivers which contain large lakes seldom flood because the lakes can absorb the run-off when rainfall is heavy, while in times of drought the water in the lakes helps to maintain a fairly steady flow of water. The Poyang and Tung Ting Lakes regulate the water-flow of the Yangtze-kiang while the Tonlé Sap regulates the water-flow of the Mekong. Many large rivers do not contain lakes but any liability of these to flood can be prevented by constructing lakes. This has been done on the Tennessee River (U.S.A.) and on the Sutlej (India), etc.

As a source for fish Some lakes are rich in fish, e.g. the Caspian Sea (U.S.S.R.) for sturgeon, the Tonlé Sap (Cambodia) and the Great Lakes (N. America), and in China and Japan man has built lakes for breeding fish.

For water storage Most urban settlements obtain their water supplies from lakes both natural and man-made. Examples of such lakes can be found in almost every country.

For providing irrigation water Some of the lakes whose water is used for generating hydro-electric power, are also used for providing irrigation water. These lakes therefore are multi-purpose. The Burrinjuck Dam on the Murrumbidgee (Australia), the Sennar Dam on the Blue Nile (Sudan) and the dams on the Tennessee (U.S.A.) all create multi-purpose lakes of this type.

For developing a tourist industry A large number of countries have used their lakes for this purpose. Switzerland, Finland, and Canada are just three examples.

As moderating influences on climate Large lakes in temperate latitudes have a moderating influence on the climate of nearby regions. In winter these lakes exert a warming influence by releasing heat which was stored up in the summer. In summer they exert a cooling influence by absorbing part of the heat. The eastern shores of Lakes Ontario, Huron and Erie have milder winters than the western shores have – because the winds, which blow from the west, are warmed by the lakes. Lakes also supply water vapour to winds passing over them and thus the rainfall pattern of nearby regions can be affected.

EXERCISES

Choose *three* of the following types of lake: ox-bow lake, crater lake, corrie lake and oasis, and for *each*:
   (i) state how it has been formed
   (ii) draw a diagram to show its main characteristics
   (iii) name *one* example or a region where an example may be found.

2 Choose *three* of the following lakes: Lake Victoria, Lake Toba, Lake Manzala, Lake Garda and the Sea of Galilee, and for *each*:
   (i) state its location
   (ii) explain its origin by means of well-labelled diagrams
   (iii) name *one* region where a lake of similar origin may be seen.

3 Choose *four* of the following features: haff, delta lake, basin of inland drainage, karst lake, tectonic lake, and a playa, and for *each* chosen feature:
   (i) briefly explain how it has originated
   (ii) draw a well-labelled diagram to show its characteristics
   (iii) name *one* example or name a region where an example may be seen.

4 A lake may be created for one or more of the following reasons:
   (i) to generate hydro-electric power
   (ii) for irrigation
   (iii) to control flooding
   (iv) to supply water to settlements.
   Choose *two* of these and for *each*:
   (a) briefly explain how it operates
   (b) name *one* specific example.

Objective Exercises

1 Which of these lakes occupies a rift valley?
   A Lake Toba
   B Lake Superior
   C Lake Eyre
   D Lake Tanganyika
   E Crater Lake

   A B C D E

2 The levels of some lakes are below sea level. Such lakes most probably owe their origin to
   A the process of faulting
   B the damming of a glaciated valley by moraine
   C the formation of a caldera
   D the growth of a spit across a river's mouth
   E a rise in the level of the land relative to sea level

   A B C D E

3 Which one of the following lakes owes its origin mainly to volcanic activity?
   A Lake Chad
   B Lake Baikal
   C Lake Huron
   D Lake Toba
   E Lake Superior

   A B C D E

4 Lake Toba, in Sumatra, occupies a
   A structural depression
   B moraine-dammed valley
   C depression in a limestone plateau
   D a caldera
   E man-made lake

   A B C D E
   :: :: :: :: ::

5 In which one of the following types of physical land-
   scape would lakes **not** develop through the process of
   silting?
   A on a deltaic plain
   B on a limestone plateau
   C on a flood plain
   D in a coniferous forest
   E on a coastal plain

   A B C D E
   :: :: :: :: ::

6 Some large lakes in temperate latitudes have a mode-
   rating influence upon the climate of adjacent land sur-
   faces. Such an influence may develop when
   A the lake is high above sea level
   B the water is shallow
   C prevailing out-blowing winds cross the lake in winter
   D the lake is a basin of inland drainage

   A B C D
   :: :: :: ::

7 All of the following types of lakes owe their origin to de-
   positional factors **except**
   A ox-bow lakes
   B cirque lakes
   C delta lakes
   D haffs

   A B C D
   :: :: :: ::

8 Some lakes are permanent (they last for several years)
   and some are temporary (they last for a season only).
   All of the following are permanent **except**
   A cirque lake
   B playa lake
   C caldera lake
   D ox-bow lake

   A B C D
   :: :: :: ::

9 In which one of the following types of region would
   you expect there to be lakes of depositional origin?
   A an arid plateau
   B an alluvial lowland
   C an ancient shield
   D limestone plateau

   A B C D
   :: :: :: ::

10 A lake will probably be saline when
   A it is connected to the sea by a river
   B its surface is high above sea level
   C rivers drain into it but not out of it
   D it has a tropical latitude

   A B C D
   :: :: ::

11 Some lakes are seasonal, that is, they dry up for a par
   of the year. Such lakes are likely to occur
   A on a deltaic plain
   B in a caldera
   C in a basin of inland drainage
   D in a glaciated valley

   A B C D
   :: :: ::

12 Weather and Climate

When we say it is hot, or wet, or cloudy, we are saying something about the *weather*. Weather refers to the state of the atmosphere: its temperature, pressure and humidity for a place for a short period of time. If we want to find out what the weather is like we must examine:

1 temperature	2 humidity
3 pressure	4 rainfall
5 wind direction and strength	6 cloud cover
7 sunshine	

MEASURING AND RECORDING WEATHER ELEMENTS

A weather station is a place where all the elements of weather are measured and recorded. Each station has a Stevenson Screen (diagrams on the right) which contains four thermometers which are hung from a frame in the centre of the screen. They are:

1 Maximum thermometer
2 Minimum thermometer
3 Wet bulb thermometer
4 Dry bulb thermometer

The screen is so built that the shade temperature of the air can be measured. It is a wooden box whose four sides are louvered to allow free entry of air. The roof is made of double boarding to prevent the sun's heat from reaching the inside of the screen, and insulation is further improved by painting the outside white. It is placed on a stand, about 121 cm (48 in) above ground level.

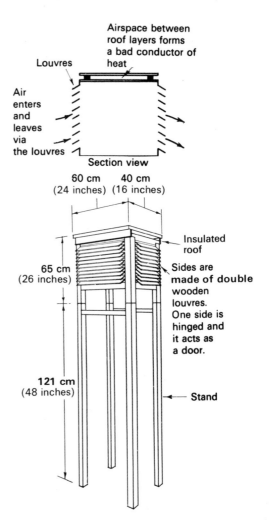

To measure maximum and minimum temperature

Maximum Thermometer

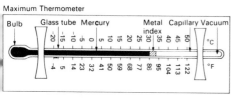

When the temperature rises the mercury expands and pushes the index along the tube.

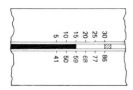

When the temperature falls the mercury contracts and the index remains behind. The maximum temperature is obtained by reading the scale at the end of the index which was in contact with the mercury. In the diagram this is 30°C (86°F). The index is then drawn back to the mercury by a magnet.

Minimum Thermometer

When the temperature falls the alcohol contracts and its meniscus pulls the index along the tube.

Minimum Thermometer

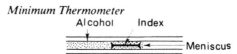

When the temperature rises the alcohol expands. The index does not move but remains in the position to which it was pulled. The minimum temperature is obtained by reading the scale at the end of the index which is nearer the meniscus. In the diagram this is 15°C (59°F). By raising the bulb of the thermometer the index is returned to the meniscus.

Six's Thermometer

This thermometer can also be used for measuring maximum and minimum temperatures. When the temperature rises the alcohol in the left-hand limb expands and pushes the mercury down this limb and up the right-hand limb. The alcohol in this limb also heats up and part of it is vaporised and occupies the space in the bulb. The maximum temperature is read from the scale on the right-hand limb. When the temperature falls the alcohol in the left-hand limb contracts and some of the alcohol vapour in the conical bulb liquefies. This causes the mercury to flow in the reverse direction. The minimum temperature is read from the scale on the left-hand limb. Note this scale is reversed.

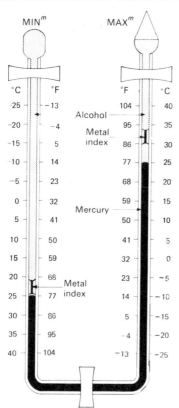

vapour in the air and the amount of vapour the air could hold at that particular temperature. This amount is called the *Relative Humidity* (R.H.). Thus if the R.H. is 80 per cent at a temperature of 30°C (86°F), then the air is holding eight-tenths of the water vapour it could hold at that temperature. When air can hold no more vapour we say the air is *saturated* and its R.H. is 100 per cent. Now the amount of vapour air can hold is dependent upon its temperature. When this rises the air is able to hold more vapour, and when it falls it cannot hold as much. When the temperature falls it may well be that the air contains more vapour than it can hold. The excess vapour then *condenses*, i.e. turns into water droplets which form either clouds, rain, mist or fog.

Measurement of Humidity

The hygrometer shown below consists of two ordinary thermometers. The bulb of one is wrapped in a piece of muslin which dips into a container of water. This thermometer is called the wet bulb thermometer. The other is called the dry bulb thermometer. When the air is not saturated, water evaporates from the muslin and this cools the wet bulb and causes the mercury to contract. The

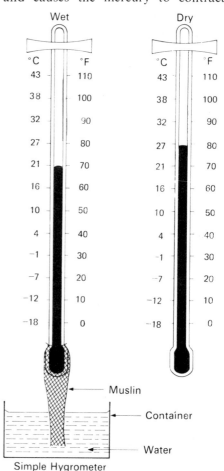

Simple Hygrometer

HUMIDITY OF THE AIR

No air is absolutely dry although some air, such as that over tropical deserts, contains very little water vapour. Humidity refers to the amount of water vapour in the air, but it is more important to know the relationship between the actual amount of

ry bulb thermometer is not affected in the same
ay, and so the two thermometers show different
eadings. When the air is saturated, there is no
vaporation and hence no cooling. The two thermo-
neters therefore show the same reading. The
ifference between the two readings is, therefore,
n indication of the humidity of the air. Remember
hese statements:

Thermometer readings
   No difference   – air is saturated.
   Small difference – humidity is high.
   Large difference – humidity is low.

ATMOSPHERIC PRESSURE AND ITS MEA-SUREMENT

Air has weight and thus it exerts a pressure on the
earth's surface. At sea level this averages 1·034
gf/cm (14.7 lb per sq in). Pressure varies with
both temperature and altitude, and the instrument
which measures pressure is called a *barometer*.
There are two principal types of barometer:
   Mercury barometer; 2 Aneroid barometer

Mercury Barometer

Although this is a large and cumbersome instrument,
it is very accurate and is used in many weather
stations. Atmospheric pressure is read in millibars.
Thus 29·92 inches of mercury are equivalent to
1013 millibars (mb) at sea level.

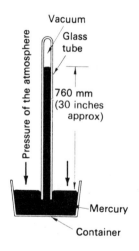

Vacuum
Glass tube
760 mm (30 inches approx)
Mercury
Container
Pressure of the atmosphere

The pressure of the air on the mercury in the container supports a column of mercury about 760 mm high. This amount of mercury has the same weight as a column of air about 18 kilometres high (almost 11 miles) having the same cross-sectional area as the mercury column

Aneroid Barometer

The heart of this instrument consists of a small
metal box which contains very little air. The top of
this box bends slightly under the influence of any
change in atmospheric pressure. The movement of
the box top is conveyed by a system of levers to a
pointer which moves across a graduated scale.
When the pressure rises the box top bends in and
when the pressure falls the spring pushes the box
top outwards.

RELATIVE HUMIDITY (per cent)

Dry bulb	Wet bulb							
	20 °C	22 °C	24 °C	26 °C	28 °C	30 °C	32 °C	34 °C
20 °C	100%	–	–	–	–	–	–	–
25 °C	65%	80%	95%	–	–	–	–	–
30 °C	40%	50%	60%	80%	90%	100%	–	–
35 °C	24%	30%	35%	45%	57%	70%	82%	95%

The actual value of relative humidity is given by
tables, a page of which is shown here. If the dry
bulb temperature is 12°C and the wet bulb is 11°C,
i.e., depression of the wet bulb is 1°C, the relative
humidity is 88%.

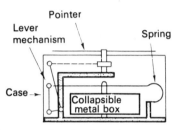

Pointer
Lever mechanism
Spring
Case
Collapsible metal box

Aneroid Barometer (sectional view)

THE MEASUREMENT OF RAINFALL

All weather stations have a rain gauge. Rain falling
in the funnel trickles into the jar below and at the
end of a 24-hour period this is poured into a
graduated measuring cylinder. The graduation of
the cylinder is such that the reading obtained is
the depth of rain that has fallen over an area
equivalent to that of the top of the funnel. Rainfall
is measured in millimetres or inches, and the cylinder
is tapered at the bottom (see inset) to enable very
small amounts to be measured accurately.

Position of the Rain Gauge

It must be placed in an open space so that no
run-off from buildings or trees, etc., enters the
funnel. It must also be sunk into the ground so
that about 30 centimetres of it sticks up above
ground level. This prevents rain from splashing into
it from the ground, and also prevents the sun's

rays from causing excessive evaporation of the water already collected in the jar.

Rain Gauge

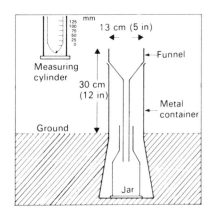

To Measure the Direction and Speed of the Wind

Wind direction is measured by a wind vane which consists of a rotating arm pivoted on a vertical shaft. The arrow of the wind vane always points in the direction from which the wind blows and the wind is named after this direction. Thus the diagram of a wind vane indicates that the wind is a north-east wind.

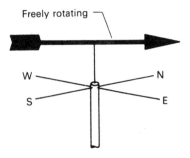

Weathercock

114

The speed of the wind is measured by an instrument called an *anemometer*. This instrument has three or four horizontal arms pivoted on a vertical shaft. Metal cups are fixed to the ends of the arms so that when there is a wind the arms rotate. This movement operates a meter which records the speed of the wind in kilometres (or miles) per hour.

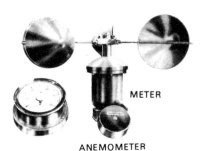

Wind velocity can also be obtained, though only approximately, by observing the way some objects are moved by the wind. This is done by referring to the *Beaufort Wind Scale.*

Beaufort Scale No.	Wind Description	Effect on land features	Speed (km/h)	Arrow Indication
0	calm	Smoke rises vertically	less than 1	
1	Light air	Direction shown by moving smoke but not by wind vane	1 - 5	
2	Light breeze	Wind felt on face; leaves rustle; wind vane moved	6 - 11	
3	Gentle breeze	Leaves and twigs in constant motion; flag moved	12 - 19	
4	Moderate breeze	Raises dust and paper; small branches moved	20 - 29	
5	Fresh breeze	Small trees begin to sway	30 - 36	
6	Strong breeze	Large branches in motion, whistling heard in telegraph wires	37 - 49	
7	Moderate gale	Whole trees in motion	50 - 60	
8	Fresh gale	Twigs broken off trees	61 - 73	
9	Strong gale	Slight structural damage occurs, especially to roofs	74 - 86	
10	Whole gale	Trees uprooted, considerable structural damage	87 - 100	
11	Storm	Widespread damage	101 - 120	
12	Hurricane	Widespread devastation experienced only in some tropical regions	above 121	

Recording of Winds
I On a Map
Winds are shown by arrows on a weather map

The shaft of an arrow shows wind direction, and the feathers on the shaft indicate wind force or speed. A half feather stands for a speed of 5 knots and a full feather a speed of 10 knots. A black pennant stands for a speed of 50 knots. Combinations of these can thus be used to show any speed. In practice the feathers do not have a definite value but indicate a speed between two limits. These values are given in brackets below the numbers 1, 5, and 10 and so on. Arrows are inserted at the positions of the weather stations on the map. The tip of the arrow away from the feathers points to the direction in which the wind is blowing.

II On a Wind Rose

The main purpose of a wind rose is to record wind direction for a specific place. A simple wind rose is shown in the above diagram. It consists of an octagon, each side of which represents a cardinal point. Rectangles are drawn on each side and each day when there is a wind a line is ruled across the rectangle representing the direction from which the wind is blowing. This is done for one month. The number of days when there is no wind is recorded in the circle in the centre of the octagon. The diagram shows the type of wind pattern Singapore often gets during December.

CLOUDS

When air is cooled some of its water vapour may condense into tiny droplets of water. The temperature at which the change takes place is called the *dew-point temperature*. Clouds are made of water droplets or ice particles, and so are mists and fogs which are really low-level clouds.

The shape, height and movements of clouds can indicate the type of weather which is about to occur, and because of this they are carefully studied by meteorologists who prepare weather forecasts. The following symbols are used on weather maps to indicate the amount of cloud cover. Lines drawn through places having the same amount of cloud are called *isonephs*.

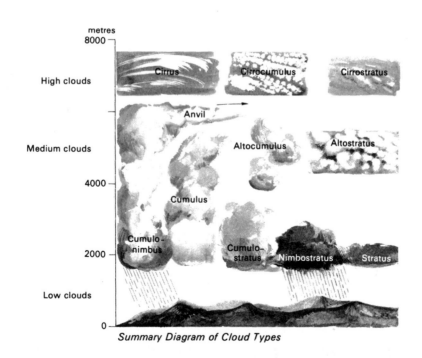

Summary Diagram of Cloud Types

Cloud Amount (oktas)	Symbol	Cloud cover description
0	○	Clear sky
1	◔	One-eighth cover
2	◑	Two-eights cover
3	◕	Three-eighths cover
4	◑	Half of sky covered
5	◕	Five-eighths cover
6	◕	Three-quarters cover
7	◕	Seven-eighths cover
8	●	Complete cloud cover
	⊗	Sky obscured (fog)
	⊗	Missing/doubtful data

Cloud Types

Clouds can be classified according to their appearance, form and height. There are four basic groups.

1 **High Clouds**: 6000 to 12 000 metres (19 600 to 39 250 feet) above sea level
   (a) *Cirrus (Ci)*: a wispy, fibrous-looking cloud which often indicates fair weather
   (b) *Cirrocumulus (Cc)*: a thin cloud, often globular and rippled
   (c) *Cirrostratus (Cs)*: looks like a thin white sheet which causes the sun and moon to have 'halos'

2 **Medium Clouds**: 2100 to 6000 metres (6900 to 19 600 feet) above sea level
   (a) *Altocumulus (Alt. Cu)*: globular, bumpy-looking clouds which have a flattened base; usually indicate fine weather
   (b) *Altostratus (Alt. St)*: greyish, watery-looking clouds.

3 **Low Clouds**: below 2100 metres (6900 feet)
   (a) *Stratocumulus (St. Cu.)*: low rolling, bumpy-looking clouds which have a pronounced wavy form
   (b) *Nimbostratus (Ni. St.)*: a dark, grey layered cloud which looks rainy, and which often brings rain
   (c) *Stratus (St.)*: fog-like low cloud (near to ground level); brings dull weather which is usually accompanied with drizzle

4 **Clouds of great vertical extent**: 1500 to 9000 metres (4900 to 29 500 feet)
   (a) *Cumulus (Cu.)*: a round-topped, and flat-based cloud which forms a whitish grey globular mass; indicates fair weather.
   (b) *Cumulonimbus (Cu. Ni)*: a special type of cumulus cloud which may reach up to 9000 metres (29 500 feet) and which forms white or black globular masses whose rounded tops spread out to form an anvil or cauliflower shape; thunder clouds indicating convectional rain, lightning and thunder.

   Cloud types can also be shown on maps by using special symbols which are similar to those used for showing the amount of cloud cover.

Cumulonimbus Cloud

Cirrus Cloud

Cirrocumulus Cloud

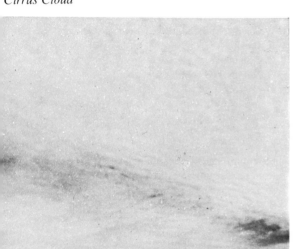

Cirrostratus Cloud

Altocumulus Cloud

Nimbostratus Cloud

Cumulus Cloud

SUNSHINE

We have already discussed the various factors which affect the amount of sunshine a region receives. Latitude and position of the earth in its revolution around the sun are the more important factors.

The number of hours of sunshine a place receives can be measured by using a sensitised card, which is graduated in hours and on to which the sun's rays are focussed. Lines drawn through places having the same amount of sunshine are called *isohels*.

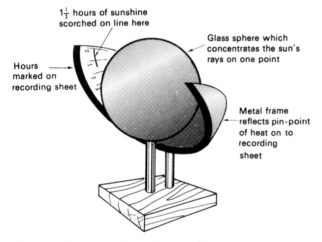

Diagram of Campbell-Stokes Sunshine Recorder

EXERCISES

1 Name the most common instruments which are contained in a weather station. Choose *two* of these, and for *each* describe: (i) how the instrument works, (ii) how its readings are taken and recorded, and (iii) the purpose of taking these readings.

2 Briefly explain the meanings of *each* of the following statements:
  (a) A Stevenson Screen should be at least 1·2 metres (or 4 feet) above the surface, its sides should be louvered and it should be placed in the open.
  (b) A Six's Thermometer is a combination of maximum and minimum thermometers.
  (c) When air is saturated wet and dry bulb thermometers should read the same.

3 Describe and name the instrument that is used for measuring rainfall. State where you would keep the instrument in order to measure rainfall accurately, and say how and when you would use it for measuring rainfall.

4 What instruments would you use to measure the following:
  (i) wind velocity
  (ii) relative humidity
  (iii) atmospheric pressure.
  Choose *two* of these instruments and for *each* draw an annotated diagram to explain how it works.

5 Study the weather data given below and then: (i) state the diurnal temperature range, (ii) state the daily mean temperature, and (iii) describe the weather conditions for the whole day giving reasons to support your answer.

Local Time	(hrs)	00.00	02.00	04.00	06.00	08.00	10.00
Temp	(°C)	12	10	10	11	12	20
	(°F)	54	50	50	51	54	68
Rel Humidity	(%)	55	58	60	59	52	49
Rainfall	(mm)	–	–	–	–	–	–
	(in)	–	–	–	–	–	–

Local Time	(hrs)	12.00	14.00	16.00	18.00	20.00	22.00
Temp	(°C)	22	25	25	21	20	16
	(°F)	72	78	78	70	68	60
Rel Humidity	(%)	44	38	32	34	39	42
Rainfall	(mm)	–	–	–	–	7	5
	(in)	–	–	–	–	0·3	0·2

Objective Exercises

1 Which one of the following is **not** an element of weather?
  A sunshine
  B cloud cover
  C height above sea level
  D rainfall
  E fog

  A B C D E
  :: :: :: :: ::

2 Minimum and maximum temperatures are obtained from an instrument called
  A a barometer
  B a Six's thermometer
  C an anemometer
  D a clinical thermometer
  E a hygrometer

  A B C D E
  :: :: :: :: ::

3 The relative humidity of a region is low when
  A the wet and dry bulb thermometers read the same
  B the difference between the readings of the wet and dry bulb thermometers is large
  C the temperatures are high
  D the temperatures are low
  E the wet and dry bulb thermometers read differently

  A B C D E
  :: :: :: :: ::

4 Weather elements can be measured by instruments. Which one of the following pairs is **incorrect**?
  A Maximum and minimum temperatures-Six's Thermometer
  B Atmospheric pressure-Barometer
  C Wind direction-Wind Vane
  D Humidity-Rain Guage
  E Wind speed-Anemometer

  A B C D E
  :: :: :: :: ::

5 Which of the following is **not** a form of precipitation?
  A snow
  B haze
  C rain
  D dew
  E hail

  A B C D E
  ·· ·· ·· ·· ··

6 A Stevenson Screen usually contains all of the following **except**
  A maximum thermometer
  B wet and dry bulb thermometers
  C minimum thermometer
  D rain gauge

  A B C D
  ·· ·· ·· ··

7 In what order do the processes of saturation, evaporation and condensation take place during the formation of clouds?
  A evaporation, condensation, saturation
  B condensation, saturation, evaporation
  C saturation, condensation, evaporation
  D evaporation, saturation, condensation

  A B C D
  ·· ·· ·· ··

8 Which one of the following types of cloud rarely forms at heights lower than 5000 metres?
  A stratus
  B cumulus
  C cirrus
  D nimbostratus

  A B C D
  ·· ·· ·· ··

9 Water vapour is turned into water droplets by the process of
  A evaporation
  B liquefaction
  C convection
  D condensation

  A B C D
  ·· ·· ·· ··

10 Under which of the following conditions would the influence of aspect on temperature be most noticeable?
  A a flat sandy surface in the Sahara Desert during July
  B hilly country in the Amazon Basin in December
  C the south-facing side of a hill in Central France in April
  D the north-facing side of a hill on the Equator in June

  A B C D
  ·· ·· ·· ··

13   Climate

When we say that Malaysia is hot and wet all the year, or that Central Chile has hot dry summers and warm mild winters, we are in fact saying something about the state of the atmosphere over a long period of time. What we are really describing is the average state of the atmosphere, and such descriptions refer to the state of the *climate*. Before we make a study of the main types of climate, we must examine temperature, pressure and rainfall, etc., and find out what it is that causes these to vary from region to region. We must also know how to find the average state of these elements.

TEMPERATURE

Insolation and how Air is Heated
The sun's energy is called *insolation* and this is turned into heat at the earth's surface. Only about 45 per cent of the incoming insolation reaches the surface. The heat generated at the surface warms the air by *radiation* (heat waves sent out by the earth's surface), *conduction* (passing of heat by contact), and *convection* (passing of heat by air currents).

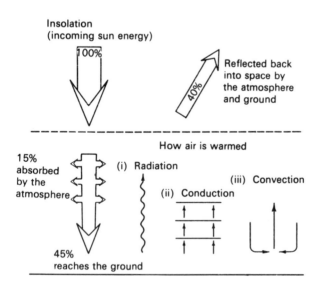

Heating and Cooling of Land and Water Surfaces
It takes over five times as much heat to raise the temperature by 2°C for a given volume of water as it does for the same volume of land. This is what the common statement 'land heats up more quickly than water' implies. The reverse of this, i.e. 'land cools more quickly than water', is equally true. In N.W. Europe, this condition causes the coastal regions to be warmed by the seas in winter and cooled by them in summer.

Water is fairly transparent and hence the sun's rays penetrate to considerable depths. Therefore, the heat which develops is more widely distributed than it is on land. Also, water movements in the seas cause the heat to be further distributed. All this means that a much greater volume of sea is heated than is the case with land.

Factors Influencing Temperature
The temperature of a place is dependent upon some or all of these factors:
1 Latitude
2 Altitude
3 Ocean currents
4 Distance from the sea
5 Winds
6 Aspect
7 Cloud cover
8 Length of day
9 Amount of dust and other impurities in the air

Latitude
The altitude of the mid-day sun is always high in the tropics and hence temperatures are always high. Outside the tropics the altitude is lower and temperatures are correspondingly lower. In general temperatures decrease from the equator to the poles. The diagram below explains this. Bands marked X contain equal amounts of sun energy, but, because area B is smaller than area A, the temperature at B will be higher than that at A. Notice also that the sun's rays at A have passed through a greater thickness of atmosphere than have those at B, and hence more sun energy arrives at B than at A.

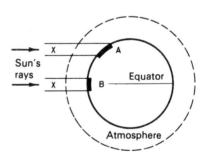

Altitude
We have already seen that the sun's rays heat the earth's surface which then passes on its heat to the air. Water vapour and dust in the air prevent this heat from rapidly escaping back into space, but at high altitudes, e.g. on the top of a high mountain, air is rarefied and it contains very little vapour or dust. The heat from the earth's surface therefore rapidly escapes and the air remains cold (see

diagram). In tropical arid regions such as hot deserts the almost complete absence of water vapour results in the earth's surface becoming intensely hot in the day. During the night most of this heat rapidly passes back into space with the result that night temperatures drop appreciably. Such regions have a large *diurnal* (daily) range of temperature.

In general, temperature falls by 6·5°C for every 1000 metres ascent, that is, 1°F for every 300 feet. A mountain which is 3000 metres (9800 feet approx.) high will have a temperature of 10·5°C (about 51°F) at its top if the temperature at its bottom is 30°C (86°F). See the diagram.

Distance from the Sea

The sun's heat is absorbed and released more slowly by water than by land. This becomes very noticeable in temperate latitudes in the winter season when sea air is much warmer than land air. Hence on-shore winds bring warmth to coastal regions. This warming influence is confined to a narrow coastal belt because the sea air rapidly loses its heat to the colder land. Air temperatures decrease from the coast inland (middle diagram). In the summer season land surfaces are warmer than sea surfaces and the air over the land is therefore warmer than that over the sea. Therefore coastal regions are cooler than inland regions (bottom diagram). Climates whose temperatures are influenced greatly by the sea are called *maritime*, or *oceanic*, or *insular*

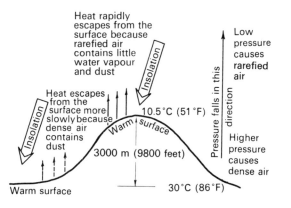

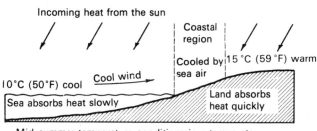

Mid-summer temperature conditions in a temperate latitude (temperatures are approximate)

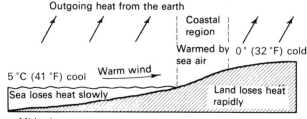

Mid-winter temperature conditions in a temperate latitude (temperatures are approximate)

Ocean Currents

Warm and cold currents often raise or lower the temperatures of land surfaces if the winds are on-shore.

(i) Warm currents moving polewards carry tropical warmth into the high latitudes, and this warming influence is very marked in latitudes 40° to 65° on the west sides of continents, especially along the seaboard of Western Europe (page 102). The warmth is conveyed to the land by the prevailing Westerly Winds. This action is almost entirely confined to the winter season. Between latitudes 0° and 40° on the eastern sides of continents warm currents raise coastal temperatures.

Note In tropical latitudes on-shore winds crossing warm currents do not raise the temperature of the air over the land because this is already high.

(ii) Cold currents have less effect upon temperatures because they usually lie under off-shore winds (page 104). There are exceptions, e.g. the coast of Labrador, when summer temperatures are lowered by on-shore winds which blow over the cold Labrador Current.

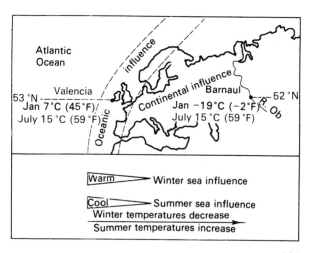

climates. These occur in coastal regions which lie under prevailing on-shore winds. Climates whose temperatures are greatly influenced by remoteness from the sea are called *continental* climates. These occur in the hearts of temperate continents.

Winds

In temperate latitudes prevailing winds from the land lower the winter temperatures but raise the summer temperature. Prevailing winds from the sea raise the winter temperatures but lower the summer temperatures. In tropical latitudes on-shore winds modify the temperatures of coastal regions because they have blown over cooler ocean surfaces. *Local winds* (see page 137) sometimes produce rapid upward or downward temperature changes.

Cloud Cover and Humidity

Clouds reduce the amount of solar radiation reaching the earth's surface and the amount of earth radiation leaving the earth's surface. When there are no clouds both types of radiation are at a maximum. The heavy cloud cover of the equatorial regions explains why the day temperatures rarely exceed 30°C (86°F) and why the night temperatures are not much lower. In hot deserts the absence of clouds and the presence of dry air result in very high day temperatures of over 38°C (about 100°F) and much lower night temperatures of 21°C (about

70°F) or below. Very humid air absorbs heat during the day and retains it at night. It also helps to prevent the loss of heat from the lower layers of the air. Thus in the humid tropics the air remains warm at night even on days when there is little or no cloud.

Hot Desert Regions

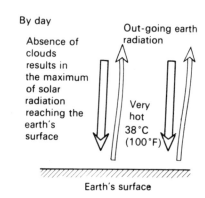

Aspect

The influence of aspect on temperature is only noticeable in temperate latitudes. In the tropics, the mid-day sun is always high in the sky and aspect is of little significance. South-facing slopes are warmer than north-facing slopes in the Northern Hemisphere, whilst in the Southern Hemisphere the reverse is true (see diagrams below).

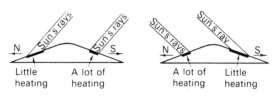

Equatorial Regions

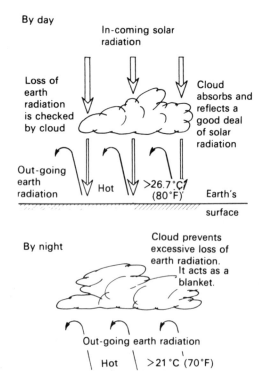

In the high latitudes, the mid-day Sun is at a low angle in the winter. Blocks of flats are usually built far apart to enable all the flats to receive some

sunshine, but in lower latitudes, flats can be built close together because the angle of the mid-day Sun is high.

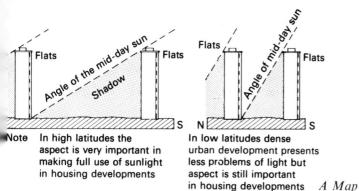

Flats Flats

Angle of the mid-day sun

Shadow

Flats Flats

Angle of mid-day sun

Note In high latitudes the aspect is very important in making full use of sunlight in housing developments

In low latitudes dense urban development presents less problems of light but aspect is still important in housing developments

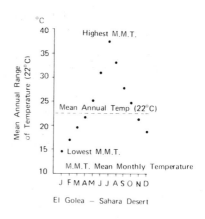

°C
Highest M.M.T.

Mean Annual Temp. (22°C)

Lowest M.M.T.

M.M.T. Mean Monthly Temperature

J F M A M J J A S O N D

El Golea — Sahara Desert

Temperature Readings and Recording

1. $\left(\dfrac{\text{Daily Max. Temp.} + \text{Daily Min. Temp.}}{2}\right)$ gives MEAN DAILY TEMPERATURE. See the diagram below.

Mean Daily Temperature is 25 C $\left(\dfrac{30° + 20°}{2}\right)$

2. (Daily Max. Temp. – Daily Min. Temp.) gives DAILY RANGE OF TEMPERATURE. See the diagram below. Daily Range of Temperature is 10°C (30 – 20)

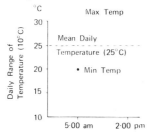

°C

Max Temp

Mean Daily

Temperature (25°C)

• Min Temp

5·00 am   2·00 pm

3. $\left(\dfrac{\text{Sum of Mean Daily Temps. for 1 month}}{\text{Number of days in month}}\right)$ gives MEAN MONTHLY TEMPERATURE.

4. $\left(\dfrac{\text{Sum of Mean Monthly Temps. for 1 year}}{12}\right)$ gives MEAN ANNUAL TEMPERATURE. See the diagram on the top right of this page. Annual Temperature is 22°C.

5. (Highest Mean Monthly Temp. — Lowest Mean Monthly Temp.) gives MEAN ANNUAL RANGE OF TEMPERATURE. See the diagram on the top right of this page. Mean Annual Range of Temperature is 22°C (37° – 15°).

Temperature Graphs and Maps

A Graph is used to show the temperature for a place. If the points in the diagram on the top right of this page were joined by a smooth curve, this would be a temperature graph.

A Map is used to show the temperatures for a region. The positions of all weather stations must first be plotted by dots on the map. The temperature for each station is usually adjusted to what it would be if that station were at sea level. This is done by adding 6·5°C for every 1000 metres of height for the station. The adjusted temperature values are then written alongside the dots and all places having the same temperature are joined by a smooth line. Such a line is called an *isotherm*. Isotherms rarely pass through a station and they must be inserted by using interpolation which is based on proportion.

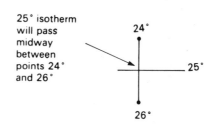

25° isotherm will pass midway between points 24° and 26°

24°

25°

26°

Isotherms in degrees C

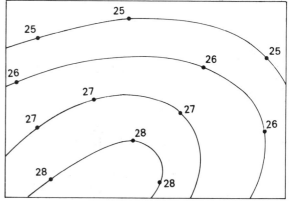

25

25

25

26

25

26

27

27

26

27

28

27

26

28

28

28

123

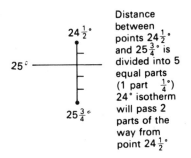

Distance between points $24\frac{1}{2}°$ and $25\frac{3}{4}°$ is divided into 5 equal parts (1 part $\frac{1}{4}°$) 24° isotherm will pass 2 parts of the way from point $24\frac{1}{2}°$

Temperature °C	A	State of the Air
Below − 10°		Very Cold
− 10° to 0°		Cold
0° to 10°		Cool
10° to 21°		Warm
21° to 30°		Hot
Over 30°		Very Hot

Annual Range of Temperature °C	B	Description
Below 3°		Negligible
3° to 8°		Small
8° to 19°		Moderate
19° to 30°		Large
Over 30°		Very Large

How to Describe the Temperature State of the Air

When we talk about the climate of a region we nearly always use such adjectives as very hot, hot, warm or cool, etc. It is important therefore to give some temperature value to each of these adjectives. In the table A on page 111 the temperature refers to daily, monthly or annual mean temperature. According to this table a month which has a mean temperature of 25°C will be referred to as a hot month. A similar table B can be used for describing the mean annual range of temperature.

World Distribution of Temperature

The following two maps show the mean temperatures for the months of January and July respectively. You should examine the isotherms of these two maps very carefully and then try to account for the following statements which can be made about these two maps.

July Temperatures in °C (°F)

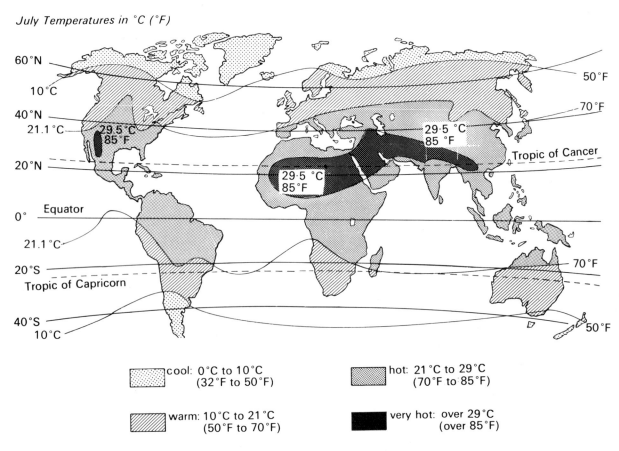

cool: 0°C to 10°C (32°F to 50°F)

warm: 10°C to 21°C (50°F to 70°F)

hot: 21°C to 29°C (70°F to 85°F)

very hot: over 29°C (over 85°F)

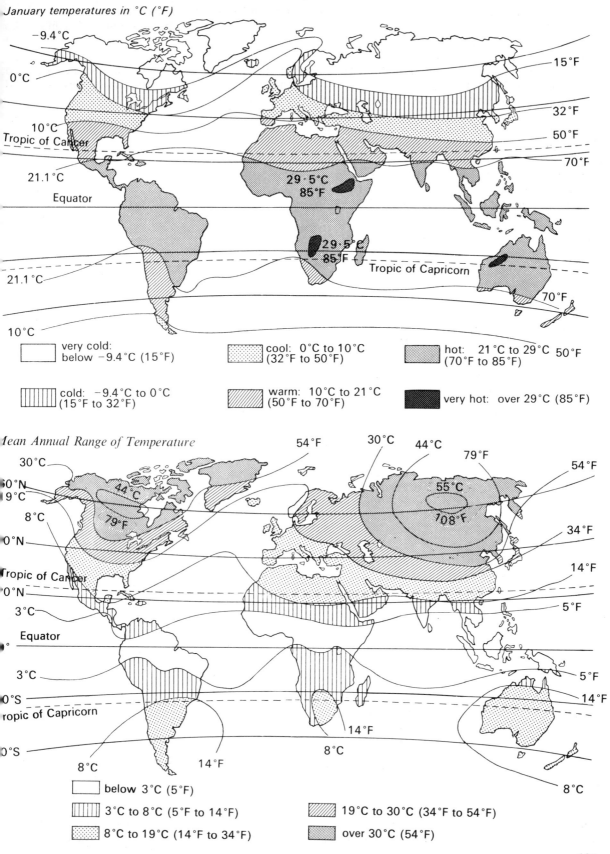

January temperatures in °C (°F)

−9.4°C

0°C

10°C
Tropic of Cancer

21.1°C

Equator

21.1°C

10°C

29·5°C
85°F

29·5°C
85°F
Tropic of Capricorn

15°F

32°F

50°F

70°F

70°F

	very cold: below −9.4°C (15°F)		cool: 0°C to 10°C (32°F to 50°F)		hot: 21°C to 29°C 50°F (70°F to 85°F)
	cold: −9.4°C to 0°C (15°F to 32°F)		warm: 10°C to 21°C (50°F to 70°F)		very hot: over 29°C (85°F)

Mean Annual Range of Temperature

54°F   30°C   44°C

79°F   54°F

30°C

60°N
19°C

8°C

44°C

79°F

55°C

108°F

34°F

0°N

14°F

Tropic of Cancer

0°N

5°F

3°C

Equator

0°

3°C

5°F

0°S

14°F

Tropic of Capricorn

0°S

14°F

8°C   14°F   8°C

8°C

	below 3°C (5°F)			
	3°C to 8°C (5°F to 14°F)			19°C to 30°C (34°F to 54°F)
	8°C to 19°C (14°F to 34°F)			over 30°C (54°F)

125

1 There is a definite northward movement of all isotherms between January and July.
2 This movement of the isotherms is greater over the land than it is over the oceans.
3 The highest temperatures for both January and July are over the continents.
4 The lowest temperatures for January are over the northern continents (Asia and North America).
5 The isotherms bend poleward over the oceans but equatorward over the continents in January.
6 The isotherms bend equatorward over the oceans but poleward over the continents in July.
7 The seasonal changes are less marked over the southern continents than over the northern continents.

The mean monthly temperatures for January and July can be used to prepare a map to show the mean annual ranges of temperatures. This has been done in the map on page 125 (bottom). The map indicates the importance of maritime and continental influences which will be discussed later. In the meantime study the map and, as with the previous two maps, try to account for these statements.

1 The range of temperature in general increases from the equator to the poles.
2 The greatest range of temperature occurs not at the poles but over Asia and North America in latitude 60°N (approx.).
3 Coastal regions have a smaller range of temperature than do continental interiors.
4 The range of temperature on the eastern sides of Asia and North America is greater than it is on the western sides in the *same latitude*.

PRESSURE
We have already seen that air has weight and therefore it exerts a pressure, called *atmospheric pressure*, on the earth's surface. This pressure is not the same in all regions nor is it always the same in one region all the time. Atmospheric pressure depends primarily on three factors:
1 Altitude
2 Temperature
3 Earth Rotation

Influence of Altitude on Pressure
The pressure of air at ground level is higher than that of air at the top of a high mountain. The air at ground level has to support the weight of the air above it and the molecules in the bottom air must push outwards with a force equal to that exerted by the air above it. The molecules of air at the top of a mountain are pushing outwards with much less force because the weight of the air above it is less. It therefore follows that *when air sinks its pressure increases*. In sinking, the volume of the air decreases but the number of molecules in it remains the same. The outward pressure of these molecules is spread over a smaller area. Similarly, *when air rises* its volume increases and the outward pressure of its molecules is spread over a larger area and *its pressure decreases*.

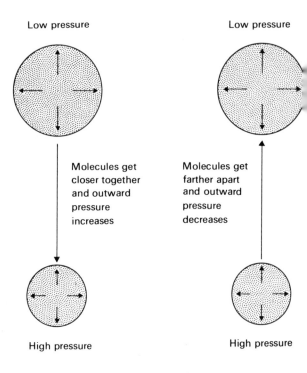

Molecules get closer together and outward pressure increases

Molecules get farther apart and outward pressure decreases

Very little air above this height

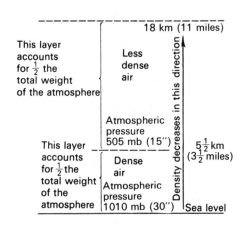

Influence of Temperature on Pressure
1 We have seen that when air sinks its pressure increases because it becomes compressed. And when it becomes compressed its molecules move

more quickly and heat is produced. *The temperature of air therefore rises when its pressure rises.*

2 When air rises its pressure decreases because it expands. When air expands its molecules move more slowly and heat is used up. *The temperature of air therefore falls when its pressure falls.*

3 When air is heated it expands and when this happens the outward pressure of its molecules is spread over a larger area. This means that the pressure of the air decreases. *The pressure of air therefore falls when its temperature rises.*

4 When air is cooled it contracts and when this happens the outward pressure of its molecules is spread over a smaller area. This means that the pressure of the air increases. *The pressure of air therefore rises when its temperature falls.*

Effects of temperature on pressure

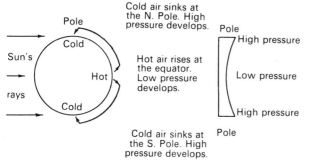

Effects of Earth rotation on pressure

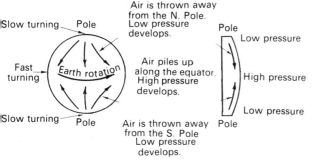

Other factors affect pressure

Supposing that only temperature affected pressure, then the pressure pattern of the atmosphere would be something like that shown in the diagram above (top). There would be a belt of low pressure around the earth at the equator, and two belts of high pressure, one over each pole. But because altitude,

temperature and earth rotation all affect pressure, the pressure pattern is not as simple as this.

Influence of Rotation on Pressure

The rotation of the earth causes air at the poles to be thrown away towards the equator. In theory this should result in air being piled up along the equator to produce a belt of high pressure, whilst at the poles low pressure should develop as shown in the diagram (bottom). But what actually happens is much more complicated and we must now try and find out how temperature and rotation together affect the pressure pattern.

Combined Influence of Rotation and Temperature on Pressure

First, the effect of temperature

1 Low temperatures at the poles result in the contraction of air and hence the development of high pressure (see page 128)

2 High temperatures along the equator result in the expansion of air and hence the development of low pressure (see page 128). This is called the *Doldrum Low Pressure.*

Second, the effect of rotation

1 Air blowing away from the poles crosses parallels which are getting longer. It therefore spreads out to occupy greater space, that is, it expands and its pressure falls. These low pressure belts become noticeable along parallels 60°N and 60°S. As air moves away from the poles, more air moves in from higher levels to take its place. Some of this comes from the rising low pressure air in latitudes 60°N and 60°S.

2 Air rising at the equator spreads out and moves towards the poles. As it does so it crosses parallels which are getting shorter and it has to occupy less space. It contracts and its pressure rises. This happens near to latitudes 30°N and 30°S and in these latitudes the air begins to sink thus building up the high pressure belts of these latitudes. These are called the *Horse Latitude High Pressures.*

Note Some of the high pressure air in latitudes 30°N and 30°S moves over the surface towards the equator, and some of it moves towards the poles. The air moving towards the equator replaces the air which is rising there. The air moving towards the poles reaches latitudes 60°N and 60°S where it replaces the air which is rising there.

If the earth had a uniform surface, i.e. if it was all land or all water, then the pressure pattern would be as shown in the diagrams on page 128. Winds have been inserted because they are produced by

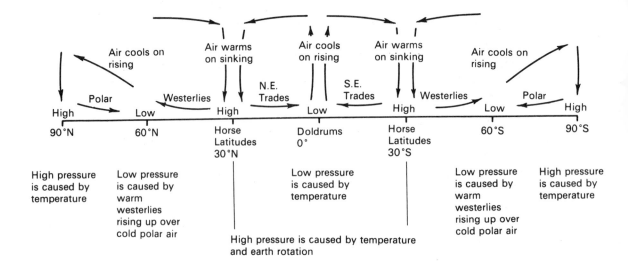

High pressure is caused by temperature | Low pressure is caused by warm westerlies rising up over cold polar air | High pressure is caused by temperature and earth rotation | Low pressure is caused by temperature | Low pressure is caused by warm westerlies rising up over cold polar air | High pressure is caused by temperature

pressure systems. You can see from the diagrams that at the surface, winds blow from highs to lows, whilst at high levels they blow from lows to highs. You can also see that there are three wind systems in each hemisphere: one operates between the pole and latitude 60°, one between latitude 60° and latitude 30° and one between latitude 30° and the equator. These three systems are called the *Polar*, the *Tropical* and the *Equatorial systems*.

Horse Latitudes (Sub-Tropical H.P.)
Doldrums (Equatorial L.P.)
Horse latitude (Sub-Tropical H.P.)

☐ High pressure   ▨ Low pressure

The Actual Pressure Systems on the Earth as it is

The earth's surface is not uniform, but is composed of land and water, and its axis is tilted at an angle of $66\frac{1}{2}°$. It has already been shown how land and water surfaces heat and cool at different rates, and

also how temperatures in regions outside the tropics and especially over land surfaces vary considerably from season to season. All of this results in changes in the pressure systems as given in the diagram above. Study the pressure maps on page 129 which are for January and July and try to account for the differences.

Pressure in January

(i) The Equatorial Low Pressure Belt extends well into the Southern Hemisphere where it is the summer season.

(ii) Low pressure is particularly well developed over Australia.

(iii) The low temperatures over the hearts of the northern continents produce strong high pressure systems. These link up with the sub-tropical high pressure systems.

(iv) The sub-tropical high pressures of the Southern Hemisphere are formed only over the oceans.

(v) Low pressure systems are well developed over the North Atlantic and the North Pacific Oceans.

Pressure in July

(i) The Equatorial Low Pressure Belt extends well into the Northern Hemisphere where it is the summer season, and it links up with the low pressure belts over north-west India and Pakistan and south-west U.S.A. Pressure in these areas is very low.

(ii) The sub-tropical high pressure belt in the Northern Hemisphere is no longer continuous, and it now exists as separate cells of high pressure only over the oceans.

(iii) The sub-tropical high pressure cells of the Southern Hemisphere combine to form one belt of high pressure which extends across the three continents.

(iv) The low pressure cells over the North Atlantic and North Pacific Oceans are poorly developed and have moved north.

Points to remember about the Seasonal Changes in Pressure

1 The revolution of the earth and the permanent tilt of its axis result in the overhead sun 'moving' between the Tropics as shown in the section dealing with the seasons. This causes the Doldrums to move north and south of the equator.

2 The Doldrums is the key to the pressure systems and hence these also move north and south of the positions they occupy when the sun is overhead along the equator (equinoxes).

3 Seasonal temperature changes over the continents in the Northern Hemisphere cause seasonal pressure changes over these continents.

Recording Pressure on Maps

The atmospheric pressures recorded at the weather stations in a region, at approximately the same time, are plotted on a map of the region at the positions of the stations. Lines are then drawn through places having the same pressures. These lines are called *isobars*.

Note: The pressures are usually 'reduced' to sea level before they are plotted. The pressures are expressed in millibars (mb).

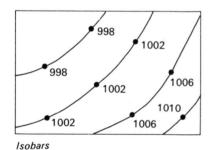
Isobars

WINDS

In our work on pressure we have studied the relationship between temperature and pressure, and we have seen that:

1 A rise in temperature causes air to expand and its density to decrease.

2 A fall in temperature causes air to contract and its density to increase.

These statements mean that *high temperatures give rise to low pressure at sea level, and low temperatures give rise to high pressure at sea level.*

Land and Sea Breezes

When a region is hotter than a neighbouring region air moves into the hot region from the cooler region to take the place of the hot air which has expanded

January Pressure

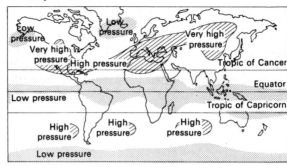

July Pressure

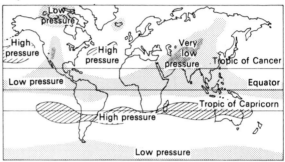

and risen. The air which moves in is a *wind*. The diagram below explains how this comes about. During the day the land gets warmer than the sea and hence air pressure is lower over the land than the sea. Air blows from sea to land as a *sea breeze*. During the night the land cools more quickly than the sea and the reverse process sets in. Land and sea breezes illustrate this process particularly well.

Earth Rotation Influences Wind Direction

In an earlier part of this book (page 10) we saw that

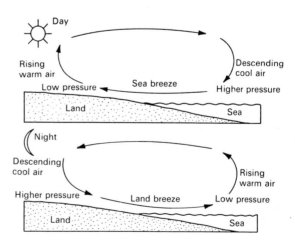

129

the rotation of the earth causes freely moving water and air masses to be deflected from their original courses. This is summarised by *Ferrel's Law* which states that freely moving bodies are deflected to their *right* in the Northern Hemisphere, and to their *left* in the Southern Hemisphere. In the diagram below dotted lines represent the paths which the winds would take if the earth was not rotating. Winds therefore do not follow straight paths, but curving ones as shown in this diagram.

In this diagram the earth is regarded as being uniform, i.e. all land or all water.

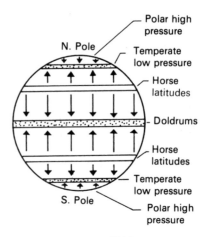

How the Winds would blow on a non-rotating earth

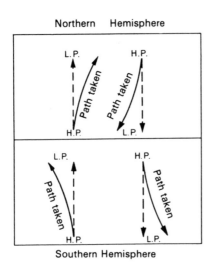

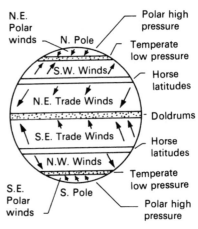

How the Winds blow on a rotating earth

Prevailing Winds

A wind which blows more frequently than any other wind in a particular region is called a *prevailing wind*. The bottom right diagram shows the prevailing winds of the world. You will notice from this diagram that winds are named by the direction from which they blow.

The maps on the top of page 131, which show the wind patterns for January and July respectively, should be carefully studied. Notice the following points:

(i) Over North America and Asia the winds are *out-blowing* during the winter when pressure is high, but are *in-blowing* during the summer when pressure is low.

(ii) The Westerly Winds in the Northern Hemisphere blow more persistently in winter than in summer. The Westerly Winds in the Southern Hemisphere blow persistently throughout the year.

(iii) The North East Trades over Asia are streng-

thened by the Asian High Pressure in winter, but completely disappear in summer because of the Asian Low Pressure.

(iv) The North East Trade Winds in eastern U.S.A. disappear in the summer because of the American Low Pressure.

(v) All the wind belts in July shift slightly northward of their January positions.

We have seen that because the earth's surface is part land and part water, changes take place in the pressure belts from season to season (page 129). This is accompanied by changes in the pattern of winds. The basic wind pattern as shown in the above diagram can be recognised in the wind patterns for both January and July. It only really breaks down over Asia and North America where winds change direction from season to season.

130

January Wind Pattern

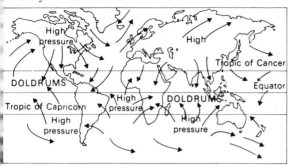

July Wind Pattern

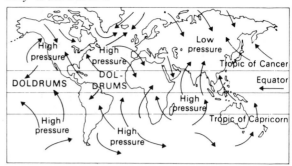

Monsoon Winds

Monsoon is derived from '*mausim*' (Arabic) which means *season*, and this word is applied to winds whose direction is reversed completely from one season to the next. Monsoon winds are best developed over Asia (Indian sub-continent, S.E. Asia, China and Japan).

In the map on the right and in that below, the full black arrows represent monsoon winds. The arrows in broken lines represent non-monsoon winds.

July

1 The Himalayas separate the Asian Low Pressure from the Punjab Low Pressure.
2 Winds blow from the **Australian High Pressure** across the **Equatorial Low Pressure to the more intense Asian Low Pressure.**
3 Winds blow from the Horse Latitudes High Pressure across the Equatorial Low Pressure to the more intense Punjab Low Pressure.

July

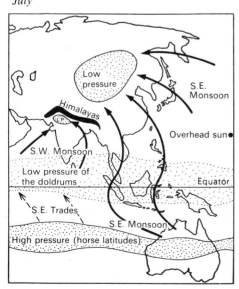

January

1 The Himalayas separate the Asian High Pressure from the Punjab High Pressure.
2 Winds blow from the Asian High Pressure across the Equatorial Low Pressure to the more intense Australian Low Pressure.
3 Winds blow from the Punjab High Pressure to the Equatorial Low Pressure.

January

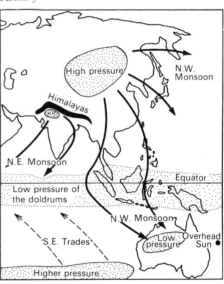

Characteristics of the Prevailing Winds

Polar Winds

1 They blow from the Polar High Pressures to the Temperate Low Pressures.
2 They are better developed in the Southern Hemisphere than in the Northern Hemisphere.
3 They are deflected to the right to become the N.E. Polar Winds in the Northern Hemisphere and to the left to become the S.E. Polar Winds

in the Southern Hemisphere.

4 They are irregular in the Northern Hemisphere.

Westerlies

1 They blow from the Horse Latitudes to the Temperate Low Pressures.

2 They are deflected to the right to become the S. Westerlies in the Northern Hemisphere and to the left to become the N. Westerlies in the Southern Hemisphere.

3 They are variable in both direction and strength.

4 They contain depressions.

Trades

1 The word 'trade' comes from the Saxon word *tredan* which means to tread or follow a regular path.

2 They blow from the Horse Latitudes to the Doldrums.

3 They are deflected to the right to become the N.E. Trades in the Northern Hemisphere and to the left to become the S.E. Trades in the Southern Hemisphere.

4 They are very constant in strength and direction.

5 They sometimes contain intense depressions.

Depressions and Anticyclones
Depression

This is a mass of air whose isobars form an oval or circular shape, where pressure is low in the centre and increases towards the outside. Depressions are well developed in the Westerly Wind and sometimes in the Trade Wind Belts. Depressions are rarely stationary and they tend to follow definite tracks. They are most influential in maritime areas because they weaken over land areas.

Anticyclone

This is a mass of air whose isobars form a similar pattern to that of a depression, but in which pressure is high at the centre, decreasing towards the outside. Anticyclones often remain stationary before gradually fading out. They often affect whole continents.

The wind circulation in depressions and anticyclones is shown in the following diagrams. You will see that *air moves anti-clockwise in a depression and clockwise in an anticyclone in the Northern Hemisphere, and clockwise in a depression and anti-clockwise in an anticyclone in the Southern Hemisphere.*

Types of Depressions

There are two types:

1 Depressions (temperate cyclones).

2 Tropical cyclones (hurricanes and typhoons).

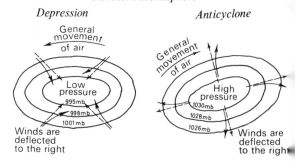

Northern Hemisphere

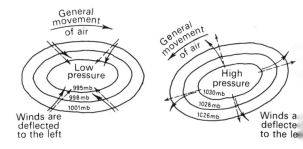

Southern Hemisphere

Depressions

These arise in the belt of Westerly Winds and are caused by the mixing of cold air from polar regions with warm, humid air from tropical regions. They consist of swirling masses of air (anti-clockwise in N. Hemisphere and clockwise in S. Hemisphere). They usually bring prolonged rain to coastal regions and often very windy weather.

Tropical Cyclones

These arise in the belt of Trade Winds where these winds begin to disappear in the Doldrums. They move in a general westerly direction and they have very low pressure. Because of this they give rise to winds of great force which are extremely destructive. Their circulation is the same as that of the depressions. In Asia they are called *typhoons*; in the West Indies they are called *hurricanes*.

How depressions and cyclones develop

The earth's atmosphere can be divided into several air masses each of which has distinctive temperature and humidity characteristics. Two such air masses are the polar air mass which originates over unfrozen land and water in high latitudes and which moves towards tropical latitudes, and the tropical air mass which originates in the low latitudes and which moves poleward.

The zone along which two contrasting air masses meet is called a *front*. The best known fronts are the *polar front* (between polar and tropical air

...asses) and the *inter-tropical front* (between the ...de wind belts of the Northern and Southern ...emispheres).

...hen contrasting air masses lie adjacent to one ...other, indentations develop along the front se-...rating them either because the warmer air of one ... mass slowly rides up over the colder air of the ...her, or because the colder air pushes into the ...armer air and forces it to rise. The positions of ...e two main fronts, the polar front and the inter-...opical front varies from season to season.

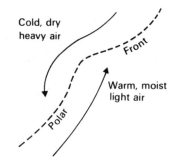

(i)

Note: Width of developing
depression: 320 to 800 kilometres
(200 to 500 miles)

...e development of a depression

...age 1   Along the polar front, cold polar air moves
... a general westerly direction, and warm tropical
...r moves in a general easterly direction. The
...ictional effects of the two air flows causes a wave
... develop. See diagram (i).

...age 2   The wave bulges into the colder air and
...ts larger. Pressure falls at the tip of the wave and
... anti-clockwise circulation of winds blows around
...is low pressure point in the Northern Hemisphere
... shown in diagram (ii). The circulation is clockwise
... the Southern Hemisphere.

...age 3   As the bulge develops, the warm air rises
... over the colder air at the front of the bulge. This
...ont is called the *warm front*. At the rear of the
...lge, the colder air forces its way under the warm
...r. The rear is called the *cold front*. The warm air
...tween the two fronts is called the *warm sector*. See
...agram (iii). You can see from diagram (iv) that the
...arm front is much more gently sloping than the
...ld front. Eventually the cold front catches up with
...e warm front and lifts it off the ground. It then
...comes an *occluded front* and it soon dies out.

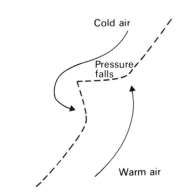

(ii)

...he weather associated with a depression

...a diagram (v) the depression is approaching A
...nd the sequence of weather at A will be a follows:
Clear sky except for a little high cirrus cloud. The
wind is from the south-east. After a while a
definite cloud cover develops and light showers
of rain occur, which get progressively heavier.
The warm front passes.
The rain stops and the wind changes direction
from south-east to south-west. Temperatures rise
and the air is humid (warm sector).
As the cold front passes the weather changes
very rapidly. The wind now blows from the
north-west and the temperature falls. Short but
often heavy falls of rain occur. With the passage
of the depression the sky clears and it remains
cool.

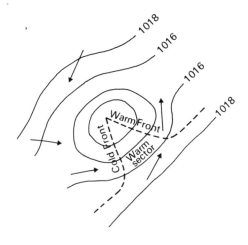

(iii)

133

(iv)

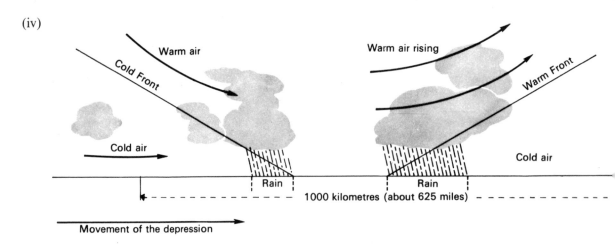

(v)

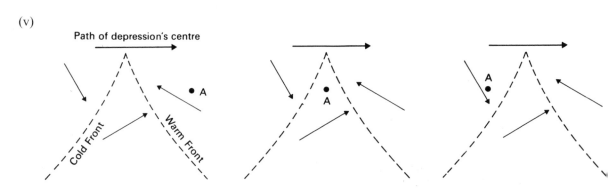

Weather Maps and Depressions

The following two maps show how warm and cold fronts appear on a weather map.

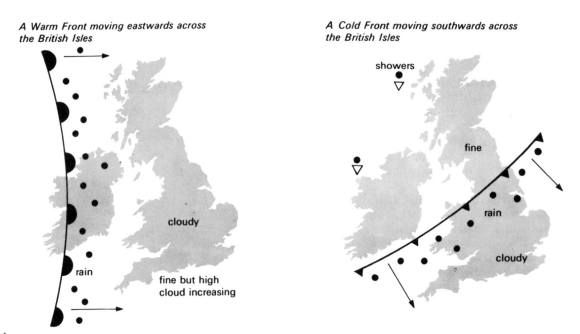

A Warm Front moving eastwards across the British Isles

cloudy

rain

fine but high cloud increasing

A Cold Front moving southwards across the British Isles

showers

fine

rain

cloudy

This diagram shows the birth of a depression. Cold air meets warm moist air moving in the opposite direction. The cold air forms a gently sloping wedge underneath the warm air and a front develops.

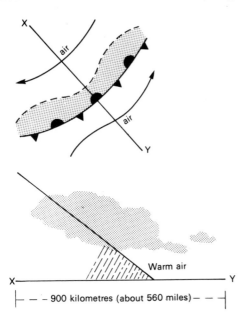

The next two diagrams show a mature depression as it appears on a weather map, and in three-dimensional form.

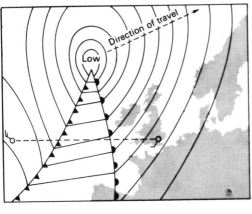

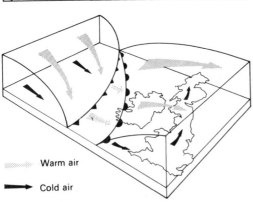

This weather map shows a depression and gives information on all aspects of weather associated with it. This type of map is often called a *synoptic chart*.

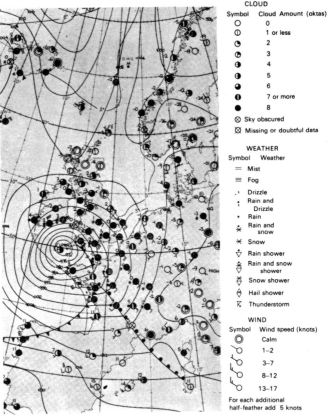

CLOUD	
Symbol	Cloud Amount (oktas)
○	0
◍	1 or less
◒	2
◒	3
◓	4
◑	5
◕	6
◕	7 or more
●	8
⊗	Sky obscured
⊠	Missing or doubtful data

WEATHER	
Symbol	Weather
=	Mist
≡	Fog
∙,	Drizzle
;	Rain and Drizzle
•	Rain
⁂	Rain and snow
✳	Snow
▽	Rain shower
⩔	Rain and snow shower
⩓	Snow shower
⬙	Hail shower
�	Thunderstorm

WIND	
Symbol	Wind speed (knots)
◎	Calm
◯‒	1–2
◯⟍	3–7
◯⟝	8–12
◯⟞	13–17
For each additional half-feather add 5 knots	

How a tropical cyclone develops
Tropical cyclones develop along the inter-tropical front where the trade wind air masses converge. These air masses over the oceans have moist lower layers but drier upper layers. When two such air masses meet, one of them will tend to be lifted up. This causes *instability*. Large cumulus clouds develop which produce intense thundery conditions. The actual origin of a tropical cyclone is not fully clear.

Location and Movement of Depressions and Tropical Cyclones

Tropical Cyclone

The diagram on page 137 shows that a tropical cyclone is funnel-shaped. The air on the outside is rapidly rising and swirling in an anti-clockwise direction in the Northern Hemisphere. The air inside the funnel is relatively calm because it is descending.

The Nature of a Tropical Cyclone

1 Before the cyclone arrives the air becomes very still, and temperature and humidity are high.
2 As the front of the vortex arrives gusty winds

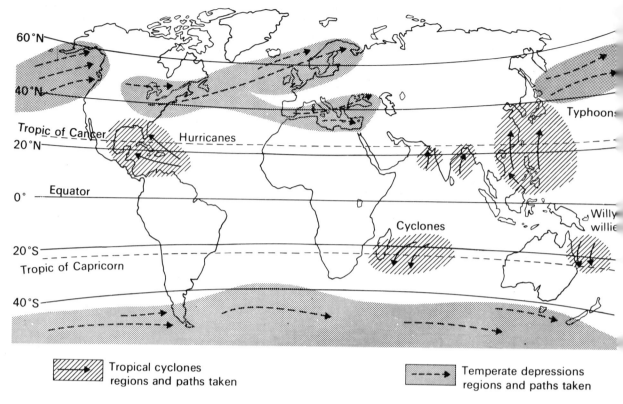

| Tropical cyclones regions and paths taken | Temperate depressions regions and paths taken |

develop and thick clouds appear.

3 When the vortex arrives, the winds become violent (upward surges) and often reach 250 kilometres (160 miles) per hour. Dense clouds and torrential rain reduce visibility to a few metres.

4 Calm conditions return when the eye of the cyclone arrives.

5 The arrival of the rear of the vortex brings in violent winds, dense clouds and heavy rain. The wind is now blowing from the opposite direction to that of the front of the vortex.

Where Tropical Cyclones form and what happens to them

They always form over the oceans in the tropics where the northerly and southerly Trade Winds meet. They follow an easterly course and soon lose their strength when they cross the coast and penetrate inland.

Section through a Tropical Cyclone

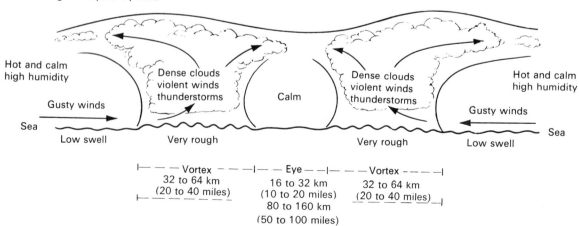

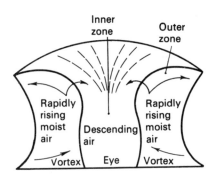

Section through a Tropical Cyclone

The different names given to tropical cyclones appear on the map on page 136.

Tornadoes, which occur in the Mississippi Valley of U.S.A., have not been shown on this map. A tornado differs from a tropical cyclone in that it forms over land. It is more destructive than a cyclone because its winds often exceed 320 kilometres (200 miles) per hour. Fortunately tornadoes are only a few hundred yards across.

Note The air moves into the depression from all directions.

Local Winds (affect only limited areas and blow for short periods of time)

Most local winds are developed by depressions. The air circulation in these is such that air is drawn in from tropical regions in the front of the depression (this gives rise to hot winds), and from polar regions in the rear of the depression (this gives rise to cold winds). See the diagram below.

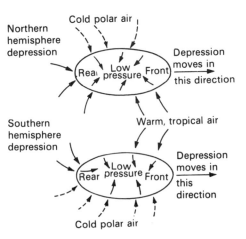

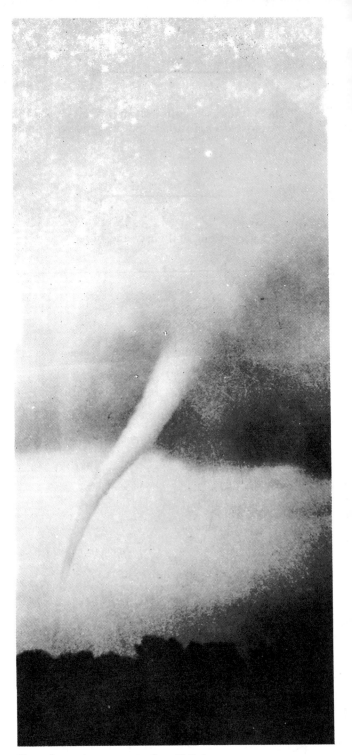

Tornado passing across southern U.S.A.

The uprising funnel of air is clearly visible. Although this is only a few hundred metres wide at the most it can cause tremendous damage by its 'vacuum' effect.

A Depression Winds

Hot Winds

Usually these are both hot and dusty, but if they have crossed a sea surface they become very humid.

Cold Winds

Often very strong and gusty and bitterly cold.

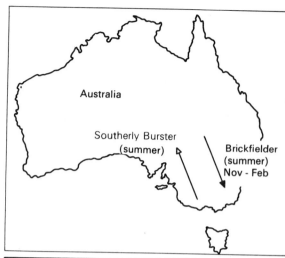

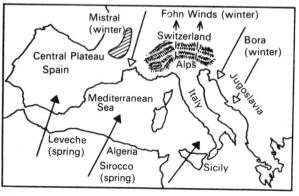

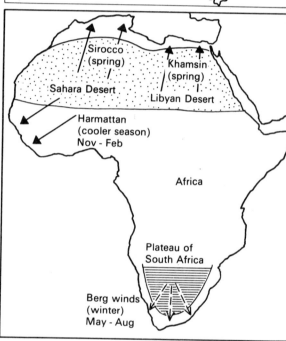

B Descending Winds

These are warm winds which descend mountain slopes onto the lowlands. As the air rises up the windward side of a mountain it cools at the rate of 6·5°C per 1000 metres. When condensation occurs, heat is given out, and the air cools at 4·5°C per 1000 metres. After crossing the mountain the air descends and it warms as it does so. Warming takes place at the rate of 6·5°C per 1000 metres. In the left diagram (see page 139) air on the windward side of the mountain at sea level is 7°C whereas on the leeward side at the same level it is 9°C.

138

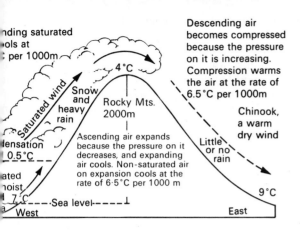

nding saturated
ols at
: per 1000m

Snow
and
heavy
rain

Saturated wind

4°C

Rocky Mts.
2000m

Ascending air expands
because the pressure on it
decreases, and expanding
air cools. Non-saturated air
on expansion cools at the
rate of 6·5°C per 1000 m

densation
0·5°C

ated
noist

7°C

West ----Sea level-----⊥ East

Descending air
becomes compressed
because the pressure
on it is increasing.
Compression warms
the air at the rate of
6.5°C per 1000m

Chinook,
a warm
dry wind

Little
or no
rain

9°C

Summary of Local Winds
Depression Winds

Hot Winds	Cold Winds
Sirocco	Mistral
Leveche	Bora
Khamsin	Pampero
Harmattan	Southerly Burster
Santa Ana	Buran
Zonda	
Brickfielder	

Descending Winds
Chinook
Föhn
Berg
Nor'wester
Samun (Persia)
Nevados (Ecuador)

RAINFALL
Under certain conditions water vapour, which is a
gas, takes the form of tiny droplets of water. These
appear as a cloud, mist, fog, hail, dew or rain. All
of these forms are referred to as *precipitation*.

Conditions necessary for Precipitation to Occur
The air must be saturated.
The air must contain small particles of matter
such as dust around which the droplets form.
The air must be cooled below its dew-point.

How Air is Cooled
Air is cooled in two main ways:
By being made to rise – (most of the world's rain
results from this type of cooling)
  (i) Hot air rising by convection currents (top
     diagram)
  (ii) A wind blowing over a mountainous region
     (middle diagram)
  (iii) Warm air rising over cold air.

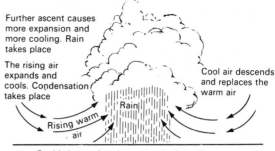

Further ascent causes
more expansion and
more cooling. Rain
takes place

The rising air
expands and
cools. Condensation
takes place

Rain

Cool air descends
and replaces the
warm air

Rising warm
air

Earth's hot surface heats the air above it.
The heated air expands and becomes lighter
than the surrounding air. It therfore rises.

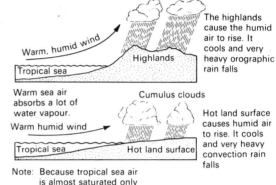

Warm, humid wind

Highlands

Tropical sea

The highlands
cause the humid
air to rise. It
cools and very
heavy orographic
rain falls

Warm sea air
absorbs a lot of
water vapour.

Warm humid wind

Tropical sea

Cumulus clouds

Hot land surface

Hot land surface
causes humid air
to rise. It cools
and very heavy
convection rain
falls

Note: Because tropical sea air
is almost saturated only
a little cooling is
required to condense the
water vapour into rain

2 *By passing over a cold surface* – (most of the
world's mist and fog results from this type of
cooling)
  (i) A warm wind blowing over a cold current
  (ii) A warm wind blowing over a cold land
     surface.

Oceans and ocean currents influence rainfall
The importance of oceans as a source of rainfall
is well known. In tropical latitudes the air over the
oceans contains much more water vapour than does
the cooler air over the oceans in temperate latitudes.
But remember not all tropical regions have heavy
or even fairly heavy rainfall.
  (i) Winds blowing over a warm current on to a
     cooler land surface usually bring heavy rain.
  (ii) Winds blowing over a cool current on to a
     warmer land surface usually bring little or no
     rain.
  (iii) Winds blowing over a warm current and then
     over a cold current usually produce fog.

Types of Rainfall
I Convection Rain
Convection rain is often accompanied by lightning
and thunder. In tropical latitudes this type of rain
is usually torrential. It is the most common type of
rain to fall in equatorial regions and in regions
having a Tropical Monsoon Climate.

A Thunderstorm

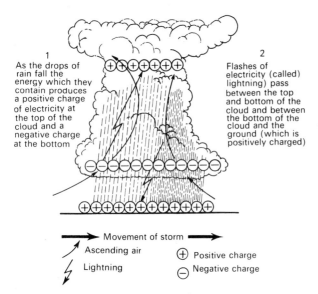

1
As the drops of rain fall the energy which they contains produces a positive charge of electricity at the top of the cloud and a negative charge at the bottom

2
Flashes of electricity (called lightning) pass between the top and bottom of the cloud and between the bottom of the cloud and the ground (which is positively charged)

Movement of storm
Ascending air
Lightning
⊕ Positive charge
⊖ Negative charge

Thunderstorms can occur whenever land surfaces become greatly heated. In humid tropical regions like Indonesia, Malaysia, Central and West Africa, the Amazon Basin and Central America, thunderstorms are very common. They usually occur in the afternoon and are especially frequent in the season of heavy convectional rains.

The thunder of these storms is caused by the rapid expansion and contraction of the air. The electrical discharges (lightning) produce intense heat which causes the air to expand. Cooling soon takes place and the air contracts.

II Depression or Cyclonic Rain

Depression rain occurs when large masses of air of different temperatures meet. The warm air is forced up and over the cooler air. In tropical cyclones the rainfall is often very heavy but lasts for only a few hours. In temperate depressions it is much lighter but lasts for many hours or even days. Cyclonic rain is common throughout the Doldrums where the trade winds meet.

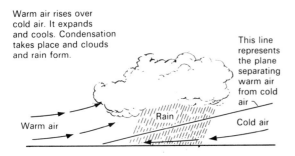

Warm air rises over cold air. It expands and cools. Condensation takes place and clouds and rain form.

This line represents the plane separating warm air from cold air

Warm air
Rain
Cold air

III Orographic or Relief Rain

Whereas convection rain only occurs in region whose surfaces are greatly heated by the sun, an cyclonic rain only occurs where masses of air c different temperatures meet, orographic rain occur in all latitudes. It is most common where on-shor winds rise up and over hilly or mountainous region lying parallel to the coasts.

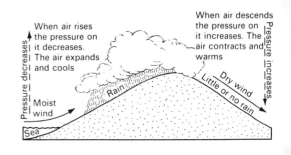

When air rises the pressure on it decreases. The air expands and cools

Pressure decreases

Moist wind

Sea

Rain

When air descends the pressure on it increases. The air contracts and warms

Pressure increases

Dry wind
Little or no rain

Fogs

Extensive fogs develop where warm and cold currents meet and where warm moist winds blow over cold surfaces. Fogs are very frequent off the mouth of the St. Lawrence (map below) and round th shores of Japan where the warm Kuro Siwo and th cold Oya Siwo (Okhotsk) meet. They often develop off the west coasts of hot deserts where warm sea breezes pass over the cold off-shore current. The sea breezes are only local winds but they give rise to belts of mist or fog just off the coasts. Examples of this type occur off the coasts of California, Peru N.W. Africa and S. Africa (lower map).

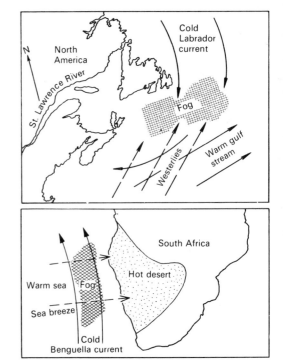

North America
Cold Labrador current
St. Lawrence River
Fog
Westerlies
Warm gulf stream

South Africa
Warm sea
Fog
Hot desert
Sea breeze
Cold Benguella current

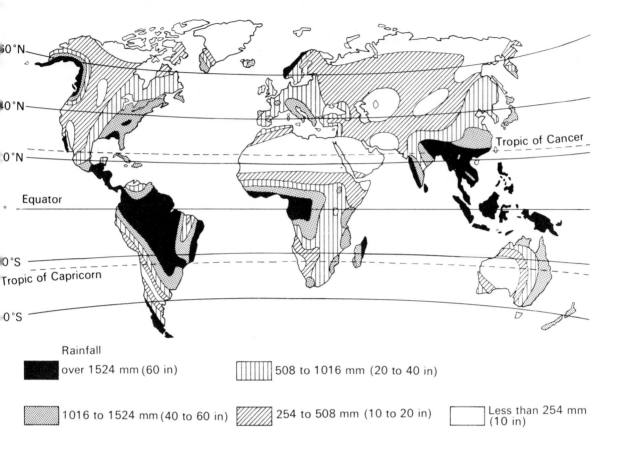

Rainfall

▮ over 1524 mm (60 in) | ⫼ 508 to 1016 mm (20 to 40 in)

▦ 1016 to 1524 mm (40 to 60 in) | ▨ 254 to 508 mm (10 to 20 in) | ☐ Less than 254 mm (10 in)

The World Pattern of Rainfall

We have already studied how the revolution of the earth and the tilt of its 'axis' result in a movement of some pressure belts and a change in others, and how this in turn causes the belts of prevailing winds to move northwards and southwards, and other wind belts to change direction during the year. Let us now see how this affects the rainfall pattern of the world. First, we will examine the map at the top of this page which shows the mean annual rainfall. In this map the world has been divided into five types of rainfall region on a basis of the amount of rain received annually. Notice the following points.

1 The wettest regions (over 1500 mm about 58 in) occur in:
  (i) Equatorial Latitudes (Amazon and Zaire Basins, Indonesian Islands, Malaysia and New Guinea)
  (ii) Tropical Monsoon Regions (S. China, Peninsular S.E. Asia, Bangladesh (formerly E. Pakistan), N.E. India and W. India, Sri Lanka (Ceylon) and the Philippines)
  (iii) Regions receiving on-shore Westerly Winds (British Columbia, N.W. Europe, S. Chile, Tasmania, South Island of New Zealand).

2 The driest regions (less than 250 mm about 10 in) occur in:
  (i) Hearts of N. America and Asia
  (ii) Regions lying permanently under off-shore Trade Winds (Sahara and Arabian Deserts, Australian Desert, Kalahari and Atacama Deserts and the deserts of S.W. States of U.S.A.)
  (iii) Arctic Lowlands (N. America, Greenland and Asia).

Now this map only tells us how much rain a region receives each year. It does not tell us at what time of the year the rain comes. When we study agriculture, i.e. the types of crops grown on the earth's surface, it is necessary to know both the amount of rain falling in a region in one year and the time of year when it comes. The maps on page 142 and 143 show the distribution of rain for the world during the summer and winter seasons.

May 1st to October 31st
(See map on page 129)

1 The sun is overhead in the Northern Hemisphere and most of the rain falls in this hemisphere.

2 The belt of equatorial convection rains is chiefly located north of the equator.

3 Southern and eastern Asia, eastern N. America and E. America receive heavy rain from on-shore winds.

4 Extensive areas in S.W. Asia, N. Africa, N. and Western Australia, S. Peru, N. Chile, S.W. States of N. America, and the Namib Desert receive little or no rain because they lie under off-shore trade winds.

5 The Arctic lowlands receive little rain because of low temperatures which prevent the air from absorbing much water vapour.

November 1st to April 30th

(see map on page 143)

1 The sun is overhead in the Southern Hemisphere and most of the rain falls in this hemisphere.

2 The belt of equatorial convection rain is chiefly located south of the equator.

3 N. and E. Australia, S.E. Africa and S.E. Brazil and E. Argentina receive rain from on-shore trade winds. (In N. Australia rain is brought by monsoon winds which are modified N.E. Trade Winds.)

4 Extensive areas of S.W. Asia, N. Africa, Centra and E. Australia, S. Peru, N. Chile and S.W Africa receive little or no rain because they li under off-shore trade winds.

5 The Arctic Lowlands as before have little or n rain.

The Migration of the Overhead Sun and th Rainfall Pattern

A *Some regions are NOT influenced*

1 Regions permanently in a belt of prevailing wind:
   (i) N.W. Europe, W. Canada, S. Chile, Tasm; nia, South Island (N.Z.).
   They lie in the belt of on-shore Westerlic (rain all the year).
   (ii) S. California, N. Chile, Sahara, S.W. Asi; W. Australia and the Namib Desert.
   They lie in the belt of off-shore Trades (littl or no rain throughout the year).

2 Regions permanently in the doldrums belt, e.; Zaire and Amazon Basins. They receive co vection rain throughout the year.

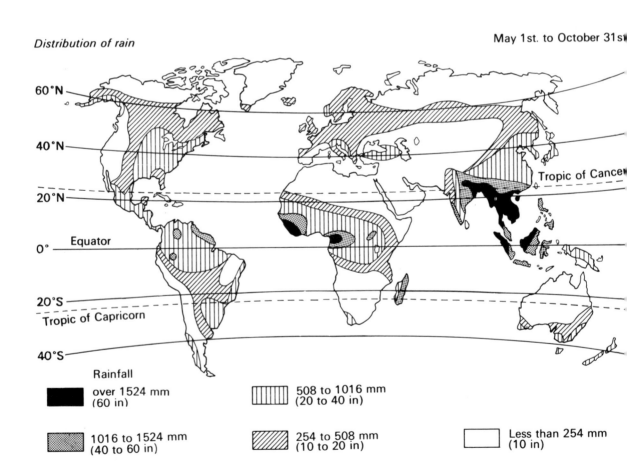

Distribution of rain

May 1st. to October 31s

60°N

40°N

Tropic of Cance

20°N

0° Equator

20°S

Tropic of Capricorn

40°S

Rainfall

■ over 1524 mm (60 in)

▨ 1016 to 1524 mm (40 to 60 in)

▥ 508 to 1016 mm (20 to 40 in)

▧ 254 to 508 mm (10 to 20 in)

□ Less than 254 mm (10 in)

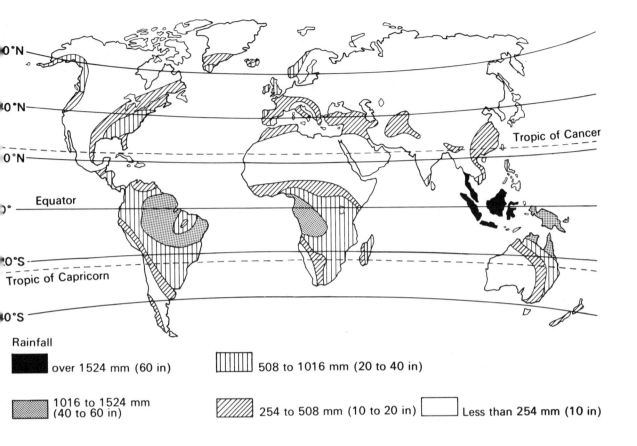

Rainfall

■ over 1524 mm (60 in) |||||| 508 to 1016 mm (20 to 40 in)

▨ 1016 to 1524 mm
(40 to 60 in)   ▨ 254 to 508 mm (10 to 20 in) □ Less than 254 mm (10 in)

B *Some regions ARE influenced*

1 Regions lying between two belts of prevailing winds:
Central California, Central Chile, Mediterranean Lowlands, S.W. Australia.
They lie between the Westerly and Trade Wind belts. The Westerlies bring rain in the cool season; in the warm season the Trades blow and there is no rain.

2 The interiors of Asia and N. America. In these regions atmospheric pressure is low in summer. In-blowing winds give rain and intense heat causes convection rain. In winter atmospheric pressure is high and winds are out-blowing. There is little or no rain.

3 Regions bordering the permanently wet equatorial regions lie under the doldrums once a year and this results in heavy convection rain. They also lie under the Trades once a year. Where these are off-shore there is little or no rain. These regions lie in S. America, Africa and Australia and they have a *Sudan climate* (see page 150).

4 Monsoon Regions:
Japan, China, Peninsular S.E. Asia, India, Bang-

ladesh, Sri Lanka and N. Australia.
For one season they lie under on-shore winds which bring rain. For the other season they lie under off-shore winds which bring little or no rain.

In the map on page 144 the world has been divided into four types of rainfall region. The four types are:
 (i) Regions receiving rain throughout the year
 (ii) Regions receiving rain chiefly in the hot season (summer)
 (iii) Regions receiving rain chiefly in the cool season (winter)
 (iv) Regions receiving little or no rain.
The following statements can be made about this map:
1 *Regions having rain all the year are located in*:
 (i) Equatorial and some tropical latitudes
 (ii) Coastal regions where winds are on-shore for most of the year.
2 *Regions having most rain in the summer are located in*:
 (i) Most of Asia excluding Malaysia and In-

143

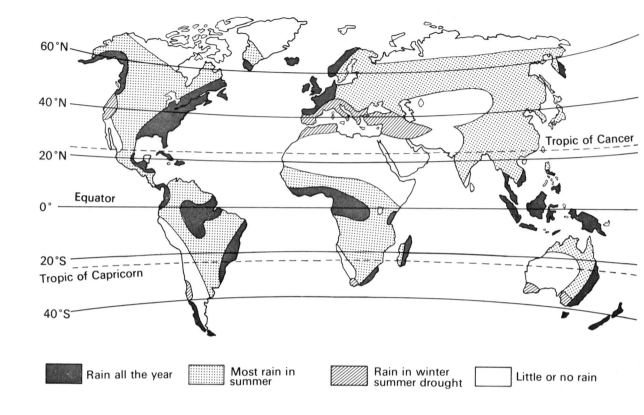

donesia and New Guinea
(ii) E. Europe
(iii) N. Australia
(iv) Regions bordering the equatorial latitude of Africa and S. America
(v) Central N. America.

3 *Regions having most rain in the winter are located in*:
(i) Central California and Central Chile
(ii) Mediterranean Lowlands
(iii) S.W. Africa and S.W. Australia.

4 *Regions having little or no rain are located in*:
(i) S. California and S.W. States of N. America
(ii) The Atacama and Kalahari Deserts
(iii) The Sahara, Arabian and Asian Deserts
(iv) The Australian Desert
(v) The Polar Deserts.

Seasonal Rainfall and Type of Rain

On page 145 is a diagrammatic representation of Eurasia and N. Africa and it shows the seasonal distribution of winds and rain together with the type of rain. The winds on the left-hand side operate over the western part of the region and those on the right-hand side operate over the eastern part of the region. Examine this figure and pay particular attention to the following:

(i) The north-south shift of some wind belts, and the change in wind direction of other belts from season to season.
(ii) The location of on-shore winds (which usually bring rain) and of off-shore winds (which usually bring no rain).
(iii) The distribution of rain in relation to the lines of latitude which are given.

Recording of Rainfall on a Map

Lines are drawn on the map through all places having the same rainfall. Such lines are called *isohyets*. They are drawn at a uniform interval (in the lower left diagram on page 145 this is 25). A scale of colours or line shading (see lower right diagram on page 145) is then worked out and applied to the map. It is usual to start with light colours or open lines for low rainfall and to use darker colours or closer lines for heavy rainfall.

Rainfall maps may show seasonal or annual values.

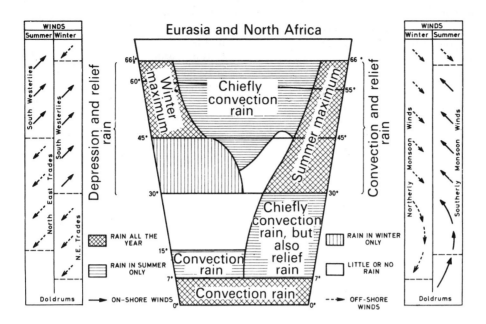

Eurasia and North Africa

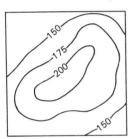
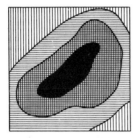

Rainfall in millimetres

EXERCISES

Carefully explain the meanings of the following:
  (i) sea breezes are day winds, and land breezes are night winds
  (ii) water surfaces gain and lose heat more slowly than land surfaces
  (iii) temperature decreases as altitude increases
  (iv) the height of the mid-day sun in the tropics is always higher than it is in temperate latitudes. Illustrate your answer with well-labelled diagrams.

By using well-labelled diagrams, describe *three* ways by which rain may be caused, and for each, name a specific region where this type of rain commonly occurs.

Explain each of the following statements:
  (i) most of the hot deserts are located on the western sides of continents between 20°S. and 30°S. and 20°N. and 30°N.
  (ii) heavy fog frequently occurs over the waters around Newfoundland

  (iii) the surface waters of the north-east Atlantic are warmer than those of the north-west Atlantic
4 Carefully explain the meaning of each of the following:
  (i) mean monthly temperature
  (ii) annual temperature range
  (iii) mean annual temperature
  (iv) diurnal temperature range
5 Briefly discuss the different factors which affect the world temperature distribution pattern, and explain the part played by temperature in climate.
6 Explain the following statements:
  (i) water vapour and dust influence the amount of insolation a region receives
  (ii) descending winds are warm winds
  (iii) temperate cyclones give changeable weather
  (iv) extremes of temperature often occur in the hearts of continents
7 Name *four* local winds and for *each*:
  (i) describe why it develops
  (ii) explain why it is warm or cold, dry or damp
  (iii) name a region where it operates
8 Name the instruments used for measuring wind direction and velocity and state how these two characteristics can be shown in diagrammatic form.
9 Carefully explain, with the aid of labelled diagrams, the following statements:
  (i) equatorial regions receive rain throughout the year
  (ii) the highest daily temperatures are recorded outside equatorial latitudes
  (iii) the hearts of the northern continents receive little rain.

Objective Exercises

1 The approximate amount of the Sun's energy (called insolation) which reaches the Earth's surface is
   A 90%
   B 75%
   C 45%
   D 30%
   E 20%

   A B C D E
   .. ..

2 Four towns, all at about the same altitude and on the same line of latitude, have mean January temperatures as given below. Which town is farthest from the sea?
   A   7°C
   B −18°C
   C −10°C
   D   3°C
   E   10°C

   A B C D E
   .. ..

3 The difference between the maximum and minimum temperatures recorded for a place during a period of one day is called
   A daily mean temperature
   B diurnal temperature range
   C daily average temperature
   D mean monthly temperature
   E mean annual temperature

   A B C D E
   .. ..

4 Perpendicular rays are usually more heating than oblique rays. It can therefore be said that
   A a south-facing slope is warmer than a north-facing slope in the summer in the Northern Hemisphere
   B a north-facing slope is warmer than a south-facing slope in the summer in the Northern Hemisphere
   C the seasons in the tropics are short
   D winter in latitude 35°S is colder than winter in latitude 35°N
   E summer temperatures in central Asia are lower than they are in central Africa

   A B C D E
   .. ..

5 Which of the following winds are predominantly seasonal winds?
   A depression(cyclonic) winds
   B prevailing winds
   C monsoon winds
   D local winds
   E descending winds

   A B C D E
   .. ..

6 All of the following are cyclones **except**
   A hurricane
   B Zonda
   C Willy Willy
   D typhoon
   E tornado

   A B C D E
   .. ..

7 Which one of the following is **not** a necessary condition for the formation of heavy rainfall?
   A air must be warm near the surface
   B moist air must rise to great heights
   C air must become saturated
   D air pressure must be high
   E humidity must be high

   A B C D E
   .. ..

8 Which one of the following statements relates to both land and sea breezes?
   A Air blows from the sea to the land during the day.
   B Air blows from the land to the sea during the night
   C Air generally moves from a cool region to a warmer region.
   D The land cools more quickly than the sea during the night.
   E The sea warms up more slowly than the land during the day.

   A B C D E
   .. ..

9 All of the following statements are true **except**
   A Relative humidity of a mass of air falls if the temperature of the air rises.
   B Air is saturated when its relative humidity is 100%.
   C When air subsides its relative humidity decreases.
   D The relative humidity of a mass of air remains constant when the air crosses over a cold land surface from a warm water surface:
   E Condensation takes place in a mass of air if there is further cooling after the relative humidity of the air reaches 100%.

   A B C D E
   .. ..

10 A line on a weather map joining all places of equal pressure is called an
   A isotherm
   B isohyet
   C isohel
   D isobar

   A B C D
   .. ..

11 Which one of the following statements most accurately describes the relationship between air pressure and air temperature?
A When air is heated it becomes lighter and it rises.
B Air which is cooled contracts and the outward pressure of its molecules is spread over a smaller area.
C When air rises its pressure decreases.
D The pressure increases in a body of air which is descending.

A B C D

12 All of these statements are true about anticyclones in general but only one is true of anticyclones in the Southern Hemisphere. Which is that statement?
A The air moves in a circular manner.
B Pressure increases from the outside to the centre.
C Anticyclones often form over the continents.
D The air moves in an anti-clockwise direction.

A B C D

13 The land on the leeward side of mountain ranges which are at right angles to on-shore winds is often dry. This is because
A the winds are descending on the leeward side
B pressure is high to the leeward side
C the air on the leeward side is cool and is therefore relatively dry
D the leeward side lies under dry land winds

A B C D

14 Equatorial lowlands usually experience
A a large diurnal temperature range
B heavy thunder rain in the afternoon
C strong winds
D cold nights

A B C D

15 The seasonal rainfall pattern of India is caused by
A the large annual range of temperature
B the tropical location of India
C the monsoon winds operating over southern Asia
D the Himalayas blocking winds from interior Asia

A B C D

16 A line on a map which joins places having the same rainfall is called an
A isohyet
B isobar
C isotherm
D isohel

A B C D

17 Altitude is one of several factors which influence the temperature of a place. Which one of the following statements best explains the influence of altitude on temperature?
A A place which has a high altitude is nearer the Sun and it therefore has a higher temperature than it would have at sea level.
B Because the Sun's rays meet the top levels of the atmosphere first, high altitudes tend to be warmer than low altitudes.
C The atmosphere is heated primarily from below and therefore temperatures at high altitudes are lower than temperatures at low altitudes.
D Air is rarefied at high altitudes and it contains very little vapour or dust. Heat from the Earth's surface therefore rapidly escapes and temperatures at high altitudes are lower than they are at low altitudes.

A B C D

18 Which one of the following factors can have the greatest influence on the temperature of a place in Equatorial latitudes?
A aspect
B distance from the sea
C altitudes
D ocean currents

A B C D

19 Which one of the following statements about winds and atmospheric pressure is **not** correct?
A Winds blow in a clockwise direction around a centre of low pressure in the Northern Hemisphere.
B Winds blow in a clockwise direction around a centre of high pressure in the Northern Hemisphere.
C Winds blow in an anti-clockwise direction around a centre of high pressure in the Southern Hemisphere.
D Winds blow in a clockwise direction around a centre of low pressure in the Southern Hemisphere.

A B C D

20 The windward slopes of coastal mountains which are at right angles to winds blowing from the sea are wetter than the leeward slopes. This is because
A they are nearer the sea
B the winds have to rise to cross them
C descending winds are warm
D the sea is warmer than the land

A B C D

21 If the temperature at sea level is 7°C, the temperature of the air at a height of 2000 metres will be about
A 15°C
B 16°C
C −6°C
D 10°C

A B C D

147

14   Types of Climate

All of us know that some regions are hot and others are cold; some are wet and others are dry; some have rain all the year and others have rain for part of the year only.

The world can be divided into climatic regions, each of which has a distinct temperature and rainfall pattern. This can be done by dividing the world into five temperature zones:

Hot Zone; Warm Zone; Cool Zone; Cold Zone; Very Cold Zone.

Each of these zones is very large, and, with the exception of the very cold zone, the rainfall* distribution is not even, i.e. one part of a zone may be wet whilst another part of the same zone may be dry. All except the very cold zone are now subdivided into rainfall regions. The resulting regions all have a distinct climate and these are shown in the map below. The only climatic type shown on this map which is not based on the temperature and rainfall division is the mountain climate. This map shows 18 climatic types in all and you must become familiar with all of them.

The main features of each climatic type are summarised in three or four statements, and a climatic diagram is drawn for a particular town for each type. At least two towns have been selected for each of the tropical climates. The climatic diagrams show:

1 Monthly Rainfall
2 Total Annual Rainfall
3 Annual Temperature Range

TROPICAL CLIMATES

EQUATORIAL CLIMATE (Padang and Singapore)

Location

1 The best examples occur in the lowlands between

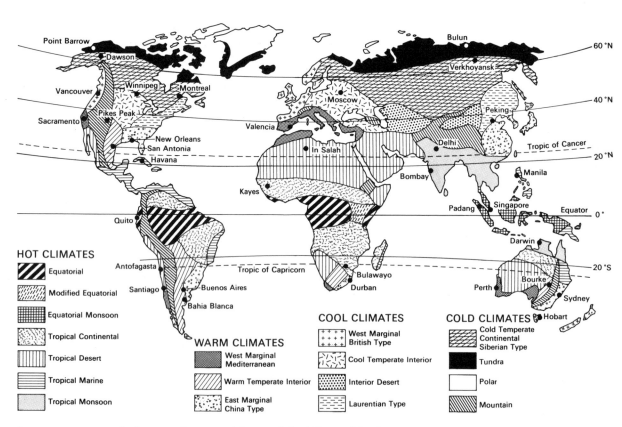

HOT CLIMATES

- Equatorial
- Modified Equatorial
- Equatorial Monsoon
- Tropical Continental
- Tropical Desert
- Tropical Marine
- Tropical Monsoon

WARM CLIMATES

- West Marginal Mediterranean
- Warm Temperate Interior
- East Marginal China Type

COOL CLIMATES

- West Marginal British Type
- Cool Temperate Interior
- Interior Desert
- Laurentian Type

COLD CLIMATES

- Cold Temperate Continental Siberian Type
- Tundra
- Polar
- Mountain

Precipitation covers both rain and snow, and this word should be used for the very cold climates and other climates where snowfall is significant.

5°N. and 5°S., e.g. the Amazon and Zaire Basins.

2 The highlands which occur between these latitudes, e.g. East African highlands, have a much modified equatorial climate. Altitude in these regions 'reduces' temperatures to about 15°C (59°F).

3 A part of the Guinea coast of West Africa receives low annual rainfall, e.g. Accra, 700 mm (28 in). This region really has an equatorial climate which is modified by monsoon winds.

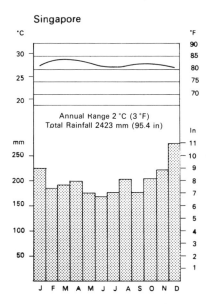

Singapore

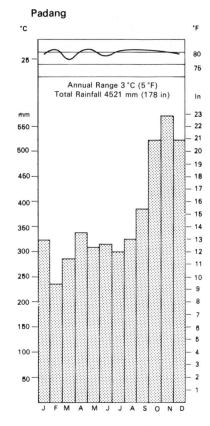

Padang

Climatic characteristics

1 This latitudinal belt lies under the Doldrums Low Pressure all the year and, therefore, there are no seasons.

2 The mid-day sun is always near the vertical and it is overhead twice a year, at the equinoxes.

3 Average daily temperatures are 26°C (about 79°F) throughout the year. These are well below the average daily temperatures of other types of climate occuring outside equatorial latitudes. Extensive cloud cover and heavy rainfall prevent temperatures from rising much over 26°C (about 79°F).

4 The diurnal temperature range is between 6°C (11°F) and 8°C (15°F) which is greater than the annual temperature range of about 3°C (5°F).

5 Rainfall is heavy and is usually convectional. Rains often come in the afternoons and are generally accompanied by lightning and thunder. Annual rainfall is about 2000 mm (80 in) though some regions get higher falls.

Note: Malaysia, Singapore and Indonesia are generally said to have an equatorial climate. Although these countries show some of the characteristics of this climate, e.g. the temperature and humidity patterns, others are not as well developed. This is because they lie under the monsoon winds which sweep across South-East Asia from the N.E. from November to February and from the S.W. from July to September. These winds moderate humidity and give a seasonal pattern to the rainfall regime of many regions.

6 Humidity is always high.

Agricultural development

1 In some of the more heavily forested parts of equatorial regions, populations are sparse and hunting and food collecting form the basic economies of the more primitive people, while shifting cultivation is practised by the more advanced forest dwellers.

2 Plantation agriculture for the cultivation of rubber (Malaysia and Indonesia), cacao (West Africa, especially Ghana) and oil palm (Nigeria and Malaysia) and other crops like sugar, bananas, and pineapples, etc. has become established in many coastal regions.

3 Equatorial forests contain valuable trees such as mahogany, ebony and greenheart, and the cutting down of these in some regions has given rise to important timber industries.

Extensive development is often made difficult by many factors, the most important of which are:

(i) diseases and insect pests which attack man, his animals and his crops, (ii) the difficulty of establishing communications in the very densely forested regions, (iii) the generally poor, thin soils which soon lose what little goodness they contain when the forest cover is removed.

TROPICAL CONTINENTAL (SUDAN) CLIMATE
(Kayes and Bulawayo)
Location
1 It occurs between 5°N. and 15°N., and 5°S. and 15°S.
2 It is best developed in Africa and east central South America.

Climatic characteristics
1 The latitudinal belt which has this climate comes under the Trade Winds for a part of the year (winter) and lies under the doldrums for the rest of the year (summer).
2 Summers are hot with temperatures around 32°C (about 90°F). Winters are cooler, 21°C (about 70°F). The annual temperature range is therefore about 11°C (20°F).
3 Heavy rains, mainly of convectional type, fall in in the summer: winters are usually dry.
4 The Trade Winds in north Africa blow from the Sahara Desert and are dry, hot winds. These are particularly noticeable in West Africa where they are called the *harmattan*. Besides being dry and hot, the harmattan are also dusty. In South Africa and in South America, south of the equator, the Trade Winds blow from the sea and they bring rain to coastal regions.
5 The annual rainfall is often around 762 mm (30 in) but it may be more, e.g. for regions near to the equatorial latitudes, or less e.g. regions near the hot deserts.

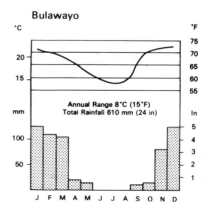

Bulawayo

Annual Range 8°C (15°F)
Total Rainfall 610 mm (24 in)

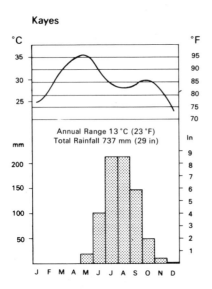

Kayes

Annual Range 13°C (23°F)
Total Rainfall 737 mm (29 in)

6 The highest temperatures occur just before the rainy season begins, i.e. April in the Northern Hemisphere and October in the Southern Hemisphere
7 Humidity is high in the summer.
8 The climate can be said to have hot, wet summers and cooler, drier winters, but note the exceptions referred to in (4).

Agricultural development
The natural vegetation consists of tall grass, often 3 m (10 ft) or more, with scattered clumps of trees. In Africa this type of vegetation is called *savana*, and it is the home of a great variety of animals such as elephant, giraffe, zebra, deer, leopard, tiger jaguar, etc.

1 Agricultural development is not far advanced in the African savanas. The Masai, who are nomadic herdsmen, maintain large herds of zebu cattle, goats and sheep which graze on the grasslands of Kenya, Tanzania and Uganda. The cattle are reared for their milk and blood both of which form important items in the diet of the Masai. The blood is taken from the neck of the animal which does not appear to suffer any ill effects. The cattle are not killed for meat.
2 Crops are grown in the African savanas, especially in northern Nigeria where the Hausa have practised crop cultivation for a long time. They grow food crops such as Guinea corn, millet, maize, bananas, groundnuts and beans. They also grow non-food crops such as cotton and tobacco which are sold for cash. The Hausa practise crop rotation which enables soil fertility to be maintained. They also keep herds of goats and cattle which supply them with milk and meat.
3 Similar crops are cultivated by the Kikuyu in East Africa and some of their produce is sold to the Masai.
4 Commercial plantation farming has become established in some parts of the African savana, especially in East Africa, Uganda, Malawi, Kenya

and Tanzania. The main crops of these plantations are sugar cane, tobacco, sisal and cotton.

5 The two savana regions of South America, known as the *llanos* and the *campos*, are both used for rearing cattle. The latter is the more extensive and the more important for animal farming. Its potentials are considerable.

6 The savana of Australia is very sparsely populated and as yet there has been no significant agricultural development.

The main factors which at present prevent any further expansion to agricultural development are (i) the unreliable nature of the rainfall (droughts are common), (ii) diseases and insect pests which attack animals and crops, (iii) loss of soil fertility through the removal of such minerals as phosphates, potash and nitrates by heavy rain in the west season, (iv) poor communications, and (v) a sparse population in relation to the extent of the savana lands. It is possible that all of these conditions will be put right or counteracted eventually, and if and when this happens, the savana lands will become major producers of animal products, and food and non-food crops.

TROPICAL MARINE CLIMATE (Havana, Durban and Manila)
Location
1 It occurs in east coastal areas of regions having a tropical continental climate, and in the coastal regions of Central America.
2 It is developed best in the lowlands of Central America, the West Indies, coastal lowlands of Brazil and East Africa, including east Malagasy, north-east Australia and the Philippines.

Climatic characteristics
1 On-shore trade winds blow throughout the year and they bring rain almost every day with rather heavier falls in the hot season.

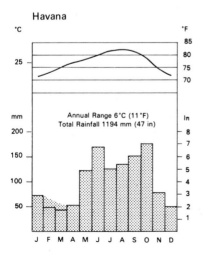

Havana

2 Annual rainfall varies from 1000 mm to 2000 mm (40 in to 80 in) depending upon location. Rainfall is both convectional and orographic.

3 Temperatures are similar to those of tropical continental climates. The annual temperature range is about 8°C (15°F) with hot season temperatures of 29°C (about 85°F) and cool season temperatures of 21°C (about 70°F).

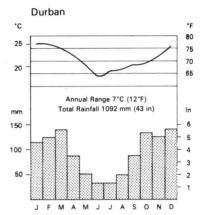

Durban

Annual Range 7°C (12°F)
Total Rainfall 1092 mm (43 in)

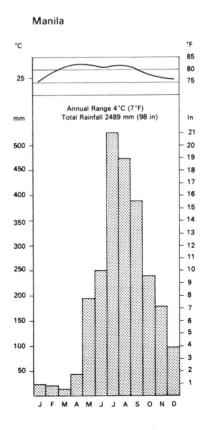

Manila

Annual Range 4°C (7°F)
Total Rainfall 2489 mm (98 in)

Agricultural development

1 The cultivation of crops, especially food crops is important in almost all tropical marine regions. Sugar-cane is grown annually on large plantations in most parts of the West Indies, East Africa (especially Natal) and Queensland (north-east Australia). Other crops include coffee in eastern Brazil, and Manila hemp in the Philippines.

TROPICAL MONSOON CLIMATE (Delhi, Bombay and Darwin)

Location

In some tropical and temperate latitudes seasonal land and sea winds operate on a huge scale affecting both continents and oceans. These seasonal winds, called *monsoon winds,* are best developed over an area extending from south-eastern and eastern Asia to northern Australia. Eastern Asia has a temperate monsoon climate (refer to the maps on page 131).

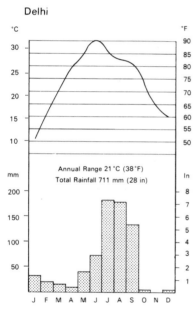

Delhi

Annual Range 21 °C (38 °F)
Total Rainfall 711 mm (28 in)

Climatic characteristics

1 Seasonal reversal of winds is the chief feature of this climate. For one season the winds blow from the sea to the land bringing heavy rainfall to coastal regions, and for another season the winds blow from the land to the sea and these give little or no rain.

2 Annual rainfall varies greatly, the amount falling depending mainly on relief and the angle at which the on-shore winds meet this. The south-facing slopes of the Khasi Hills in Assam receive as much as 12 500 mm (500 in) annually. In contrast the region around Delhi receives only 620 mm (25 in) annually.

3 Temperatures range from 32°C (about 90°F) in the hot season to about 15°C (59°F) in the cool

season, thus giving an annual range of about 17°C (31°F). But these values vary very much (compare Bombay and Delhi temperature graphs).

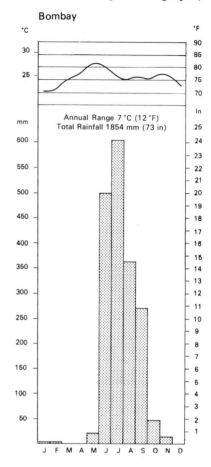

Bombay

Annual Range 7 °C (12 °F)
Total Rainfall 1854 mm (73 in)

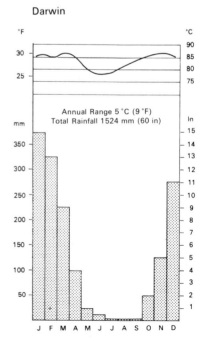

Darwin

Annual Range 5 °C (9 °F)
Total Rainfall 1524 mm (60 in)

4 A typical tropical monsoon climate consists of three seasons. For the Indian sub-continent and Burma these are:
   (i) Cool, dry season (November to February) when the off-shore north-east monsoon winds blow.
   (ii) Hot, dry season (March to May) when temperatures are high because of the near overhead mid-day sun, and when winds are almost absent.
   (iii) Hot, wet season (June to October) when the on-shore south-west monsoon winds blow and when rainfall is very heavy in regions receiving the full force and fury of on-shore winds. During this season cloudy skies cause temperatures to fall a little, but humidity rises to its maximum for the year.

Agricultural development

1 The intensive cultivation of food crops takes place in all parts of tropical monsoon regions where soils are adequately fertile and rainfall or the availability of water in the form of lakes and rivers, is sufficient to meet crop requirements.
2 There is also extensive cultivation of non-food crops usually on plantations or estates which in the main have been established on cleared forest land.
3 Padi is the most commonly cultivated food crop and it forms the staple diet of most of the people in the wetter parts of India, Pakistan, Burma, southern China, and the countries of northern South-East Asia. Wet padi is grown on well-watered lowlands and hill slopes (which have to be terraced). Dry padi, which does not require as much water, is grown on drier hill slopes.
4 Wheat, millet, maize, sorghum, etc. are grown in the drier areas where padi cannot be grown. These are particularly important food crops in northern India and Pakistan.
5 Other important lowland crops are sugar cane, especially in the wetter regions, cotton, a most important crop in India and Pakistan, and jute, mainly in the Ganges Delta.
6 In the highland areas of many tropical monsoon regions tea forms an important plantation crop. This crop is extensively grown in Sri Lanka and the Himalayan foothills of India and Bangladesh.

The cultivation of food crops, especially padi is entirely dependent upon the on-shore monsoon bringing the right amount of rain at the right time. If the monsoon arrives late it can result in widespread famine. Alternatively, monsoon rains may be unusually severe, resulting in widespread flooding and crop destruction, thus again causing famine.

The main obstacle to better farming in many tropical monsoon regions is the farmers' ignorance of modern farming methods involving maintenance and improvement of soil fertility by using fertilisers, manure and crop rotation. Before this obstacle can be overcome, the practice of land division on the death of the owner will probably have to be stopped, and money will probably have to be made available to farmers for the purchase of good seed, fertilisers and labour-saving farming equipment.

TROPICAL DESERT CLIMATE (Antofagasta and In Salah)

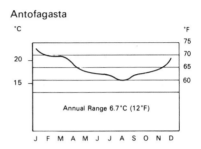

Antofagasta

Annual Range 6.7°C (12°F)

Location

1 It occurs on the western sides of land masses in the belt of permanent trade winds. The only exception is the desert belt in north Africa which extends from the west coast right across the continent and into south-west Asia. This is because the trade winds affecting the eastern part of north Africa are blowing in from the land mass of south-west Asia and they are therefore dry winds.
2 The most important regions having this type of climate are the Sahara Desert, the Arabian, Iranian and Thar Deserts, the Australian Desert, the Kalahari and Namib Deserts, the Atacama Desert and the Californian and Mexico Deserts.

Climatic characteristics

1 Rain rarely falls and the average annual fall is usually below 12 cm (5 in). Sometimes there are sudden torrential downpours which give rise to temporary flooding.
2 The hot deserts occur in the tropical high pressure belts where air is subsiding. Such air absorbs rather than yields moisture. Also, most of the winds blowing into the hot deserts originate in cooler regions. When crossing the desert these winds get hotter and again this prevents condensation. Note also that on-shore winds which sometimes affect the west coasts of desert belts cross cold currents which occur along these coasts. The winds are cooled and condensation takes place giving fog or light showers. The winds are therefore dried and on reaching the warm land surface they are dried still further.

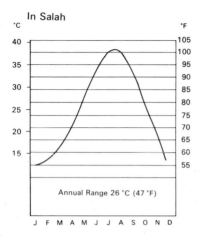

In Salah

Annual Range 26 °C (47 °F)

3 Temperatures vary from 29°C (about 85°F) in the hot season to 10°C (50°F) in the cool season.

4 Because there are no clouds, day temperatures often go over 38°C (about 100°F) and temperatures of over 49°C (120°F) are not uncommon. At night, again because there is no cloud cover, radiation is rapid and temperatures can fall to 15°C (59°F) in the hot season to 5°C (about 40°F) in the cool season. Diurnal temperature ranges are therefore very high.

Agricultural development

1 Regions having this type of climate can be cultivated and lived in if there is a constant supply of water.

2 The lowlands of the valleys of the Nile, the Tigris-Euphrates and the lower Indus are intensively cultivated for both food and non-food crops. By developing irrigation canals, all of these regions have extended the area of cultivation beyond the limits of the river valleys.

3 The cultivation of crops in desert areas where there are no rivers can only take place if water is available in some other form. Water does occur in oases and it is at these sites that settled agriculture takes place. Some oases are small but some are very extensive. Dates, wheat, vegetables and fruit are usually grown. Oases occur mainly in the Sahara and Arabian Deserts.

4 Nomadic herding takes place in the Sahara and Arabian Deserts where Bedouin Arabs own large flocks of sheep and goats. These animals graze on the scant pastures which occur in some parts of the desert, especially in the higher parts where there is heavy condensation,. in the form of dew at night. The animals provide the Bedouin with most of his basic requirements: the rest he obtains by barter trade with the oases farmers.

The main factor preventing further development in the deserts is the very limited quantities and distribution of available fresh water. Some deserts, especially the Sahara Desert, are known to have large supplies of water below the surface. One day it may be possible to make this available on a huge scale which could then see an extension of farming activities. It may also be possible to tow ice-bergs from the Antarctic to the hot desert areas.

WARM TEMPERATE CLIMATES
WARM TEMPERATE WESTERN MARGIN CLIMATE – MEDITERRANEAN CLIMATE
(Sacramento, Santiago and Perth)

Location
1 This occurs between 30°N. and 45°N. and 30°S. and 40°S., on the western sides of continents.
2 The climate is best developed around the shores of the Mediterranean Sea, in south-west Africa, central Chile, central California, and south-west and southern Australia (Adelaide to Melbourne).

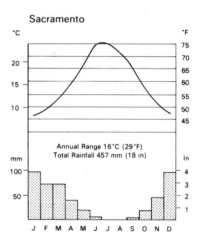

Sacramento

Annual Range 16°C (29°F)
Total Rainfall 457 mm (18 in)

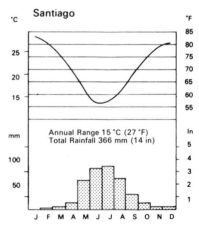

Santiago

Annual Range 15°C (27°F)
Total Rainfall 366 mm (14 in)

Climatic characteristic
1 Temperatures range from 21°C (about 70°F) in the summer to 10°C (50°F) or below in the winter.
2 Off-shore trade winds blow in the summer. These are dry and give no rain. The sky is cloudless and humidity is low.
3 On-shore westerly winds blow in the winter

bringing cyclonic rain. This usually amounts to about 500 mm (20 in) though considerably more may fall on steep slopes lying at right angles to the wind. The rain often comes in heavy showers which sometimes cause floods.

4 The annual rainfall ranges from 500 mm (20 in) to 760 mm (30 in).

5 Mediterranean climates experience both hot and cold local winds. Some examples are the *sirocco*, a hot dusty, dry wind which blows in the summer across the Mediterranean Sea from the Sahara Desert; the *mistral*, an intensely strong, cold wind, which blows in the winter down the Rhône Valley from the north and often reaching the Mediterranean coast; the *bora*, another winter

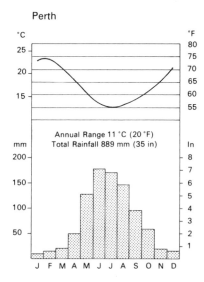

Perth

Annual Range 11 °C (20 °F)
Total Rainfall 889 mm (35 in)

wind which develops because of pressure differences between central Europe and the Mediterranean Sea and blowing south across Yugoslavia to the Adriatic.

6 The climate can be summarised as having bright, sunny, hot and dry summers and mild rainy winters.

Agricultural development

1 The climate permits a large range of crops to be grown. These include fruits and cereals.

2 Citrus fruits (oranges, lemons, grapefruit and limes) are extensively cultivated in many regions, very often by means of irrigation. The semi-desert regions of central California, central Chile and Israel are now very important producers of citrus fruits made possible by the use of irrigation. Other important fruits are peaches, apricots, plums, cherries and pears.

3 The olive tree is especially important to Mediterranean regions. Its fruit is rich in oil which is used in cooking in much the same way as coconut oil. The fruit is also eaten fresh and forms an

important item in the diet of Mediterranean people.

4 Nuts such as chestnuts, walnuts and almonds are grown.

5 The cultivation of grapes, which is called *viticulture*, takes place in most regions having a Mediterranean climate. A large part of the fruit is used for making wine, though some is dried (sultanas, currants and raisins) and some is eaten fresh.

6 The two most important cereals grown are wheat (the more important) and barley. Both crops are grown in the winter though in some regions, e.g. the Po Valley of Italy and the Ebro Valley of Spain they are grown in summer if irrigation water is available.

7 Other crops which are cultivated in some regions are figs, tobacco and cotton.

8 Agriculture has given rise to specialised industries such as wine-making, flour-milling, fruit-canning and food processing.

WARM TEMPERATE INTERIOR CLIMATE (Bourke and San Antonia)

Location

1 It occurs in the interior of continents, excluding Asia, between latitudes 20°N. and 35°N. and 20°S. and 35°S. The interior of Asia in these latitudes has very low winter temperatures and very low annual rainfall. Its climate is therefore better described as Temperate Desert.

2 The climate is best developed in the southern continents and is especially well developed in the Murray-Darling Lowlands (Australia) and in the High Veldt (Africa). It also occurs in the U.S.A. in western Oklahoma, Texas and northern Mexico.

Climatic characteristics

1 Temperatures range from 26°C (about 79°F) in the summer to 10°C (50°F) in the winter.

2 The annual rainfall varies from 380 mm (15 in) to 700 mm (30 in) depending on location. A good deal of the rain is brought by South-East Trade

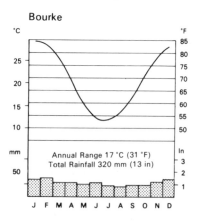

Bourke

Annual Range 17 °C (31 °F)
Total Rainfall 320 mm (13 in)

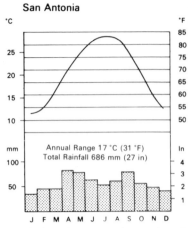

San Antonia

Annual Range 17 °C (31 °F)
Total Rainfall 686 mm (27 in)

Winds to the southern continental regions in the summer months. This decreases from east to west which means that the western parts of these climatic regions are fairly dry, sometimes semi-arid. There is also convectional rain which is caused by the low pressure systems over these areas in summer. Most of the rain occurs in summer.

3 Evaporation is high in summer and this often makes irrigation essential for crop cultivation.

Agricultural development

1 Grass is the natural vegetation of regions having this climate. The grasslands have specific names, e.g. the *Downs* (Australia), the *Veldt* (South Africa) and the *Pampas* (Argentina).

2 Vast herds of beef cattle and flocks of sheep are reared on these grasslands. Special fodder crops are grown for the animals.

WARM TEMPERATE EASTERN MARGIN CLIMATE – CHINA TYPE (New Orleans, Buenos Aires, Sydney)

Location

1 It is located on the eastern sides of continents between latitudes 23°N. and 35°N. and 23°S. and 35°S.

2 It is best developed in central China, south-eastern U.S.A., southern Brazil and the eastern part of the Pampas (Argentina), south-eastern Africa and south-eastern Australia.

Climatic characteristics

1 As in the Mediterranean climate, the trades and westerlies are the dominant seasonal winds. But notice the contrasts: in this climate the trades are on-shore winds and they bring rain whilst the westerlies are off-shore winds. These bring lighter rain.

2 Summers are hot 26°C (about 79°F) and winters are mild 13°C (about 55°F). Temperatures in the winter can be dramatically lowered suddenly

when local winds caused by depressions develop, e.g. *pampero* (Argentina) and southerly *burster* (Australia), and blow strongly.

3 Monsoonal winds tend to develop in both south-eastern U.S.A. and in China. In China the development is marked and there is a definite seasonal wind reversal.

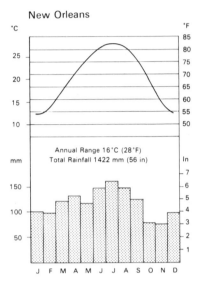

New Orleans

Annual Range 16°C (28°F)
Total Rainfall 1422 mm (56 in)

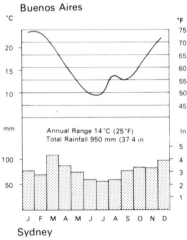

Buenos Aires

Annual Range 14°C (25°F)
Total Rainfall 950 mm (37.4 in)

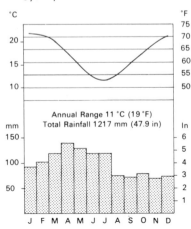

Sydney

Annual Range 11°C (19°F)
Total Rainfall 1217 mm (47.9 in)

4 Most of the rain falls in summer and it is convectional. The ligher rains of winter are caused by depressions developing in the off-shore westerly winds. Total annual rainfall is about 1000 mm (40 in).

5 Typhoons (south China) and hurricanes (south-east U.S.A.) are common in summer.

Agricultural development

1 This climate is especially well suited to intensive and continual crop cultivation. The temperature and rainfall patterns enable crops to be grown throughout the year.

2 Padi (often two crops a year) is cultivated extensively in China and it forms the basis of the daily diet of most Chinese in central China. Padi farming relies heavily upon human labour and it is grown on a subsistence basis.

3 A great variety of crops are grown in south-eastern U.S.A. Of these, cotton, maize and tobacco are the most important. Most of the maize is used for fattening pigs and cattle and hence animal farming is also important.

4 In south-east Africa, especially in Natal, maize is extensively grown as a basic food crop, and sugar cane, tobacco and cotton are grown as cash crops.

5 Cattle and sheep are reared in vast numbers in that part of South America which has this climate. Some wheat and maize are grown as well.

6 In south-east Australia the main activity is dairy farming and this region produces large quantities of milk, butter and cheese.

COOL TEMPERATE CLIMATES

COOL TEMPERATE WESTERN MARGIN CLIMATE — THE BRITISH TYPE (Vancouver, Valencia, Hobart)

Location

1 It occurs on the western sides of continents between 45°N. and 60°N. and south of 45°S.

2 It is best developed in north-west Europe, western Canada (British Columbia), coastal southern Chile, Tasmania and South Island of New Zealand.

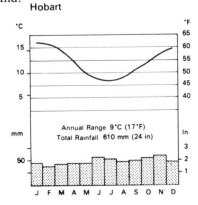
Hobart
Annual Range 9°C (17°F)
Total Rainfall 610 mm (24 in)

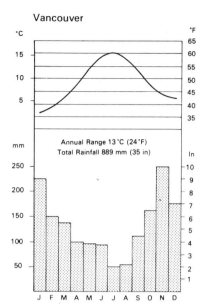
Vancouver
Annual Range 13°C (24°F)
Total Rainfall 889 mm (35 in)

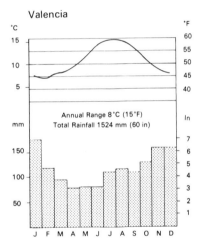
Valencia
Annual Range 8°C (15°F)
Total Rainfall 1524 mm (60 in)

Climatic characteristics

1 Winter temperatures range between 2°C (about 35°F) and 7°C (about 45°F), while summer temperatures range from 13°C (about 55°F) to 15°C (59°F). The annual temperature range is between 8°C (14°F) and 11°C (24°F).

2 Prevailing winds throughout the year are from the west, but they blow more strongly and persistently in the winter.

3 The on-shore westerlies, especially in the Northern Hemisphere, cross warm ocean currents (North Atlantic and North Pacific Drifts) which prevent winter temperatures from falling very low. It is the combined effect of ocean currents and winds which results in a small annual temperature range.

4 Sub-tropical and sub-polar air meet in these latitudes and this gives rise to depressions and

anticyclones which move in from the west. These pressure systems produce changeable weather.

5 Rain falls throughout the year, though there is a maximum in winter. Cyclonic and orographic rains both occur. The total annual rainfall is about 760 mm (30 in) though in mountainous regions this may be as high as 2500 mm (100 in).

Note Westerly winds and the warming influence of the warm ocean currents affects western coastal regions of Alaska and Norway which lie in the cold temperate latitudes. Winter temperatures in these regions are lower, often dropping to just below freezing point. However, these are still much higher than for places in the same latitudes and which are a few hundred miles to the east. Summer temperatures are usually below 15°C (59°F).

Agricultural development

1 A large part of the natural vegetation of deciduous forest has long since been cut down to make way for farming.
2 Extensive areas, especially in the wetter regions, are under grass and cattle and sheep farming is very important.
3 In north-west Europe the following types of farming take place:
 (i) beef cattle and dairy farming
 (ii) sheep farming for wool and meat
 (iii) cereal farming, especially for wheat, barley, and oats
 (iv) mixed farming – cattle, root crops and cereals on a crop rotation basis
 (v) market gardening, especially near to the urban conurbations
 (vi) fruit farming.
4 Agricultural activities in British Columbia are similar to those of north-west Europe except that fruit farming, especially apple farming is one of the major activities.
5 Vast flocks of sheep are reared in South Island of New Zealand and sheep farming for wool and meat dominates the economy.
6 Sheep farming is also important in Tasmania and southern Chile, but the latter region is the least developed of all regions having this type of climate. Fruit farming, especially for apples, is a major activity in Tasmania.

COOL TEMPERATE EASTERN MARGIN CLIMATE – LAURENTIAN TYPE (Montreal, Peking and Bahia Blanca)

Location
1 It occurs on the eastern sides of North America and Asia between 35°N. and 50°N. and on the eastern side of South America, south of 40°S.
2 It is best developed in the Maritime Provinces

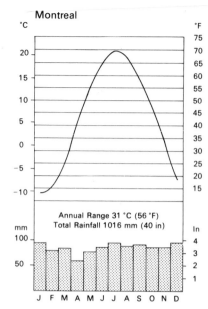

of eastern Canada, and the New England states of the U.S.A., northern China, Manchuria, Korea and northern Japan.

Climatic characteristics
1 Winter temperatures range from −9°C (about 15°F) to −7°C (about 20°F), and summer temperatures 15°C (60°F) to 24°C (about 75°F). The annual temperature range is therefore high and it averages about 24°C (45°F). Compare this with the temperature range of 11°C (20°F) for Western Margin climates.
2 Cold winds blow seawards from the interiors of North America and Asia in the winter and this causes the temperatures to be low. In North America these winds pick up moisture from the Great Lakes which gives rise to heavy falls of snow. In Asia the out-blowing winds are dry.
3 Off-shore currents, the Labrador Current (N. America) and the Kuril Current (Asia) reduce the winter temperatures along the coasts still further. Compare with the warm currents off the western coasts in the same latitude.
4 Precipitation (as rain and snow in N. America and N. Japan) occurs throughout the year and is fairly evenly distributed. In north-east Asia (except N. Japan and N. Korea) however, rainfall is confined to the summer (see (2) above). The total annual rainfall varies from 530 mm (25 in) to 1000 mm (40 in). The rain is both convectional and cyclonic.
5 North-east Asia has a typical monsoon wind pattern, i.e. seasonal wind reversal. Dry cold out-blowing winter winds give no rain to the mainland but heavy snowfalls occur in northern Japan and northern Korea. Moist, in-blowing summer winds bring rain to all parts.

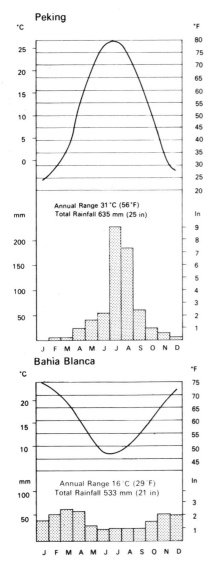

Peking

Annual Range 31 °C (56°F)
Total Rainfall 635 mm (25 in)

Bahia Blanca

Annual Range 16 °C (29°F)
Total Rainfall 533 mm (21 in)

COOL TEMPERATE INTERIOR CLIMATE
(Moscow and Winnipeg)

Location
1 It occurs in the interior of North America and Eurasia between latitudes 35°N. and 60°N.
2 It is best developed in the provinces of Alberta, Saskatchewan and Manitoba, in Canada, the north-central and mid-west of the U.S.A., in central and eastern Europe, and western U.S.S.R.

Climatic characteristics
1 Winter temperatures often fall to as low as −19°C (about 0°F) with summer temperatures rising to

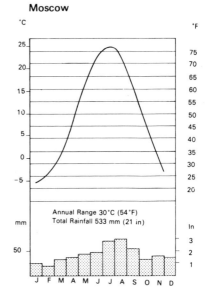

Moscow

Annual Range 30°C (54°F)
Total Rainfall 533 mm (21 in)

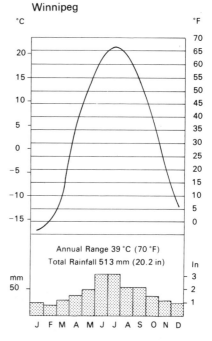

Winnipeg

Annual Range 39 °C (70 °F)
Total Rainfall 513 mm (20.2 in)

Agricultural development
1 Mixed farming with cattle, hay, oats and wheat is a major activity in the lowlands of north-east America. Dairy farming and market gardening are important near the towns. Fruit farming, especially for apples, is very important in some areas, e.g. Nova Scotia (Canada).
2 Crop farming is extensive in Manchuria and other parts of mainland north-east Asia which have this type of climate. Important crops are wheat, maize, kaoliang (a millet) and soya bean. The latter is rich in proteins and provides oil which is used in much the same way as coconut oil or olive oil. The plant also enriches the soil and is used in crop rotation.
3 Sheep farming is about the only important agricultural activity in south-eastern South America (Patagonia). The rainfall here is very low and grasslands are poor. Both mutton and wool are produced.

18°C (about 65°F) which gives an annual temperature range of 37°C (65°F).

2 Rainfall occurs mainly in the summer and it is convectional. The annual total rarely exceeds 513 mm (20.2 in). Rainfall decreases towards the west in N. America and towards the east in Eurasia. Rainfall is so low east of the Caspian Sea that the climate is of the Desert type and it is better described as Interior Desert (see the map on page 148).

Agricultural development

1 The cultivation of wheat, using large farm units, is of major importance in both the American and Eurasian regions.

2 Mixed farming, usually cattle with wheat and other temperate cereals, takes place in the European and Soviet regions.

COLD TEMPERATE CONTINENTAL CLIMATE – SIBERIAN TYPE (Verkhoyansk and Dawson)

Location

1 This is located between the Cool Interior Climate and the Tundra Climate in both N. America and Eurasia.

2 It is best developed in Canada and the U.S.S.R.

Climatic characteristics

1 Winter temperatures range from −34°C (about

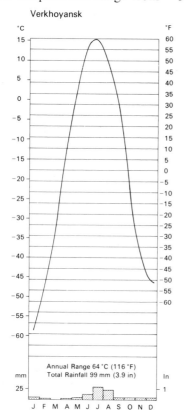

Verkhoyansk

Annual Range 64°C (116°F)
Total Rainfall 99 mm (3.9 in)

−30°F) in parts of Canada to −45°C (about −50°F) in parts of the U.S.S.R. Summer temperatures average about 21°C (about 70°F). The annual temperature range is normally over 55°C (100°F).

2 Total annual rainfall rarely exceeds 380 mm (15 in) and most of this occurs in the summer. These rains result from the entry of moist sea air. Snow falls in winter in eastern Canada are produced by depressions which move eastwards along the St. Lawrence Valley.

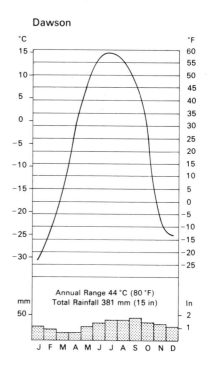

Dawson

Annual Range 44°C (80°F)
Total Rainfall 381 mm (15 in)

Agricultural Development

The subsoil is frozen for most of the year and this prevents most types of agriculture from taking place.

TUNDRA CLIMATE (Barrow Point and Bulun)

Location

1 It occurs in the northern continents north of the cold temperate continental climate.

2 It is best developed in northern Canada, and northern Asia.

Climatic Characteristics

1 Winter temperatures range from −29°C (about −20°F) to −40°C (−40°F) and summer temperatures are about 10°C (50°F). The annual range varies from 39°C (70°F) to 50°C (90°F).

2 Winter nights are long with hardly any daylight and summer days are long with hardly any night.

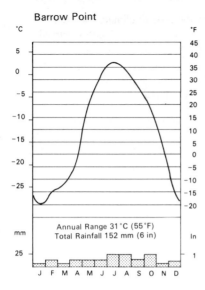

Barrow Point

Annual Range 31°C (55°F)
Total Rainfall 152 mm (6 in)

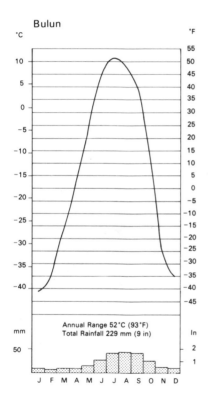

Bulun

Annual Range 52°C (93°F)
Total Rainfall 229 mm (9 in)

POLAR CLIMATE

Location
This occurs in Greenland, interior Iceland and in Antarctica.

Climatic characteristics
1 Temperatures are permanently below 0°C (32°F).
2 Blizzards are frequent.
3 Winters are really one continuous night and 'summers' one continuous day.

MOUNTAIN CLIMATE (Pike's Peak and Quito)

Location
This type of climate is best developed in regions of young fold mountains, e.g. the Rocky Mountains, the Andes, and the Himalayas, and associated mountains.

Climatic characteristics
1 In general, pressure and temperature decrease with altitude while precipitation increases. How-

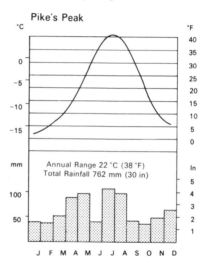

Pike's Peak

Annual Range 22°C (38°F)
Total Rainfall 762 mm (30 in)

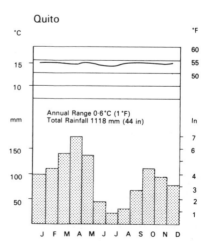

Quito

Annual Range 0·6°C (1°F)
Total Rainfall 1118 mm (44 in)

3 The total annual precipitation is about 250 mm (10 in) some of which falls as rain in the summer and some as snow in the winter.
4 Humidity is always low because of the low temperatures.

Agricultural development
Subsoils are permanently frozen and there is no agriculture of any type.

ever, if mountains are high enough, there is a height at which maximum precipitation occurs and above which it decreases as the moisture content of the air falls because it becomes rarefied and its temperature falls.

2 Because the air in high mountain regions is rarefied and relatively dust-free, it cannot absorb much heat. The air is therefore always cool and the daily temperature range is small. In comparison the ground absorbs much more heat in the summer days but loses it rapidly during the night. The daily ground temperature range is much greater.

3 There is usually a succession of temperature belts, not unlike those extending from the equator to the poles, in very high mountain regions. Such a succession is particularly well developed in the Andes.

4 Local winds develop in most mountain regions. These tend to blow up the valleys and up the mountain slopes in the day, whilst the winds blow down the valleys and mountain slopes in the night. Other local winds like the Chinook (in the Rockies) and the Föhn (in Switzerland) are caused by air crossing mountain ranges and descending down the lee side.

EXERCISES

1 Briefly describe the main characteristics of an Equatorial climate and its vegetation, and state what difficulties these present to agricultural development.

2 What are the main temperature and rainfall differences between Cool Temperate Western and Eastern Marginal climates?

3 Explain the following statements, using diagrams or sketch-maps, to illustrate your answers:
   (i) the characteristic aspects of a tropical monsoon climate are land and sea breezes on a continental scale
   (ii) the west coasts of continents within the tropics receive less rain than the east coasts

4 The following are the characteristics of three different types of climate:
   (a) A hot and a warm season with an annual temperature range of about $-7°C$ (about $20°F$) and with most rain falling in the hot season
   (b) Well distributed heavy rainfall with high temperatures throughout the year
   (c) A very large daily temperature range with annual rainfall usually below 120 mm (5 in)

(i) name the type of climate for each
(ii) choose *one* of the climates named, describe its temperature and rainfall pattern and account for this
(iii) for *each* climatic type named, give *one* region which has that climate.

5 Explain and account for the following:
   (a) The daily temperature range of In Salah (Libyan Desert) is higher than its annual temperature range.
   (b) Hot deserts are usually located on the western sides of continents.
   (c) Rainfall decreases from 60°N. to the Tropic of Cancer on the west side of a continent and increases, in the same direction, on the east side.
   (d) Desert plants are able to survive for long periods without water.

6   **A** Altitude 60 m (197 ft): Latitude 32°

	J	F	M	A	M	J	J	A	S	O	N	D
Temperature (°C)	22	23	22	19	16	13	12	13	14	16	18	22
(°F)	73	74	71	66	60	56	55	56	58	60	65	71
Rainfall (mm)	8	8	17	43	124	167	162	142	83	53	20	15
(in)	0·3	0·3	0·7	1·7	4·9	6·6	6·4	5·6	3·3	2·1	0·8	0·6

B Altitude 16 m (51 ft): Latitude 30°

	J	F	M	A	M	J	J	A	S	O	N	D
Temperature (°C)	12	13	17	20	24	27	28	27	25	21	16	13
(°F)	54	56	63	68	75	80	82	81	78	69	61	56
Rainfall (mm)	114	104	114	119	106	139	167	144	116	88	93	119
(in)	4·5	4·1	4·5	4·7	4·2	5·5	6·6	5·7	4·6	3·5	3·7	4·7

Note: Fahrenheit equivalents are approximate.

Study the figures given for the climate of the two places **A** and **B**. For each of these places:
   (a) State whether it is north or south of the equator and give reasons to support your answer.
   (b) Briefly describe the characteristic features of (i) the distribution and amount of rainfall, and (ii) the temperature.
   (c) State the type of climate for each of the places **A** and **B** and name *one* region where it occurs.
   (d) Choose *one* of the climatic types named in (c)
   (e) Briefly explain why it occurs in the region that you name.

7 Each of the following statements summarises a specific type of climate: Name the climate for *each* statement and then for each of these, (i) name one region where it occurs, (ii) describe the characteristic features of its natural vegetation, and (iii) briefly describe how the climate has influenced agricultural developments.
   (a) Mild winters, 0° to 6·6°C (32° to 44°F), warm summers usually over 15·5°C (60°F); rain throughout the year being brought by westerly winds.
   (b) Warm winters, over 15·5°C (60°F), and hot summers, over 26·7°C (80°F); on-shore trade winds throughout

the year with rain falling in all months but with a maximum in summer.

:) Mild winters, over 10°C (50°F), hot summers, over 26·7°C (80°F); light summer rainfall, mainly convectional of about 500 mm (20 in).

For *each* of the three towns given below (a) briefly describe the temperature and rainfall patterns, (b) state the type of climate giving reasons for your answer, (c) name one region where this type of climate occurs.

	J	F	M	A	M	J	J	A	S	O	N	D
itude 676 m 2,220 ft												
». (°C)	−15	−11	−5	5	11	14	16	15	10	5	−4	−10
(°F)	5	11	23	40	51	57	62	59	50	41	24	13
ʼall (mm)	23	15	20	23	48	29	84	58	33	18	18	20
(in)	0·9	0·6	0·8	0·9	1·9	3·1	3·3	2·3	1·3	0·7	0·7	0·8
itude 2,880 m 9,446 ft												
». (°C)	15	15	15	14	15	14	14	15	15	15	14	15
(°F)	59	59	59	58	59	58	58	59	59	59	58	59
ʼall (mm)	99	112	142	125	137	43	20	30	69	112	97	79
(in)	3·9	4·4	5·6	6·9	5·4	1·7	0·8	1·2	2·7	4·4	3·8	3·1
itude 6 m 20ft												
». (°C)	28	28	26	25	24	22	21	21	22	22	25	26
(°F)	82	82	79	77	74	72	70	70	72	72	77	79
ʼall (mm)	336	376	452	399	264	282	302	203	132	99	117	262
(in)	14·4	14·8	17·8	15·7	10·4	11·1	11·9	8·0	5·2	3·9	4·6	10·3

Describe and account for the temperature conditions experienced in the following types of climate: (i) warm temperate interior, (ii) cool temperate western margin, (iii) tropical continental, (iv) mountain types of climate as illustrated by the figures given below

mean monthly temperatures

Place	Altitude	Lowest	Highest
i) **Kimberley**	1,197 m	July 10°C	January 25°C
(29°S 25°E)	3,927 ft	(50°F)	(77°F)
ii) **Victoria**		January 4°C	July 16°C
(48°N 123°W)		(39°F)	(60°F)
ii) **Kayes**	30 m	January 25°C	May 34°C
(14°N 12°W)	198 ft	(77°F)	(94°F)
v) **Uspallata**	2,845 m	June 1°C	December 12°C
(33°S 70°W)	9,335 ft	(34°F)	(53°F)

Objective Exercises

1 A Tropical Continental (savana) climate is one which has

A a hot dry summer and a wet rainy winter

B a hot summer, cooler winter with most rain falling in the summer

C high temperatures and low rainfall throughout the year

D westerly winds in the winter and trade winds in the summer

E warm, wet summer and a cool, dry winter

A B C D E
·· ·· ·· ·· ··

2 Which one of the following stations is situated in a region which has a tropical monsoon climate?
A Durban (South Africa)
B Valencia (Ireland)
C In Salah (North Africa)
D Lagos (Nigeria)
E Bombay (India)

A B C D E
·· ·· ·· ·· ··

3 A Mediterranean Climate is characterised by very dry summers because
A relative humidity is low
B skies are cloudless
C trade winds blow from the land to the sea
D there are no on-shore winds
E temperatures are high

A B C D E
·· ·· ·· ·· ··

4 This is a description of a specific climatic type "Warm winter, over 15.5°C, and hot summers, over 26°C; on-shore trade winds blow throughout the year bringing rain in all months but with a maximum in summer." Which one of the following climatic types does this describe?
A Warm Temperate Western Margin
B Tropical Monsoon
C Warm Temperate Eastern Margin
D Tropical Marine
E Equatorial

A B C D E
·· ·· ·· ·· ··

5 The climate of Singapore differs from that of a station on the Mediterranean coast during July in that it experiences
A higher average temperatures and a higher average rainfall
B higher average temperatures and a lower average rainfall
C lower average temperatures and a higher average rainfall
D lower average temperatures and a lower average rainfall

A B C D
·· ·· ·· ··

6 There are very few low-growing plants in a tropical humid forest because there is
A little sunlight at ground level because of the canopy of tall trees
B lack of moisture at ground level because of the covering of tall trees

C a thick covering of leaves and decaying vegetation on the forest floor which prevents plant growth

D a high evaporation rate on the forest floor which causes plants to wither

A B C D

7 Which of the following statements best explains the difference between the climate of The British Isles and that of Southern Portugal?

A The British Isles are nearer the North Pole.

B The British Isles lie under westerly winds for most of the year.

C The western parts of the British Isles are mountainous.

D The British Isles are surrounded by water.

A B C D

8 What type of climate occurs on the western side of continents between latitudes 30°N and 45°N?

A British Type

B China Type

C Mediterranean

D Hot Desert

A B C D

9 All of the following types of climate occur in the Southern Hemisphere **except**

A Hot Desert

B Savana

C Equatorial

D Tundra

A B C (D)

10 The climate of Western Europe is mild for its latitude because

A the influence of altitude on temperature is minimal

B the prevailing winds are the Westerlies

C the coast is washed by the North Atlantic Drift

D the region is within the Low Pressure belt

A B C D

11 Which of the following tropical climates has the largest diurnal temperature range?

A Tropical Monsoon

B Equatorial

C Hot Desert

D Savana

A B (C) D

12 Which of the following statements best describes the climate represented by the graph given below?

A The annual temperature range is fairly high and rainfall occurs mainly in the winter.

B The summer is hot and dry.

C The summer is hot and dry and the winter is mild and rainy.

D Most of the rain falls in the winter when temperature are below 15°C

A B C D

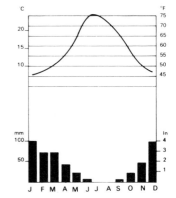

164

5   Vegetation

Geographers divide the world's vegetation into a number of types according to the appearance of the plants. The basic types, shown in the map on page 59 are:

1 Forest 2 Grassland 3 Desert; but before examining these it is important for us to know something about what a plant requires in order to maintain its growth.

PLANT REQUIREMENTS

Green plants make their own food by using:

(i) Water
(ii) Sunlight
(iii) Carbon dioxide
(iv) Mineral salts.

Water with the mineral salts in solution enters the roots of a plant from the soil and passes through the stem into the leaves, where a process called *photosynthesis* goes on. In this process carbon dioxide, which enters the leaves from the air, combines with the water in the presence of sunlight to form carbohydrates (food). The surplus water passes out of the leaves into the air and this movement of water is called *transpiration*. These two processes enable plant growth to take place. If any of the four requirements are missing or are inadequate then these processes either slow down or cease altogether.

The Influence of Temperature and Water on Plant growth

Temperature and water are the two most critical plant requirements because they do not occur in adequate amounts in all regions, and in some they vary in amount from season to season.

Plant growth normally ceases when the temperature falls below 6°C (about 42°F). In polar regions the temperature is always below 6°C (about 42°F) and there is no plant life. In other regions the temperature is always above 6°C (above 42°F) and continuous plant growth is possible. In still other regions the temperature is below 6°C (about 42°F) for one season and above it for another season. Plant growth is therefore seasonal.

Plants, like animals, have become adapted to their physical environment, in particular to the water and temperature aspects of this. Because water and temperature vary from natural region to natural region, the plants also vary in size and form. Thus hot wet regions normally have a forest vegetation whereas regions having light seasonal rains usually support a grass vegetation. Generally speaking, trees require more water than do grasses

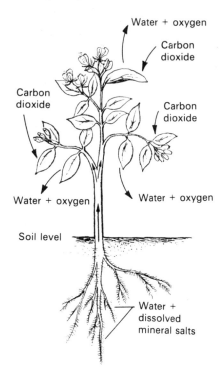

During sunlight

Water + oxygen
Carbon dioxide
Carbon dioxide
Carbon dioxide
Water + oxygen
Water + oxygen
Soil level
Water + dissolved mineral salts

which in turn require more than desert plants. We need not examine all the ways in which plants have adapted themselves to their environments, but we must have a look at the influence of drought and cold on plant adaptation.

Influence of Drought on Plants

1 Some plants develop long roots to reach water supplies far below the surface.
2 Some plants develop water storage organs, e.g. the baobab tree stores water in its trunk.
3 Some plants have special leaves which reduce transpiration, e.g. thorn-like leaves, rolled-up leaves, leaves with waxy surfaces.
4 Some plants shed most of their leaves when the dry season is also hot. If the plants do not do this then they would literally dry up as transpiration proceeds. Monsoon forests are almost leafless in the dry season.

Influence of Cold on Plants

When temperatures fall below 6°C (about 42°F) for several months of the year, many plants are unable to obtain sufficient water from the soil and many

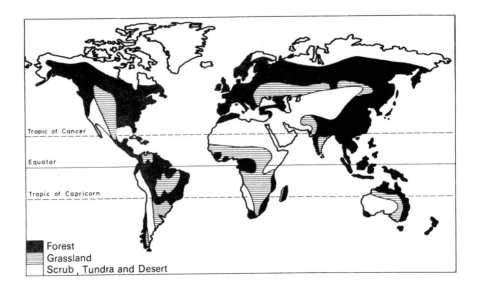

Forest
Grassland
Scrub, Tundra and Desert

of them shed their leaves. Such plants are said to be *deciduous*. This reaction to cold is particularly common in temperate forests.

Note Some trees have become adapted to cold weather and do not shed their leaves. *Coniferous trees* (except larch) keep their leaves throughout the cold season and they are called *evergreen* trees. They are able to do this because:

   (i) their leaves are rolled and little transpiration takes place from them.
   (ii) they need less water than other trees.

TYPES OF NATURAL VEGETATION

The map on page 169 shows the basic types of natural vegetation. This map gives a very simplified picture of the vegetation pattern, and so does the map on page 170, although this map shows more detail. It would appear from these maps that natural vegetation covers the entire land surfaces of the earth, but, of course, this is not so. Extensive lowland regions have long since been cleared of their vegetation to make way for man's crops and settlements. These maps are intended to show what the natural vegetation pattern would be like if man and his animals had in no way interfered with the land. Natural vegetation is only extensive today in those parts of the world where man finds great difficulty in mastering the environment, e.g. the Arctic lands and the equatorial river basins.

We will now examine these basic vegetation types.

For each type we must know the location and th names of some of the more common plants togethe with the way in which these plants have adapte themselves to the climatic characteristics.

THE FORESTS

I Tropical Evergreen Forest

Location   Amazon and Zaire Basins; Wes
              African coastlands; Malaysia
              coastal Burma, Cambodia an
              Vietnam; most of Indonesia an
              New Guinea.

Characteristics   1 Contains a great variety of plant
              which are close together.
              2 The forest consists of thre
              layers:
              (a) *top layer*: tall trees with buttres
              roots;
              (b) *middle layer*: tree ferns, liana
              e.g. rattan, and epiphytes, e.g
              orchids;
              (c) *bottom layer*: ferns, herbaceou
              plants and saprophytes.
              3 Nearly all the trees are broad
              leaved evergreens because hig
              temperatures and evenly distri
              buted rainfall permit growt
              throughout the year.
              4 Absence of seasonal climati
              change results in some plants bein

166

Equatorial Evergreen Forest in Ghana

in flower, others in fruit and others in leaf-fall at one and the same time.

5 The leaves of the tall trees form an almost continuous canopy which shuts out most of the light at ground level. There is therefore little undergrowth.

6 Mangrove trees, with stilt roots, form dense forests in coastal swamps and the lower valleys of tidal rivers.

7 When a part of a tropical evergreen forest is cleared, either for shifting cultivation or for lumbering, a less luxuriant forest growth takes over. This is called *secondary forest* and it consists of short trees and dense undergrowth. In Malaysia it is known as *belukar*.

Examples of trees   *Mahogany*, *ebony*, *rosewood*, *ironwood* and *greenheart* are common trees. *Palms* and *tree ferns* also occur in most equatorial forests.

These trees, as well as most of the other plants, occur singly or in small groups. There are rarely extensive stands of a particular tree.

II Tropical Monsoon Forest

Location   Burma; Thailand; Cambodia; Laos; N. Vietnam; parts of India; east Java and the islands to the east; N. Australia.

Characteristics   1 There is a smaller number of species than in the tropical evergreen forest.

2 Most of the trees are deciduous, losing their leaves in the hot, dry season which is a period of rest for them.

3 Heavy rain and high temperatures in the wet season result in rapid growth and the trees soon become covered with leaves.

4 The trees are tall, often as high as 30 metres (approx. 100 feet), but they are not as close together

167

Mangrove Swamp

as they are in tropical evergreen forests. Because of this, under growth is more dense (more light reaches the ground). Bamboo thickets are common.

5 Most monsoon forests contain valuable hardwoods, e.g. teak.

6 These forests merge into equatorial forests in regions where the dry season becomes short or non existent, and into grasslands in regions where the wet season becomes short and rainfall less heavy.

Examples of trees

Teak (especially in Burma, Thailand, Cambodia, Laos and E Java); *bamboo* (especially in Thailand, Cambodia, Laos and Vietnam); *sal*; *sandalwood*; and *lianas. acacia, eucalyptus* and *casuarina* occur in E. Java and N Australia.

Tropical Monsoon Forest

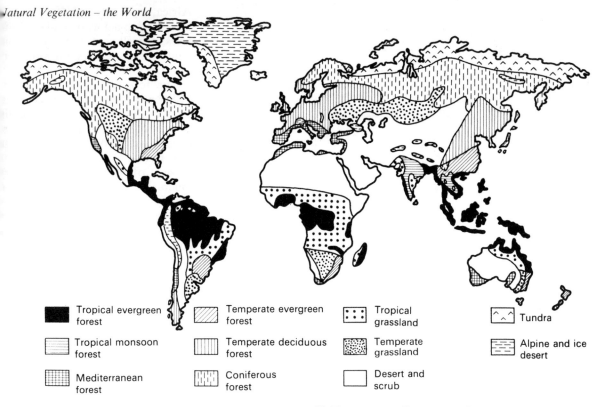

Tropical evergreen forest

Tropical monsoon forest

Mediterranean forest

Temperate evergreen forest

Temperate deciduous forest

Coniferous forest

Tropical grassland

Temperate grassland

Desert and scrub

Tundra

Alpine and ice desert

Structure of an Equatorial Forest

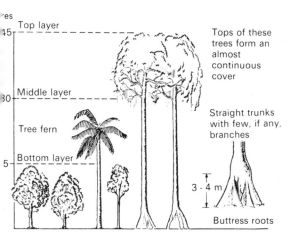

The diagram shows the three layers of vegetation in an equatorial forest. Many of the tall trees have buttress roots which give them support (remember they are of great height). Trees with buttress roots also occur in monsoon forests.

III Temperate Evergreen Forest

Location This occurs chiefly on the eastern sides of land masses in the warm temperate latitudes: S. China; S. Japan S.E. Australia; North Island (New Zealand); Natal coastlands (Africa); S. Brazil, and S.E. states of the U.S.A.

Characteristics 1 Most of these regions have rain throughout the year with winter temperatures often over 10°C (50°F) which means that plant growth can go on all the year. Most of the trees are broadleaved evergreens, although there are deciduous trees as well, especially in the Gulf States of the U.S.A.

2 Most of the evergreens are hardwoods e.g. quebracho (S.E. Brazil) and cedar.

3 The forests of China and southern Japan contain evergreen oak, magnolia and camphor trees, all of which are of economic value.

4 Walnut, oak, hickory and magnolia trees form the bulk of the forests in the Gulf States of the U.S.A.

5 The highlands of south-east

169

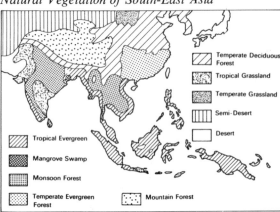

Natural Vegetation of South-East Asia

Temperate Deciduous Forest

Tropical Grassland

Temperate Grassland

Semi-Desert

Desert

Tropical Evergreen

Mangrove Swamp

Monsoon Forest

Temperate Evergreen Forest

Mountain Forest

Examples of trees

Africa, especially in Natal, contai
ironwood, blackwood and chestnu
trees, as well as wattle trees whos
trunks are used as pit props i
the coal mines of Natal.

6 The forests of south-east Aus
tralia are famous for eucalyptu
trees while those of the mountai
slopes of south-east Brazil ar
equally well known for the Paran
pine trees.

7 Most of these forests look ver
much like tropical forests in tha
the vegetation is thick and profuse
Evergreen oak; *magnolia* (especiall
China and the U.S.A.); *campho
and *bamboo* (especially China)

Temperate Evergreen Forest in North Island (New Zealand)

cedar (evergreen), *maple* and *walnut* (both deciduous); *eucalypts* (Australia); *tree fern* (New Zealand); *mulberry* (China); and *cypress* (U.S.A.).

V Mediterranean

Location   This occurs chiefly on the western sides of land masses in the warm temperate latitudes. Lowlands around the Mediterranean Sea; S.W. Australia and the Adelaide District of Australia: S.W. Africa; central Chile; central California.

Characteristics   1 Originally the vegetation was forest but much of it has been cut down and the rest has been ravaged by foraging goats so that all that now remains is a scattered woodland-type of vegetation.

2 Summers are hot and dry which make plant growth difficult. However, the plants of this region have become adapted to the summer drought by storing water obtained from the winter rains, in leaves and bark. Many plants have waxy, spiny or small leaves to cut down the rate of transpiration. Other plants, e.g. grape vine, have long tap roots which can reach down to moist rock layers well below the surface.

3 In the drier parts the vegetation becomes scrub-like and consists of sweet-smelling herbs and shrubs such as lavender, rosemary, thyme and oleander.

4 In the wetter parts, e.g. on mountain slopes, coniferous trees are common.

Examples of trees   *Evergreen oak*; *cork oak*; *eucalypts, jarrah* and *karri* (S.W. Australia); *cedar*, *cypress*, and *sequoia* or *redwood* (California), all of which are conifers. The scrub vegetation which includes *lavender, rosemary, myrtle* and *oleander* is a secondary type of vegetation, i.e. it has arisen in consequence of the destruction of the original vegetative cover. In France this is called *maquis*; and in California *chaparral*.

V Cool Temperate

Location   W. and central Europe; eastern U.S.A.; N. China; N. Japan;

Mediterranean Vegetation in Greece

Korea; S. Chile and South Island (New Zealand). The forest is poorly developed in S. Chile because of the high relief.

Characteristics   1 In most of these regions, winter temperatures fall below 6·1°C (43°F), the minimum temperature for plant growth, which results in most of the trees shedding their

171

Temperate Deciduous Forest

leaves and becoming deciduous.

2 Some of the deciduous trees are hardwood and the most common are, oak, beech, birch, hornbeam and ash.

3 Many of the trees occur in pure stands and they are of great economic value.

Examples of trees

Oak; beech; hazel; elm; chestnut poplar, and, in N. America, walnut maple; hickory; cedar and spruce (both conifers).

VI Coniferous Forest

Location

This is most extensive in high latitudes and on high mountains although it does develop on sandy soils in warmer regions. There are two main belts of this forest (i) across Eurasia extending from the Atlantic to the Pacific (ii) across N. America extending from coast to coast.

Characteristics

1 Most of the trees are evergreen and they are coniferous. They are especially well adapted to the long, cold and often snowy winters, by growing needle-shaped leaves which reduce transpiration to a minimum. Further, the leaves have a tough thick skin which protects them from winter cold. The tree is conical and has flexible branches which allow the snow to slide off. It has widely spread shallow roots to collect water from the topsoil, above the permafrost layer.

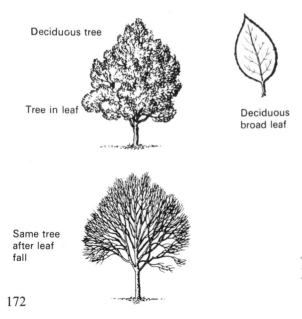

Deciduous tree

Tree in leaf

Deciduous broad leaf

Same tree after leaf fall

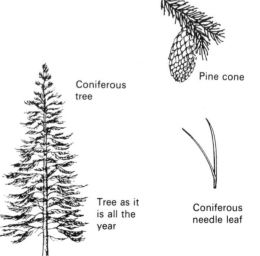

Coniferous tree

Pine cone

Tree as it is all the year

Coniferous needle leaf

Temperate Coniferous Forest

2 Coniferous trees produce soft wood which is in great demand for making paper, especially newsprint, matches, furniture and synthetic fibres such as rayon.

3 These forests have practically no undergrowth because the soil is frozen for many months each year.

*Examples
of trees*

Hemlock, spruce, pine and *fir*. The trees of coniferous forests in Mediterranean regions are chiefly *cypress* and *cedar*.

THE GRASSLANDS

I Tropical Grassland

Location

These are located mainly in the continental regions of tropical latitudes, where the rains occur in the hot season which lasts for about 5 months; north and south of the Zaire Basin, West Africa and the East African Plateaus; parts of Brazil; the Guiana Highlands; north and east of the Australian Desert; and parts of the Indian Deccan Plateau.

Characteristics

1 Although tall grasses form the dominant plant life of these regions, trees are common, especially near water courses and in the more humid areas, that is, where the region merges with equatorial forest.

2 The grasses are usually 2 metres (approx. 6 feet) high, or more, and they grow in compact tufts. Their long roots can reach down to the moist rock layers far below the surface. During the dry season the leaves of the grasses turn yellow and die but the roots remain dormant (sleeping). These grasses are therefore deciduous.

3 The trees of these regions are also deciduous, shedding their leaves in the dry season. Some

173

Baobab tree

Temperate Grassland in South Africa

trees, e.g. baobabs and bottle trees, store water in their swollen trunks and in this way they are able to survive the dry season.

4 Where these grasslands merge with the hot deserts, the vegetation changes greatly. There is no longer a continuous vegetation cover: instead, clumps of scrub-like plants scatter the surface. *Mallee*₁ and *mulga*₂ characterise this type of vegetation in Australia.

Grassland names Tropical grasslands have different names according to their location, e.g. *Campos* (Brazil); *Llanos* (Guiana Highlands); *Savana* (Africa and Australia).

1 *Mallee* – eucalyptus bushes set in a thicket of
– coarse grass.
2 *Mulga* – clumps of acacia set in a thicket of coarse
– grass.

II Temperate Grassland

Location These are best developed in the continental interiors of temperate latitudes, e.g. hearts of Asia and N. America. Less extensive areas occur in S. Africa, S. America and Australia.

Characteristics 1 These grasslands are almost tree-less and they contrast sharply with the tropical grasslands.

2 In the moister regions (rainfall over 500 mm or 20 in) the grasses are tall, though not as tall as those of tropical grasslands, and they are nutritious. These grasses are typical of the Black Earth region of the Ukraine (southern U.S.S.R.) and the moister parts of the American Prairies (now mainly under wheat).

In the drier regions (rainfall under 500 mm or 20 in), the grasses are shorter, tougher and less nutritious. These grasses are typical of the High Plains (U.S.A.) and the Asian Steppes.

3 During the heat of summer the grass begins to wither and most of it dies in the autumn. The roots do not die, and with the coming of spring, new leaves form and the land is once again covered with a green mantle.

4 A few trees, such as poplars, willows and alders, grow in the damper soils flanking water courses.

5 On the poleward side, the temperate grasslands merge with the coniferous forests while on the equatorward side, they merge with scrub of semi-deserts.

Grassland names   The following names are generally recognised: *steppe* (Eurasia); *prairie* (N. America); *pampas* (Argentina); *veldt* (S. Africa); *downs* (Australia).

Note In most of the grasslands the leaves of the grasses wither and die in the dry season (tropics) or cold season (temperate regions). The roots of the plants however do not die, and, in the following season when the rains come, the aerial parts of the plants grow afresh.

DESERT AND SEMI-DESERT

1 Tropical Desert

Location   Usually lies between 15°N. and 30°N. and 15°S. and 30°S. They lie on the western sides of land masses except for Africa where they extend from coast to coast, linking up with the Asian deserts. The chief regions having a desert vegetation are: Sahara (N. Africa); Arabia; parts of Iran, Iraq, Syria, Jordan and Israel; part of Pakistan; central Australia; the Namib Desert (S.W. Africa); Atacama (coastal Peru and N. Chile); and S. California, N. Mexico, parts of Arizona (N. America).

Characteristics   1 Only very small parts of tropical deserts are without any type of vegetation.

2 Desert plants are all very special in that they can withstand high temperatures and long periods when no rain at all falls.

3 The plants have become adapted to conditions of extreme drought in several ways. Some plants have long roots, others have few or no leaves, and what there are are tough, waxy or needle-shaped to

Giant Cactus from the Sonora Desert – Mexican Border with the U.S.A.

175

reduce transpiration to a minimum.

4 Many plants produce seeds which lie dormant for years until a little rain falls and then they germinate.

5 The most common plants are cacti, thorn bushes and coarse grasses.

II Semi-desert and Scrub

Location   This type of vegetation occurs in regions which either border the tropical deserts, or in the interiors of continents where the rainfall is just insufficient to maintain a continuous cover of vegetation.

Characteristics   In tropical scrubland the chief plant is the thorny acacia. During the period of rains a short-lived but rich mantle of grasses and flowering plants covers extensive areas. Sage bush is common in N. American scrubland. Mulga is sometimes included in this vegetation type.

III Tundra

Location   This type of vegetation is chiefly confined to the Northern Hemisphere, fringing the Arctic Ocean in the continents of Eurasia and N. America.

Characteristics   1 The growing season is very short (about two months), and at this time the surface soil thaws but the subsoil remains frozen. Water therefore lies on the surface. This influences the pattern of vegetation.

2 Where the tundra meets the coniferous forest there are stunted willows and birches.

3 Most of the tundra consists of a low cover of vegetation made up of lichens, mosses, sedges, and flowering shrubs such as bilberry and bearberry.

EXERCISES

1 Carefully describe the influences that high temperatures, low temperatures and drought can have on plants, and by choosing specific examples, show how plants have become adapted to hot, dry conditions.

2 Make a list of the main forest types and for *two* of these:
   (i) name *two* types of common tree
   (ii) state whether the trees are deciduous or evergreen

(iii) name *one* region where this type of forest occurs.

3 Carefully explain, by using appropriate diagrams, where relevant, the meanings of the following statements:
   (i) the hearts of continents have few trees
   (ii) the undergrowth of equatorial forests is not as well developed as is that of tropical monsoon forests
   (iii) vegetational zoning takes place both latitudinally and vertically.

4 Make a list of the main types of grassland and for each type:
   (i) name *two* regions where it occurs
   (ii) state any regional names it may have
   (iii) briefly describe how it is used by man.

Objective Exercises

1 All of the following are essential to the proper growth of green plants **except**
   A water
   B sunlight
   C carbon dioxide
   D high temperatures
   E high humidity

   A B C D E

2 A forest whose broad-leafed evergreen trees grow to great heights and which are festooned with parasitic plants and epiphytes, and beneath which the undergrowth is very sparse, is known as
   A Mediterranean Forest
   B Temperate Evergreen Forest
   C Equatorial Forest
   D Coniferous Forest
   E Temperate Mixed Forest

   A B C D E

3 Plants which have developed special water storage organs or which can survive long periods of hot dry weather are common to regions which have
   A an Equatorial Climate
   B a Tropical Continental (Savana) Climate
   C a Tundra Climate
   D a Warm Temperate Interior Climate
   E a Temperate West Margin Climate

   A B C D E

4 The temperate grasslands of the interior of Eurasia are called
   A prairie
   B steppe
   C veldt
   D downs
   E pampas

   A B C D E

5 A specific type of natural vegetation is known by all of the following names **except**
A veldt
B steppe
C campos
D downs
E prairie

A B C D E

6 Some plants are in flower, others are in fruit, and others lose their leaves, all at the same time, in regions which have no seasonal climate changes. This type of plant occurs in
A Coniferous Forests
B Mediterranean Forests
C Tropical Rain Forests
D Temperate Evergreen Forests

A B C D

7 A region which has a long cold winter and a short cool summer, and whose sub-soils are frozen for most of the year will have a natural vegetation of
A maquis and low scrub
B acacia and sal
C lichen and sedge
D sage bush and cacti

A B C D

8 An equatorial forest may contain all of the following trees **except**
A ebony
B iron wood
C baobab
D greenheart

A B C D

9 The trees of tropical rain forests and coniferous forests are similar in that
A they have large broad leaves
B their branches slope downwards
C they lose their leaves at the same time
D they are evergreen

A B C D

10 What type of vegetation occurs where the following conditions prevail? (a) podzol soils, (b) annual temperature range 38°C, (c) annual precipitation about 300 mm which is fairly evenly distributed through the year.
A thorn scrub
B coniferous forest

C temperate grassland
D monsoon forest

A B C D

11 Evergreen Mediterranean forest occurs
A on the eastern sides of land masses which have a warm temperate climate
B in hot wet equatorial lowlands
C on the western margin of Europe where the climate is of the cool temperate type
D between 30° and 40° N and S on the west sides of continents

A B C D

12 Which of the following is associated with a Savana Climate?
A shallow-rooted trees
B trees that grow throughout the year
C scrub
D dense evergreen forest

A B C D

16 Soils

GENERAL

Weathering processes break up the surface of a rock into small particles. Air and water enter the spaces between these and chemical changes take place which result in the production of chemical substances. Bacteria and plant life soon make their appearance. When the plants die, they decay and produce *humus* which is all-important to soil fertility. Broadly speaking, humus consists of the decayed remains of both plants and animals. Bacteria play a vital part in the decomposition of these remains. The end product of these chemical and biological processes is *soil*. From this it will be clear that the nature of any soil is influenced by:

(i) weathering
(ii) vegetation
(iii) parent rock
(iv) climate (this determines both the type of weathering and the natural vegetation).

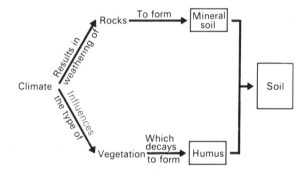

Contents of the Soil

All soils contain:

(i) mineral particles (inorganic material)
(ii) humus (organic material)
(iii) water
(iv) air
(v) living organisms, especially bacteria.

The actual amounts of each of these depend upon the type of soil. Many soils are deficient in one or more of these.

SOIL PROFILE

A soil profile is a vertical section through the soil to the underlying solid rock. Most soil profiles consist of three layers which are called *horizons*. These are lettered A, B and C.

Horizon A – the soil proper
Horizon B – the subsoil
Horizon C – the solid rock.

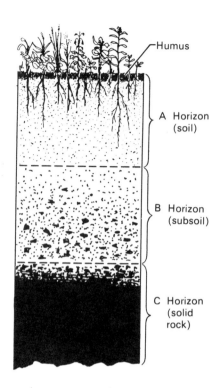

You will see from the diagram above that the top part of the A horizon is often rich in humus, and that the texture of the soil becomes coarser as the C horizon is approached.

Factors Influencing the Soil Profile

Water Movement in the Soil

When rainfall is greater than evaporation, water moves downwards in the soil and mineral matter is removed from the top layer (A horizon) and is deposited in the layer beneath (B horizon). Very often these deposits give rise to a hard layer which is called a *hardpan* and this can cause drainage to be poor. When this takes place in a soil the soil is said to be *leached*. In cold wet regions leaching helps to produce a grey soil which is known as a *podzol*; in hot wet regions it may produce a red-brown soil which is known as *laterite*.

Laterite

In humid tropical regions soil water contains very little organic matter, and such water does not dissolve iron and aluminium hydroxides. Most other minerals however dissolve and these are carried in solution to the B horizon where they are deposited. Ultimately a soil may be formed which is composed mainly of *iron* and *aluminium compounds*. This is called *laterite*. This is usually red in colour and becomes extremely sticky when wet. It is a useful material for making bricks since it sets hard on drying. Because it is the end product in the process of weathering, it is almost completely resistant to further weathering and buildings made of it last for a very long time. Some laterites are very rich in aluminium compounds and are called *bauxite*. Bauxite deposits are usually white or grey in colour. Hardpans also occur in laterites.

Note Laterite can form from any type of rock.

In hot desert and semi-desert regions there is an upward movement of water in the soil. This results in the deposition of mineral matter in the A horizon. Some of the important deposits of saltpetre have been formed in this manner.

CLASSIFICATION OF SOILS

Climate plays the greatest part in the formation of a soil (left diagram, page 178), and any classification of soils therefore tends to have a climatological basis. This means that each major soil type corresponds with a specific type of climate. These types are called *zonal soils*. They generally occur in extensive belts and they are mature soils.

Zonal Soils

A *Tropical*
   (i) Laterite
   (ii) Red Soils
   (iii) Black Soils
   (iv) Desert Soils.
B *Temperate*
   (i) Podzols
   (ii) Chernozems (Black Soils)
   (iii) Brown Soils
   (iv) Desert Soils.

For laterite, tropical red soils and tropical desert soils see notes under laterite.

Tropical Black Soils

This type of soil develops in humid tropical regions which have volcanic rocks. The soil is rich in calcium carbonate and other minerals, and when wet it becomes very sticky. It occurs extensively in N.W. Deccan (where it is cultivated with cotton), and in smaller areas in N. Argentina, Kenya and Morocco. In the Deccan it is called *regur*.

Chernozems

These are probably the richest soils in the world. They form under a natural vegetation of grass in the temperate latitudes and are rich in humus. This is because there is very little leaching. Most of the world's wheat is grown on these soils.

Temperate Brown Soils

These soils formed in climatic regions which had a

natural vegetation of deciduous forest. The soils are rich in humus, though since much of the forest has now been removed for agriculture, manures have to be applied to the soils to maintain the organic content.

Temperate Desert Soils

The soils of arid temperate regions have not been leached. They are therefore rich in plant foods, and, if irrigated, they are often very fertile. The soils are grey to brown in colour and they occur chiefly in the U.S.S.R. (between the Caspian Sea and Lake Aral), and in parts of west U.S.A.

Other Soils

There are other soils which are less influenced by climate than the zonal soils. These cannot be discussed in any detail here, but the names of some of the more important will be given.

Peat Soils: they form where drainage is very poor, e.g. swamps. Vegetation does not properly decay.

Alluvial Soils: they consist of a mixture of clay, sand and silt which has been deposited by water. They form some of Asia's most important agricultural regions.

Terra Rossa Soils: These are red soils which form in limestone regions under semi-arid conditions.

Wind-deposited Soils: the chief of these is *loess*.

SOIL EROSION

Soils and Agriculture

Farmers are not generally interested in soil profiles and soil classification such as we have just discussed. What chiefly interests them is whether the soil is fertile or not. In many regions farmers have learnt how to keep the soil fertile, or to make it more fertile by adding manures, or growing different crops in rotation. But this practice is by no means universal. Soils will deteriorate if they are not properly cultivated, and soil deterioration usually results in the removal of soil by wind or water. When this happens soil erosion is said to take place.

Causes of Soil Erosion

Soil erosion is frequently caused by the action of Man and his domestic animals. Most of it begins with the removal of the natural vegetation, be it grass or forest. So long as the vegetation cover remains, there can be little if any erosion because the roots of the plants bind the soil particles together, and the vegetation itself protects the soil from the action of wind and rain. The destruction of the natural vegetation may be caused by turning the land into crop land or by putting too many domestic animals on it which results in over-grazing, or by setting fire to it which is still a common practice of shifting cultivators.

Types of Soil Erosion
I By Water
(i) *Sheet Erosion*

This affects large areas and it occurs when rain falls on a gentle slope which is bare of vegetation. This type of erosion results in the removal of a uniform depth of soil. The middle photograph shows the nature of sheet erosion in a woodland. You will see that the land is by no means bare of vegetation in this example.

(ii) *Gully Erosion*

This is more localised and occurs when heavy rainfall rushes down a steep slope, cutting deep grooves into the land. The grooves become deepened and widened to form *gullies* which finally cut up the land to give 'badlands'. This type of erosion is especially frequent in semi-arid regions.

II By Wind

Regions having a low rainfall or a definite dry season are liable to have their soils reduced to dust and blown away by the wind if the land is bare of vegetation. We have already seen how the wind

The photograph below shows the force of soil erosion. Some idea of the depth to which erosion has taken place can be obtained from the height of the trees.

Gully Erosion in Kenya

Severe Soil Erosion at Ceara, Brazil

Sheet Erosion in South Africa

Part of the 'Dust Bowl' in the U.S.A.

removes fine particles of material in desert regions by the process of deflation. The same process also take place in the marginal zones of some of the temperate grasslands which have had the grass vegetation replaced by crops such as wheat, or which have been over-grazed by cattle. In these marginal zones rainfall is not only lower than in the grasslands proper, but is also less reliable. Sufficient rain may fall for a few years to enable the crops to thrive but this is inevitably followed by a series of drier years when the crops fail and the land is abandoned. It is at this time, when the land is bare, that wind erosion sets in. One of the worst cases of such erosion occurred in the 1930s in the mid-western States of the U.S.A. The area devastated here is called the 'Dust Bowl' (see (photograph).

Common Farming Practices which Lead to Soil Erosion

1 The cultivation of crops in regions which do not have a reliable rainfall, i.e. dry years following wet years, and which have only just sufficient rainfall for crop farming.
2 The ploughing of land up and down the slope. This provides man-made channels which can be enlarged into gullies by surface run-off.
3 In shifting cultivation (top photograph, page 165) a piece of forest is destroyed by fire, and crops are grown in the soils of the cleared patch which are now enriched by wood ash. After one or two years of such cultivation, the patch is abandoned and a new clearing is made. The abandoned patch soon experiences soil erosion by rainfall. Shifting cultivation usually takes place in the wet tropics. It is a common method of crop farming with wandering forest peoples.
4 The cultivation of the same type of crop on a piece of land year after year, leads to soil depletion if manures or fertilisers are not used. Most plants usually make a demand on a particular mineral compound and if the same type of plant is grown over a number of years then the soil will become deficient in this mineral. When this happens the soil deteriorates and soil erosion may set in.
5 The cutting down of forests, especially on the higher slopes, may result in soil erosion and the spreading out of the transported soil over the lowlands where farm land can be seriously damaged. (The middle photograph on page 182 shows the sort of damage that occurs when the transported soils are deposited in built-up areas.)

181

Shifting Cultivation in Chile

Deposition of soil after heavy rains in Hongkong in 1958

Terrace Cultivation in N. China. The low embankments around the fields protect the soil from erosion by preventing the water from rushing down the slopes

Regions where Soil Erosion takes place
Wind Erosion
Mid-western States of the U.S.A. The eroded region is called the *Dust Bowl*.

Water Erosion
The Mediterranean countries. It is caused by over-grazing by the goat especially in the Eastern countries. Destruction of forests in Spain and parts of Italy has caused erosion.

Parts of India, Sri Lanka, Sumatra, Java and Thailand. Caused mainly by shifting cultivation.

The savana regions of Africa. Caused by shifting cultivation and over-grazing.

SOIL CONSERVATION
Soil erosion has made millions of acres of land unproductive. As the world's population increases year by year, so more and more food has to be produced if famine and disease are to be eliminated, and if all people are to get an adequate and balanced diet The governments of most countries have long since realised that soil erosion is one of Man's greatest enemies, and measures are now being taken to reduce erosion to a minimum and to reclaim land which has already been eroded. International organisations like the Food and Agriculture Organization (F.A.O.) have for many years been carrying out a programme of soil conservation. This has involved the introduction of sound methods of farming which not only prevent erosion but which also keep the soil in a healthy state.

Types of Soil Conservation
1 Contour Ploughing
The furrows in which the crops are planted follow the contours. If they go up and down the slopes

they promote gullying. The photograph on the right is an aerial view of contour ploughing in Texas (U.S.A.).

2 Terracing

The slope is cut into a series of wide steps on which the crops are grown. This is very common in Asian countries in regions of rice cultivation. The rice terraces are flooded during the growing season, the water passing from one terrace to the next one below it. The flooding of the terraces is a good indication of the ability of the terraces to prevent soil erosion.

3 Planting of Shelter Belts

Belts of trees are often planted across a flat region which is liable to suffer from wind erosion. The trees break the force of the wind and thus protect the strips of land between the belts from being eroded.

4 Strip cultivation and crop rotation

Other farming methods include the cultivation of alternate strips at right angles to the prevailing winds, so that when one strip is laid bare for ploughing, the adjacent strip is under grass or is growing a crop. If the wind blows soil off the bare strip it will be caught and anchored by the vegetated strip. Crop rotation and the use of fertilisers ensures that the soil remains fertile and does not lose its 'structure'. It will therefore stick together better and be less likely to blow away.

Contour Ploughing in Texas

Shelter Belts in the Dry Steppe in the U.S.S.R. The belts are of trees.

183

The photograph above shows gullied land in North Carolina (U.S.A.). The bottom photograph shows the same area, 20 years after it was re-afforested.

The photograph above shows gullied and eroded land in Nebraska (U.S.A.). The bottom photograph shows the same area after soil conservation methods have been used. The land is now used for growing crops.

EXERCISES

1 Briefly describe the composition and formation of soil and name the essential ingredients of a good soil.

2 Make a simple classification of soils and show how climate influences the development of soil types.

3 Briefly explain what is meant by (i) soil erosion; (ii) soil conservation. Illustrate your answer with diagrams and examples.

4 (a) Briefly explain what is meant by the terms 'soil erosion' and 'soil conservation'.
   (b) Name *three* ways in which soil erosion can be produced.
   (c) Locate and name *two* regions, one in the tropics and one in temperate latitudes, where soil erosion has actively taken place, and suggest what steps have to be taken to stop it.

Objective Exercises

1 Which of the following factors does **not** contribute to the formation of soil?
   A air and water
   B decaying plants
   C rocks and temperature
   D water-logging
   E bacteria

   A B C D E

2 In cold wet regions leaching helps to produce grey soils which are called
   A laterites
   B chernozems
   C podzols
   D regur
   E loam

   A B C D E

3 Alluvial soils are usually very fertile. This is because
   A they are derived only from igneous rocks
   B they are acidic
   C they consist of fine particles derived from several types of rocks
   D they often form level plains
   E loess often forms along the edges of semi-arid regions

   A B C D E

4 In which country did wind erosion remove the top soil from a very large area, thereby creating the Dust Bowl?
   A the U.S.A.
   B Mexico
   C Canada
   D Brazil
   E Australia

   A B C D E

5 Soil erosion can be caused in several ways. Which of the following does not cause soil erosion?
   A deflation
   B overgrazing
   C deforestation
   D weathering
   E over-cropping

   A B C D E

6 All of the following, **except** one, account for the low fertility of soils in hot climates. Which is the exception?
   A they are leached
   B they are easily exhausted
   C they are acidic
   D they are poorly drained

   A B C D

7 Soil erosion often results from
   A the process of weathering
   B contour strip cultivation
   C afforestation
   D deflation

   A B C D

8 Soil erosion in a humid tropical region can be checked and corrected by
   A terracing
   B cutting down the forests
   C burning the grasslands
   D not growing crops

   A B C D

9 Which one of the following statements is **not** true with reference to soils?
   A A laterite can form from any type of rock.
   B Hardpan is a layer of hard deposits which occurs in the lower layers of some soils.
   C Peat soils develop best under hot arid climatic conditions.
   D Terra Rossa soils form under semi-arid conditions.

   A B C D

10 Soil erosion by rain wash can be caused by all of the following **except**
   A scanty vegetation cover
   B steep slopes
   C aridity
   D tropical rainstorms

   A B C D

11 Soil erosion may not be prevented by
   A terracing
   B planting trees
   C contour ploughing
   D ploughing grasslands

   A B C D

12 The following photograph shows a method of cultivation
which is intended to
A permit more than one crop a year to be grown
B increase yield per acre
C grow several different types of crop
D reduce soil erosion

A B C D

17   Introduction to the Work of Man

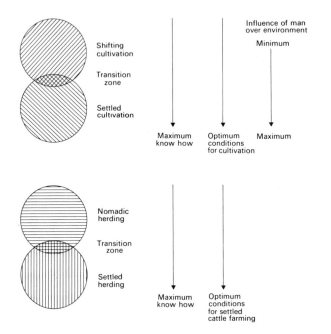

REARING OF ANIMALS

Nomadic herding: Animals wander over a large area living on what grass or other plants are available.

Settled farming: Animals are confined to specific areas, and specific crops are planted for them.

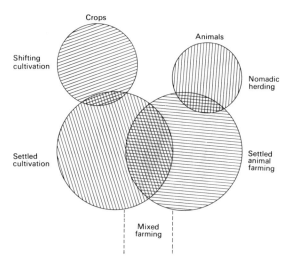

AGRICULTURAL ACTIVITIES

Agriculture involves the cultivation of crops and the rearing of animals for the purpose of producing food and materials, and making clothes and other essential things which man needs. The methods of crop cultivation and animal rearing vary from region to region.

CULTIVATION OF CROPS

Shifting cultivation: This is a primitive method which does not involve scientific knowledge. The land is cultivated for only one or two years, after which it is abandoned.

Settled cultivation: This is an advanced method. The land is cultivated year after year and crop yields are kept high by constantly using manure and fertilizers, and by regulating the use of water. Settled cultivation may be either *intensive*, as when a small piece of land is cultivated by a family, or *extensive*, as when a large piece of land is farmed or cultivated by means of machinery and hired labour.

EXTRACTIVE INDUSTRIES

These industries are concerned with the removal of natural products from the ground (on it or below it) and from the sea, and they include mining, quarrying, forestry and fishing.

Mining This is the name given to the extraction of mineral ores and oil from the earth's crust. When extracted, minerals cannot be replaced, and therefore mineral reserves in the earth's crust will eventually decrease. Throughout history, man has always been attracted to settle and work in areas where mineral wealth has existed.

Mining methods differ from region to region, depending on the location of the minerals, the depth at which they exist, and the level of technological skill available. Shaft mining is practised in places where minerals occur deep in the earth's crust, while open-pit (open-cast) mining, using either dredges or hydraulic pumps, or some other method, is practised where minerals occur at, or near the earth's surface.

Quarrying This is similar to mining, especially to open-cast mining, but it differs in that the rocks extracted are generally of lower value. These rocks are used for building and for road construction. All rock deposits which are quarried lie near to the surface and usually the quarries are located near to the populated areas which use the products of the quarries. Commonly quarried rocks are granite, slate, sandstone, limestone and gravels.

Forestry This involves the felling of trees, the wood of which is used in the construction of buildings, or the manufacture of furniture and matches, or for the making of paper pulp. Most of the trees in coniferous forests are softwoods and the wood of these is used mainly for making such products as matches and paper pulp. The trees of deciduous forests are usually hardwoods and the wood of these trees is usually used for construction work.

Fishing Although the fishing industries of the world are mainly located in the oceans and seas, fishing in rivers, and in lakes geared to fish breeding, is becoming important, especially in many Asian countries. Careful control must be exercised in fishing and forestry to ensure that the rate at which the product is extracted is balanced by its replacement, e.g. the replanting of new trees to balance those that have been felled. This control is being exercised in several countries. Controls to prevent over-fishing are also practised by most of the countries which have large fishing industries.

PROCESSING INDUSTRIES

These industries use natural products either for manufacture into goods for direct sale to consumers, e.g. the making of butter and cheese from milk, or for manufacture into an intermediate form for subsequent further manufacture into goods ready for use, e.g. the manufacture of iron and steel girders and plates from iron ore or scrap iron. Many processing industries have arisen either to preserve the natural product, e.g. the drying of fruit, or to reduce the bulk of the natural product to produce a reduction in transport costs, e.g. the concentration of mineral ores usually near to the mining sites.

The United Nations Organisation defines manufacturing as the transformation of substances into new products with the work being performed by power-driven machinery or by hand, either in a factory or in a worker's home, with the products being sold wholesale or retail. According to this definition, manufacturing takes place when an aborigine makes a blow-pipe or a factory turns out yet another car. The raw materials may be derived from animals, plants or minerals, and the

Typical agricultural landscape in Western Europe. This photograph is a view of Le Puy (France). Notice the small fields and the different uses to which they are put.

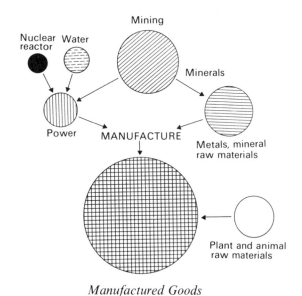

Manufactured Goods

manufactured products may be in the form of food, textiles and clothing, or light and heavy machinery. Modern manufacture usually takes place in a factory and requires advanced technological skill in manpower and machinery.

A mining landscape which is common to Europe and North America. This is a photograph of the Rhondda Valley in Wales (Great Britain).

Manufacturing landscape. Photograph of Stevenage (Great Britain)

Manufacturing can be simple or complex. Manufacturing is *simple* when only one kind of raw material is used and the finished product is simple, when, for example, an iron mill turns iron ore into iron rods. Manufacturing becomes *complex* when either (i) a wide range of different raw materials, or (ii) much scientific skill and elaborate machinery, are required to make the finished article, for example, in the manufacture of a motor car or a watch. In the more advanced industrial countries, such as the U.S.A., the U.K., the U.S.S.R. and Japan, a large part of manufacturing is complex.

To sum up Although manufacturing of one type or another takes place in almost every country, it has reached different stages of development from region to region. In some regions it involves the use of simple tools and human power: in others it involves the use of complicated tools and power-driven machines.

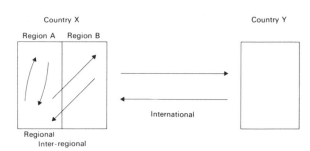

MEANING OF TRADE

Trade is the buying and selling of goods between two persons, firms or countries. There are *four* levels of trade:

1 Local trade is trade between persons or firms within one neighbourhood or community.
2 Regional trade is trade between persons or firms within one region.
3 Inter-regional trade is trade between persons or firms in different regions within the same country.
4 International trade is between different countries.

The oldest form of trade is barter trade which is the exchange of one kind of article for another without involving the use of money. Barter trade still exists in primitive and backward societies. International trade, a modern form of trade, is the outcome of the development and improvement of world land, sea and air transport and communications. International trade centres on the large ports of the world.

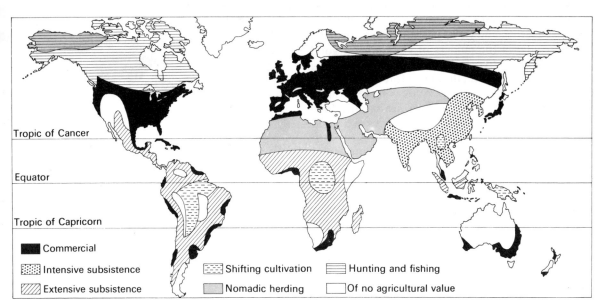

Tropic of Cancer

Equator

Tropic of Capricorn

- ■ Commercial
- ⬚ Intensive subsistence
- ⬚ Extensive subsistence
- ⬚ Shifting cultivation
- ▤ Nomadic herding
- ▤ Hunting and fishing
- ☐ Of no agricultural value

Agricultural Land Uses

CROP FARMING

Shifting cultivation

Shifting cultivation is practised by primitive peoples in many parts of the under-developed regions of South-East Asia, Africa and South America. It is known by different terms in various parts of the world, e.g. *ladang* in Malaysia, *taungya* in Burma, *caingin* in the Philippines and *milpa* in Rhodesia.

In shifting cultivation the farmer and his family cut down a small part of the forest when the dry season begins. The cut-down trees are left to dry. In the meantime, the family builds a wooden hut near the cleared plot. When the cut-down trees are dry, they are burnt and reduced to ashes, which fertilize the soil. Planting begins before the rainy season starts. The crops which are usually grown are hill rice, millet, sweet potatoes, tapioca and maize. The farmers do not use manure or fertilizers, and the same plot of land is cultivated year after year until the soil is no longer fertile. This usually lasts from two to four years. As the soil becomes less fertile, the yield from the crops decreases until, finally, the amount produced is insufficient to feed the farmer and his family. They are then forced to move to a new patch of forest and start the process again.

Meanwhile, the abandoned plot is left to revert slowly to forest. This system of cultivation which involves the shifting of the cultivated land and which uses rudimentary techniques of planting, is usually practised in areas where the population is sparse and where forest land is relatively abundant.

Subsistence farming

Subsistence farming does not aim to produce crops or animals for sale. In subsistence farming, a farmer usually owns a small piece of land on which he and his family work. They cultivate crops or rear animals for their own consumption. Any surplus products are usually stored for future use in periods when the harvest is poor. Subsistence farming sometimes becomes intensive so that a surplus of food crops is produced. This surplus is then sold. Subsistence farming is usually characterised by the use of simple techniques of cultivation, by the employment of family labour and by a low standard of living. It is generally practised by farmers who live in economically backward areas. Shifting cultivation is a primitive form of subsistence farming.

Commercial crop farming

Commercial crop farming is farming which aims at producing food or non-food crops for sale. It depends largely on a good transport system and a good marketing organization to distribute the produce to regions where it is wanted.

Commercial crop farming may be either *intensive* or *extensive*. Intensive farming usually involves the application of labour and/or capital to relatively small sized farms, and yields per unit area are usually high. Farming in the polders of Holland is a good example of intensive farming. Extensive farming involves the use of a small labour force, and sometimes considerable mechanization on very large sized farms. Yields per unit area are usually low. Wheat farming on the Canadian Prairies is a good example of extensive farming.

Commercial farming makes full use of manures and fertilizers as well as regulating water flow so that the maximum yields can be obtained without damaging the soil or lowering its fertility. Part or all of the produce obtained is sold for cash. The cultivation

Intensive Commercial Farming in Holland

Extensive Commercial Farming in the U.S.S.R. This photograph shows highly-mechanized wheat farming.

of rice and vegetables for sale are examples of intensive commercial farming.

Extensive commercial crop farming in the tropics is usually geared to producing a single crop for sale to industrial countries. The most common crops are rubber, palm-oil, cocoa and tea. Most crops grown on plantations have to be processed and this is usually done in factories on the plantations. Most of the workers live on the plantations in small settlements which are usually self-contained, that is, there are schools, shops and recreational centres.

ANIMAL FARMING

Commercial animal farming

This type of animal farming can be divided into three parts.

1 Commercial livestock farming

The products of this type of farming are meat, hides and wool, and the main producing areas are in the grassland belts of both hemispheres. Farms are very large and the number of animals per farm is high and it often runs into thousands. This type of farming is particularly important in sub-humid and semi-arid environments, and in these regions, the farm unit is called *ranch* in N. America, *estancia* in S. America, and *station* in Australia. Hired labour is usually employed and the animals are well looked after. Specially sown pastures of fodder crops such as alfalfa are grown for animal feed, and emphasis is placed on the selective breeding of animals to obtain better quality products. New grass species such as 'sudex' (Sudan grass X alfalfa) are being developed.

2 Dairy farming

This type of farming, whose product is milk, is located near to large urban centres or in regions having access to urban centres by efficient transport. Such centres may be vast distances from the dairy farming regions, for example, industrial Britain and those parts of Australia and New Zealand which

A herd of Highland Cattle in Scotland (U.K.)

supply it with dairy produce. The climate is mild and damp, so the cattle can graze outside for most or all of the winter and therefore they do not need extra fodder. The efficient transport linking the consumer to the producer, in this case, is the refrigerated cargo ship. Dairy farming for the production of milk usually occurs near to towns, while dairy farming for the production of butter and cheese may be remote from towns.

3 Mixed farming

In this type of farming, as the name implies, animals are reared and crops are grown. Grain crops such as wheat, oats, barley, rye and maize are partly grown for sale and partly for feeding to livestock. In addition root crops, especially turnips, potatoes and sugar beet may be grown. These may be for sale or for feeding to livestock. Pastures are of course important and these are sometimes specially sown to improve their quality. Many mixed farms which are near to urban centres may grow vegetables and fruit for sale to the cities. Most mixed farms are small, in comparison with ranches, and most of them can be regarded as practising intensive farming. However, in Australia, areas previously growing wheat as a mono-culture are now being run as large scale mixed farms.

The success of commercial animal farming depends on several factors.

1 A high and steady demand for livestock products.
2 Adequate pasture land for the livestock to graze on. Pastoral farming in many interior parts of Australia has been made difficult mainly because of the lack of suitable pastures as a result of the dry climate.

3 Good transport lines, particularly railways, to send livestock to local markets or to abbatoirs which are mostly located at the ports. In the abbatoirs, the animals are slaughtered for their meat which is to be exported.
4 Facilities such as cold storage plants and refrigerator ships to preserve the freshness of the meat during the journey to overseas markets.

Countries in which commercial animal farming is important are the U.S.A., the pampas region of South America, the British Isles, the Netherlands, Denmark, New Zealand and Australia. Each of these countries specializes on a different animal product. On the Canterbury Plains of New Zealand, lambs are reared for their meat; in Australia, cattle are reared in tropical parts for their beef, and sheep in temperate parts, mainly for their wool; in the Netherlands and Denmark, livestock are reared mainly for their milk which is manufactured into dairy products such as butter and cheese for export.

Nomadic animal farming

Nomadic animal farming is practised by wandering groups of people in the remote areas, particularly in the semi-desert and desert regions of the world. Each family in a group rears a number of the same kind of animals which graze on available natural pastures. When the pastures have been eaten over, the group and their animals move to fresh pastures. Thus, unlike commercial herding, nomadic herding is not confined to a large, fixed area.

Nomadic herders lead difficult lives. As they are often on the move in their search for new pastures, their amenities are reduced to a minimum. They usually live in portable tents. Almost their entire lives revolve around their herds. Much of their food,

Nomadic herdsmen of the Chinese Steppes.

and sometimes their clothes, are obtained from their animals. The chief objective of the nomadic herders is not the sale of their animals. Many nomadic herders sell only under compulsion when their animals are exchanged for necessities required by them. Usually, their aim is to increase their herds. This is because a man's social status among nomadic herders in determined by the number of animals he owns, irrespective of their condition. This is a great hindrance to efforts to improve the condition of grazing lands in Africa where many nomadic tribes herd cattle, (e.g. Masai in E. Africa). In large parts of the North African desert, nomadic herders rear camels, sheep and goats. On the steppe lands in central Soviet Asia, the Kirghiz rear sheep and cattle, while in northern Scandinavia, the Lapps rear herds of reindeer.

FACTORS INFLUENCING COMMERCIAL CULTIVATION AND LOCATION OF CROPS

There are three types of factors:
1 climate (temperature, and rainfall)
2 soil
3 economic factors (transport, market, labour and capital)

Climatic Factors

Three aspects of climate are very important. These are:
(a) temperature
(b) rainfall
(c) seasonality

Plant growth for most plants ceases if the temperature falls below 6°C (42°F). When the subsoil becomes frozen for five consecutive months, most plants cannot survive. Thus, crop cultivation is impossible in polar regions because winters are too long and too cold. Besides temperature, the total amount and distribution of rainfall per year is important because plants need both suitable temperatures and adequate rainfall. In desert regions scanty rainfall impedes plant growth and crop cultivation. The amount of rainfall needed for cultivation, however, varies from region to region and from crop to crop. The time of year when the rain falls is also important. Rain should fall when the young crops are growing rapidly. Heavy rain at harvest time can cause severe damage to mature crops. Wind can also be destructive, particularly when there are gales at the time crops, such as wheat, are ready for harvesting. Climatic variations such as unexpected late frosts, summer storms, and periodic droughts can have adverse affects on crops.

Soil Factors

Soil factors also influence crop cultivation because different types of soil are suited to different crops, e.g. wet rice grows well in clay soils, whereas coconut trees grow well in sandy soils. However, regions which have fertile soils often have climates which favour the cultivation of crops because soils are also influenced by climate. The type of crop which can be profitably grown in a region depends largely on climate and soil e.g. a region which has a constantly hot wet climate and predominantly laterised soils is unsuitable for wheat cultivation (wheat requires a warm climate, moderate rainfall and loamy soils), but it would be very suitable for the cultivation of rubber.

World Farming Regions

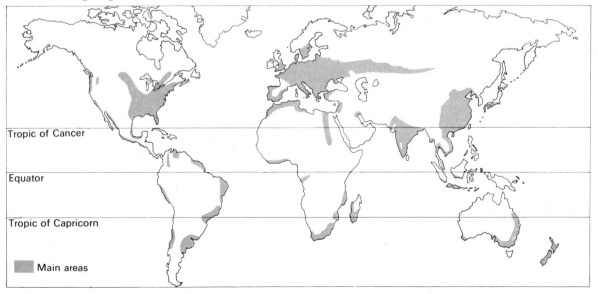

Tropic of Cancer

Equator

Tropic of Capricorn

Main areas

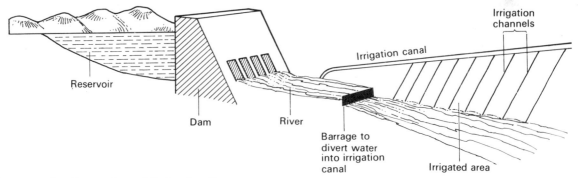

Diagram of a Reservoir and Irrigation System

Economic Factors

Besides suitable climate and soil, there must be a demand, either local or from abroad, for a crop if it is to be planted by a commercial farmer. Although many regions which have a Mediterranean climate are able to grow both grapes and olives, commercial farmers often tend to grow grapes in preference to olives since this crop has a wider demand, and yields a greater profit. When the demand for a crop which is being planted in a region drops and profits decline, then that crop will be replaced by a more suitable crop. Thus, in eastern Brazil, the history of farming is marked by one crop replacing another in successive periods. Sugar cane was extensively grown in the 17th Century and rubber trees in the 19th Century, but today coffee is the main crop produced in that region.

Besides effective demand, some other economic factors which affect crop cultivation are efficient transport, abundant human labour and easily available capital. The latter is particularly important for the cultivation of tree crops because these crops usually take several years to reach maturity, which means that the farmer has to wait a long time before he gets a return on his capital investment. The Amazon Basin, with its equatorial climate and fertile alluvial soils, is most suitable for wet rice cultivation, but little or no rice is cultivated because of a lack of the above factors.

DRY FARMING AND IRRIGATION

Dry farming

Dry farming involves preserving in the soil as much moisture as possible, and it is practised in regions having a low annual rainfall or one which varies from year to year. The purpose of dry farming is to allow sufficient moisture to remain in the soil for plant growth. The most common types of crops planted in regions of dry farming are grains, e.g. wheat, millet and corn because these can withstand drought, and they require less water from the soil than many other crops.

There are several dry farming techniques:
1 Crops are planted in alternate years, that is, crops are planted for one year, after which the land is left to lie fallow. Thus the moisture which soaks into the ground for two years is used to produce one crop.
2 Contour-strip farming is practised in areas where there is rapid soil erosion by wind and water. On sloping land crops are planted in strips which follow the contours. Strip cultivation of this type slows the rate of flow of water over the surface and allows more of it to seep into the soil. It is therefore used in some regions as a dry farming technique.
3 The top soil is loosened by deep ploughing to reduce the rate of evaporation from it and to allow rain water to soak into the sub-soil.
4 A non-porous soil is added to a porous soil to reduce its porosity so that the loss of water in the soil through percolating downwards is reduced.

Examples of regions where dry farming is practised are (i) the Central Plains of the U.S.A. and Canada; (ii) the steppe lands of Asia and Mongolia; and (iii) parts of India and Pakistan.

Irrigation

The putting of water on the land to enable crops to grow is called *irrigation*. Irrigation is used in dry regions, to which water can be taken, and in regions where the annual rainfall is too low or too variable from year to year to allow proper crop cultivation. Water has to be collected and stored before irrigation can be practised.

Irrigation is not confined to regions of low rainfall. A large part of the world's rice is grown on irrigated land in regions where the annual rainfall rarely falls below 1524 mm (60 in).

There are several types of irrigation. The main ones are:

Perennial irrigation: By this method water is supplied to cultivated areas throughout the year. It involves the construction of huge dams across rivers so that reservoirs are created to store the water. The flow of water from the reservoirs into the irrigation canals and thence onto the fields is controlled by sluice gates. Modern perennial irrigation involves the construction of major engineering works, which, though costly, provide water for vast areas.

Examples

(i) North American projects: Most of these are in the western half of the continent. The most important are: the Grand Coulee scheme, the Snake River scheme and the Sacramento and San Joaquin scheme. In the eastern half of the continent, the Tennessee Valley Authority (T.V.A.) is by far the most important.

(ii) The U.S.S.R.: There are extensive irrigation works in the semi-arid and arid areas of south central U.S.S.R. Some of the irrigation projects are on a very large scale, for example, the Kara-Kum canal, which takes water from the Amu towards the Caspian. This canal is over 960 kilometres (600 miles) long and it irrigates 1·4 million hectares ($3\frac{1}{2}$ million acres) of arable land and supplies water to 4 million hectares (10 million acres) of pasture.

(iii) Gezira scheme (Sudan): The irrigated area occupies over 40 000 square kilometres (16 000 square miles). A large dam across the Blue Nile, near Sennar, has created a reservoir whose water is led to the Gezira Plain by a network of irrigation canals.

(iv) The Aswan dam project: This project, in Upper Egypt, has enabled over 26 000 square kilometres (10 000 square miles) of land to be irrigated, but see the note below.

(v) The Indus projects: Dams constructed across the Indus Valley at Kotri, Sukkur and Gudu in Pakistan, enable over 40 000 square kilometres (16 000 square miles) of land to be irrigated.

(vi) The China projects: Extensive irrigation schemes are being constructed on the Hwang-ho, the Yang-tse-kiang and the Si-kiang. Most of these are dual purpose schemes for providing perennial irrigation and H.E.P. In Sinkiang Province the government has extended the network of irrigation canals around the edges of the deserts to allow for the establishment of pioneer settlements.

(vii) The Murray-Snowy Rivers scheme of South-east Australia: Irrigation water is drawn off the Murray, Murrumbidgee and Snowy rivers to irrigate thousands of acres of land for fruit

Perennial Canal Irrigation in south central U.S.S.R.

cultivation. The Snowy River scheme also involves the generation of H.E.P.

Note Not all irrigation schemes work entirely to man's advantage. The Aswan Dam has slowed down the flow of the Nile and very little sediment is now carried to the delta and the sea. This has resulted in:

(i) a slowing down of the building up of sandbars and banks in the delta with the result that the Mediterranean has flooded over 400 000 hectares (1 000 000 acres) of fertile delta land.

(ii) the decreasing deposition of sediments in the sea is causing the fishing industry to collapse.

(iii) Some irrigation schemes have an adverse affect on traditional types of farming which are based on the renewal of soil fertility by annual river floods. This happens when dams are constructed to control river floods. The soils of valleys below the dams rapidly lose their fertility in tropical countries which means that fertilizers have to be used if successful farming is to continue.

2 **Annual irrigation**: This is sometimes known as *basin irrigation*. It depends entirely on the annual flooding of a river during the period of heavy rainfall in the upper parts of its valley. The flood plain is usually carefully levelled and banked by mud walls into small fields, and irrigation ditches are dug to help distribute the flood waters during the season preceding the rainy season. During the rainy season, flood water moves down the valley and enters the irrigation ditches and so spreads

out over the flood plain. The silt, which is suspended in the flood water, is gradually deposited and the soil of the plain is thereby enriched. After the flood water has subsided, the fields are ploughed and planted with crops. Annual inundation, in contrast to perennial irrigation, has the disadvantage that irrigation water is available only during the rainy season. However, it has the advantage of maintaining soil fertility because of the deposition of silt.

Examples
(i) The Ganges Valley: Part of the flood plain of the Lower Ganges Valley is flooded annually, though the construction of dams is continually reducing the area affected.

(ii) The Nile Valley: At one time basin irrigatio was very widespread, especially in Lowe Egypt, but the construction of dams across th Nile has reduced its importance except in th the Nile Delta.

3 **Tank irrigation**: This method is practised in man parts of India and Sri Lanka. It involves the con struction of tanks (areas which are bunded s that water collects) which catch and store wate direct from seasonal or occasional rainfall.

4 **Well irrigation**: A well is a shaft sunk into th ground to below the water-table so that wate seeps into and collects in it. This type of irrigatio can only take place in regions made of sedimen tary rocks below whose surface there are consi derable quantities of water.

The Nile Valley

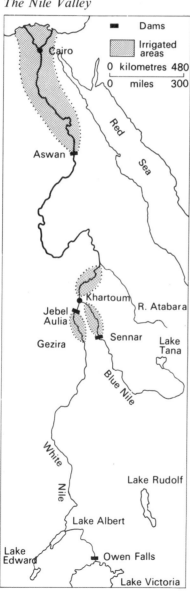

The Indus Valley

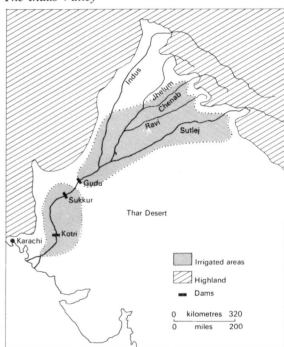

South-East Australia

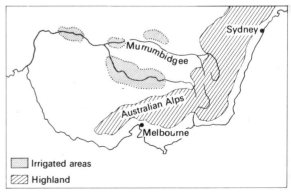

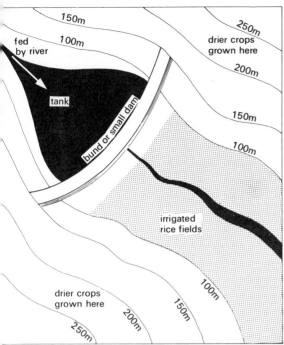

Tank Irrigation on the Deccan

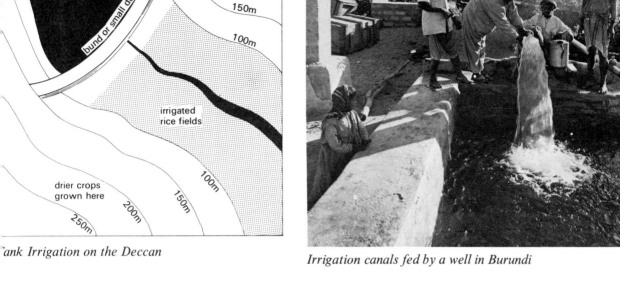

Irrigation canals fed by a well in Burundi

Examples

i) The Indo-Gangetic Plain of the Indian sub-continent.

ii) Parts of east central Australia overlying the Great Artesian Basin.

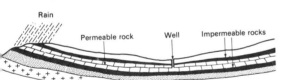

Structure of an artesian basin. Water collects in the permeable rock layer, called an *aquifer*. The impermeable rocks above and below it prevent the water from draining away. A well sunk into the aquifer is called an *artesian well*.

Note There is a limit to the number of wells that can be efficiently operated in an artesian basin because if the water is drawn up faster than it is replaced, the water table falls and the wells in the basin eventually run dry. In many countries laws are often passed to control the number of wells that can be sunk in an artesian basin.

RECLAMATION AND DRAINAGE

The turning of water-logged land or areas of shallow sea or lake into dry land is called *reclamation*. Many lowlands have been built by the deposition of silt by rivers, lakes and seas, but large areas of some

Irrigation by well in India. Skin water-bags are raised by ropes and pulleys from a deep well by a team of oxen which trudges backwards and forwards along a short sloping ramp.

199

lowlands are still in a water-logged state. And there are other regions where active deposition is taking place but which are still permanently water-covered. The manner in which such water-logged and water-covered regions are turned into dry land depends upon the nature of the regions.

Flood plains which experience seasonal or annual flooding can be made permanently dry by cutting off river meanders to straighten rivers and thus making them flow more quickly.

By this process, the flood waters are carried away to the sea quickly. Swamps and marshes can be drained and the shallow parts of coastal seas and lakes can be filled in with rubble and earth.

Examples

(i) The Netherlands: Large parts of western Holland whose coast is bordered by a chain of sand dunes, were once permanently covered with water. This area used to be called the Zuider Zee. During the last three centuries, the Dutch have built dykes (walls made of clay and reinforced with steel piling and mats of reeds and brushwood) around parts of the Zuider Zee to keep out the sea. The land inside the dykes was then drained. This used to be done by using windmills, which raised the water enabling it to be emptied into drainage canals and river. Diesel pumps are now used because they can pump out larger volumes of water and also because they are more reliable.

The drained land is called a polder. Today, there are four main polders (a fifth one is not yet completed), and these are shown on the map. After a polder has been made, its soils are treated so that it can be used for growing crops. But drainage does not cease with the formation of a polder. Because polders are low-lying they are liable to flooding, and constant work has to be done on the dykes to ensure they are in a good order. Also, the drainage pumps and drainage canals have to be kept in good working order.

The formation of a polder and the maintenance of the dykes, diesel pumps and drainage canals, costs a huge amount of money which is provided by the Dutch government.

(ii) The Fenlands: This region lies to the west of the Wash in eastern England. Until the mid-seventeenth century it was a water-logged region, except for a few 'islands' of boulder clay, despite attempts to drain it which were made by the Romans. Since 1650 drainage canals have been dug which have resulted in the

almost complete drainage of the region. Wat is pumped from the drainage canals into wid straight main canals which carry it to the se But drainage has created a problem. The fe around the Wash are made of peat. When t latter are drained they shrink, which mea that their level falls. This has, in fact, happene and all drainage water removed from the pe fens has to be pumped out as indicated abov

The Netherlands and the Fens are only two of mar drainage and reclamation schemes. Similar schem

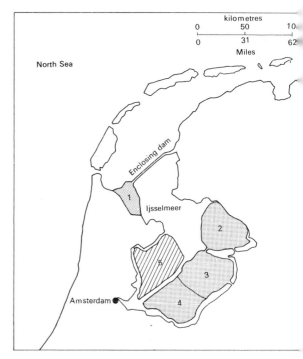

1 North-West Polder (1930)
2 North-East Polder (1942)
3 East Polder (1957)
4 South Polder (1968)
5 South-West Polder (not yet finished)

Reclamation of Ijsselmeer of The Netherlands

Dyke or embankment

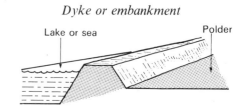

200

being worked out in many parts of the world, especially in regions where there is a shortage of land suitable for either agriculture or for building.

THE USE OF FERTILIZERS

Fertilizers are artificial manures which contain mineral compounds needed to enrich the soil. Different kinds of plants extract different mineral compounds such as nitrogen, potash and magnesia from the soil. When a piece of land is cultivated with the same crop continuously for several years, the soil will become deficient in a particular mineral compound because the crop will have extracted this mineral from it. Hence, manures or fertilizers have to be placed in the soil to replace the minerals used by the plants.

The use of fertilizers is essential in commercial agriculture because they enrich the soil and enable plants to produce high yields. Moreover, the use of fertilizers enables agriculture to be carried on in areas where the climate is suitable but the soils are poor. In parts of South Australia, scientists discovered that in large areas of poor scrubland the soil was deficient in the trace elements copper, cobalt and zinc, all of which are essential for proper plant growth. After these elements had been added, these scrublands were planted with alfalfa and they were turned into productive stock-rearing farms.

EXERCISES

Briefly outline the main differences between shifting cultivation and subsistence cultivation. For *each* (a) indicate the effects it has on soil conditions, (b) name *two* areas where it is practised.

(a) What is meant by the term 'Commercial Crop Farming'?

(b) Choose any *one* region where this type of farming is practised and (i) describe its main features indicating the types of crop grown and the main markets for them, (ii) state *two* reasons to account for the establishment of this type of farming in the region you choose.

(c) Name any other *two* regions where commercial crop farming takes place and name the crops grown.

Write a short essay on 'Animal farming' emphasising the differences among the main types of animal farming.

Make a list of the main factors which influence the cultivation of crops on a commercial scale. For each factor name one region where it is particularly important.

(a) Name *one* region of heavy rainfall, and *one* region of low rainfall where crops are grown under irrigation, and for *each*

(i) state why irrigation is used; (ii) state how it is used; (iii) state the source of the irrigation water.

(b) Name any other *two* irrigation schemes and state the importance of each to the economy of the region where it is located.

6 (a) Draw a large sketch map of a region, preferably one that you know, where land reclamation has taken place. Your map should show (i) the areas where reclamation has taken place, (ii) the type of reclamation, (iii) the uses to which the reclaimed land has been put.

(b) Beneath your map write a concise account of land reclamation as shown in your map, and name any other *two* areas where land reclamation has taken place.

Objective Exercises

1 Extensive farming implies all of these **except**
A high yield per acre
B high yield per man
C large area per farm
D small number of crop types
E high income per farm

A B C D E

2 Which of the following statements is incorrect?
A Perennial canals are superior to inundation canals.
B Perennial canals contain water in the wet season only.
C Tank irrigation is practised in parts of Sri Lanka.
D Irrigation is used in dry farming.
E Irrigation schemes are often associated with H.E.P. schemes.

A B C D E

3 Agriculture which involves the haphazard cutting and burning of vegetation, is a part of
A nomadic farming
B dry farming
C subsistence farming
D shifting cultivation
E extensive farming

A B C D E

4 The only condition that dairying does **not** require is
A regular attention
B fertile pastures
C plenty of water
D large areas of land

A B C D

5 The disadvantages of nomadic animal farming to the economy of a country include
A available natural pastures are utilised
B settlements are impermanent

C the people depend on their animals for food and clothing

D animals are kept in quantity and their quality is not important

A B C D
·· ·· ·· ··

6 Perennial irrigation differs from annual irrigation in that it
A needs less capital
B is seasonal
C covers the area with fertile silt
D involves a larger area in each scheme

A B C D
·· ·· ·· ··

7 Which of the following is **not** a characteristic of both the Fenlands of eastern England, and the Netherlands?
A the land is flat and low-lying
B drainage canals are often several metres higher than the surrounding farmland
C windmills are often used to operate irrigation works
D the land is intensively cultivated

A B C D
·· ·· ·· ··

8 Shifting cultivation is known by all of the following names **except**
A ladang
B taungya
C transhumance
D milpa

A B C D
·· ·· ·· ··

9 *Truck Farming* is carried on extensively in many parts of the U.S.A. Another name for *truck farming* is
A sericulture
B market gardening
C horticulture
D fruit farming

A B C D
·· ·· ·· ··

10 Intensive agriculture is associated with all of the following **except**
A vegetable farming in the Netherlands
B rice farming in the lowlands of South-East Asia
C fruit farming in California
D wheat farming in Canada

A B C D
·· ·· ·· ··

11 In which of the following regions is nomadic animal farming **not** a major activity?
A northern Scandinavia
B South-East Australia

C the arid steppes of Central Asia
D parts of East Africa, especially the Sudan

A B C
·· ·· ··

9 Cereals

Planting rice on an alluvial plain in Japan. Notice the bunds (low mud walls) that separate the fields.

CEREALS

Wheat, rice and maize are the three principal grain crops of the world. They grow best under specific climatic conditions. Wheat grows in the cooler, and maize in the warmer temperate regions, while rice grows in the wet tropical regions. However, all three grains are grown outside these climatic limits, e.g. wheat is grown in India and Egypt, and rice is grown in Spain and Italy. These three plants produce edible seeds which form an important part of the daily diet of the world's population.

RICE

Between one third and one half of the world's population eats rice. Rice grows very well in the warm, wet monsoon regions of tropical Asia, and in some sub-tropical regions such as lower Egypt and southern California which have an abundance of irrigation water. Although there are several varieties of rice, each having its own climatic and soil requirements, the two most common types are lowland rice and upland rice. The main difference between the two is that whilst they both need flooded conditions in the growing period, upland rice requires rather less water. Lowland rice is grown extensively on the river and coastal plains of Asia, such as the lower valley plains and deltas of the Mekong, Ganges and Irrawaddy, while upland rice is grown on the terraced slopes of hills in Sri Lanka, Java and parts of Japan.

Growing conditions for rice

The successful cultivation of rice requires:
1 A growing season of about 5 months with minimum temperatures of 21°C (about 70°F)

followed by a dry sunny period in which the crop ripens.

2 An annual rainfall of over 2000 mm (about 80 in) with at least 120 mm (about 5 in) falling in each month of the growing season.

3 Flat land to permit the soil to retain the water. Upland rice is grown on terraced slopes. Each terrace is like a broad step which means that the slopes are turned into a series of level steps.

4 Heavy alluvial soils with an impervious sub-soil of clay to retain the water.

The cultivation of rice

The rice fields are carefully levelled and earth walls, called *bunds*, are built around them so that the water is retained. This takes place before the wet season begins. Because rice requires abundant water during the growing season, but little if any during the harvest season, irrigation and drainage canals are constructed to control the flow and amount of water on the fields at all times. The sowing of the rice seed, which takes place when the wet season begins, is preceded by the ploughing, and the manuring of the fields. The rice seeds are either sown direct in the fields or in nurseries where they are allowed to germinate and grow into young seedlings which are then transplanted into the rice fields. The young plant grows quickly in the flooded fields and it reaches maturity after about six months. The fields are then drained and the rice is allowed to ripen, after which it is harvested.

The harvesting and processing of rice

Rice is harvested by hand in many parts of Asia, but machines are used in the U.S.A., in Australian rice areas and in some other rice regions. The small rice fields of Asia make the use of machines uneconomic although in some regions where rice land

Terraced slopes used for growing padi in Japan. In the foreground padi seedlings are being transplanted into flooded fields.

Common Cereals

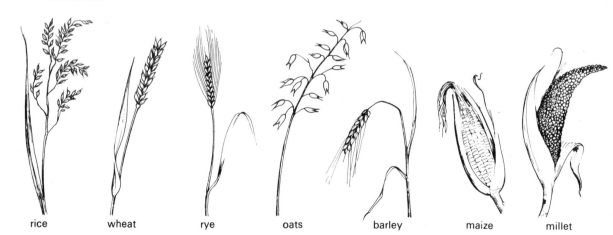

rice   wheat   rye   oats   barley   maize   millet

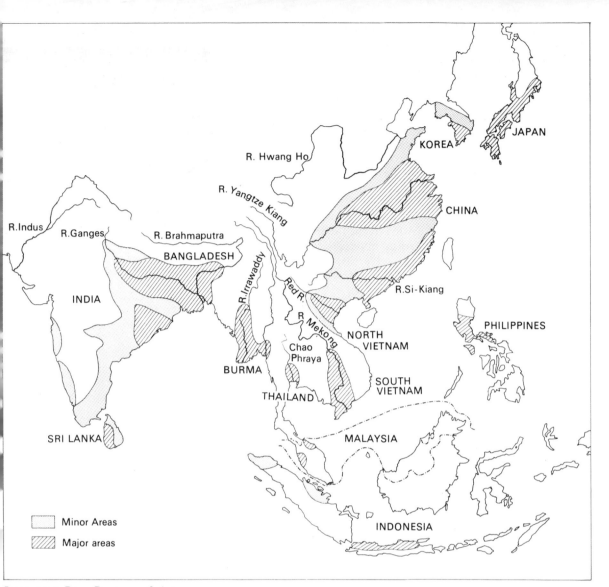

Important Rice Regions of Asia

orms large units, machines are being increasingly
sed. Most Asian rice farmers are too poor to buy
r hire machines, but cooperative organisations are
vercoming this problem in some areas, especially
Japan and China. After the rice is harvested, the
ce seeds are separated from the stalks by threshing,
fter which they are taken to the rice mills where
e seeds are dehusked and polished. In many rural
reas this process is carried out by the farmer himself
y pounding the seeds in a large wooden tub until the
usks come off. After this, the seeds are winnowed
y tossing them in a bamboo tray. When tossed,
e husks are blown away and the seeds are left
ehind in the tray. Rice cultivation and harvesting
equire a large number of workers but rice cultiva-
on produces more food per unit area than does

the cultivation of most other cereals. In many
tropical rice regions which have fertile soils, abundant water supply, and good irrigation and drainage
facilities, two crops of rice a year are common. As
a result, the rice lands of Asia have some of the
highest population densities in the world. Such an
area is the Red River Delta of North Vietnam which
has a population density of over 1500 people per
square kilometre (about 570 people per square mile).

The main rice-growing regions
Over 80% of the world's supply of rice comes from
Asia, especially from the Si-Kiang and Yangtze
basins of China, the Ganges Delta of India and
Bangladesh, the Red River Delta of North Vietnam,
the Irrawaddy Delta of Burma, the Chao Phraya

Delta of Thailand, the Mekong Delta of Cambodia, and the alluvial coastal plains of Japan and Indonesia. Outside Asia, there are important rice-producing regions in the Mississippi Delta and in parts of southern California (in the USA), the Po Basin (northern Italy), the Nile Delta (Egypt) and in parts of New South Wales (Australia).

Rice production and world trade

Because rice is the staple food of a large part of the population of Asia, most of the demand for, and the supply of, rice comes mainly from Asian countries. Rice-consuming countries which have a surplus, export rice to rice-consuming countries which have a deficit. However, owing to the rapidly increasing populations of most Asian countries, in relation to available agricultural land areas, Asia produces insufficient rice to meet its requirements. In most Asian countries, with the notable exception of Japan, the yield of rice per unit area is not as high as it could be because the farmers are too poor to buy fertilizers and efficient farming equipment. However, since the early 1960's, the International Rice Research Institute in the Philippines, has developed a number of very high-yielding rice plants, and these plants are now cultivated in some parts of Asia. These high-yielding varieties, especially IR8 and IR23, are known as *miracle rice* plants because they produce about 50% more grain than other varieties of rice. If these new varieties were extensively cultivated throughout Asia, the annual production of rice would be sufficient to meet the world demand. Unfortunately, inefficient transport and storage facilities, and inadequate trading arrangements make it difficult for some areas to benefit. Also, many farmers are unwilling to grow the new varieties because their continuing successful cultivation involves the use of fertilizers and improved irrigation facilities both of which cost money which many farmers do not have. Only a small part of the world's annual production of rice enters international trade. The three largest exporters of rice are the U.S.A., Burma and Thailand, and most of the rice these countries export is imported by Indonesia, India, Japan and Malaysia.

WHEAT

Although wheat is the staple food of most of the people in temperate latitudes, it forms an important part of the diet of the peoples of northern India and Pakistan, and it is becoming an increasingly important food for many peoples in other parts of Asia.

Wheat is grown in several types of climatic regions but the most important regions of commercial production are between the latitudes of 30°N and

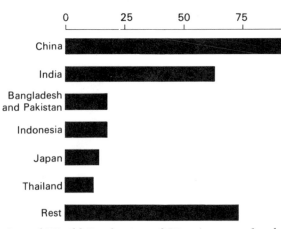

Annual World Production of Rice (average for the last five years in millions of metric tons). Approximate annual production 300 000 000 metric tons.

40°N, and 25°S and 45°S. Wheat is of two main types – *spring wheat* and *winter wheat*. Spring wheat is planted in the spring in wheat-growing regions which have severe winters, e.g. on the Canadian Prairies, in the northern parts of the U.S.A., in the U.S.S.R., in eastern Europe and in northern China. Such wheat is harvested in the following summer. Spring wheat has been specially produced by crossing different types of wheat to produce a strain which matures in the short growing season between spring and autumn. Winter wheat is sown in the autumn in regions which have mild winters. The wheat begins to germinate during the winter and is harvested in the following summer. This type of wheat is grown in Australia, in Argentina, in central Chile, and in parts of North-West Europe. In the wheat regions of India and Pakistan, the wheat is sown just after the summer monsoon rains.

The Main Wheat Regions of the U.S.A.

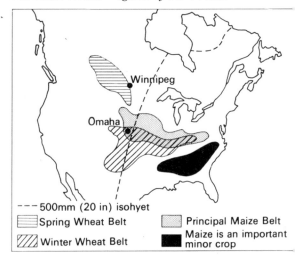

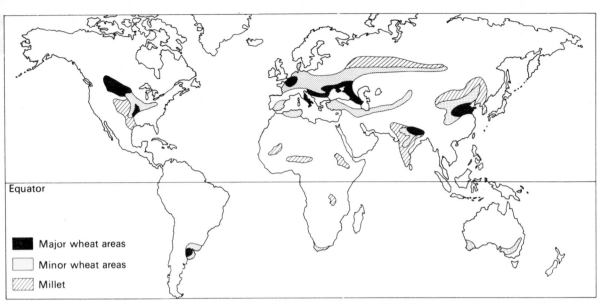

Important Wheat and Millet Regions

Legend:
- ■ Major wheat areas
- ▨ Minor wheat areas
- ▨ Millet

begin, and it is harvested before the hot wet season begins.

Growing conditions for wheat

Wheat grows best in regions which have the following conditions:

1 An annual rainfall of 380 to 900 mm (15 to 35 in), most of which falls in the growing season. However in the western parts of the Canadian Prairies where wheat is grown, the rainfall rarely exceeds 200 mm (about 8 in). Wheat cultivation is made possible here by using, dry farming techniques (see Chapter 10). Wheat is also grown in semi-desert regions, where irrigation water is available, e.g. in the Nile Valley, and in the valleys of the Upper Ganges and Indus.

2 Growing season temperatures of between 15°C (59°F) and 21°C (about 70°F)

3 A warm, dry sunny period to enable the wheat to ripen.

4 A growing season of at least three months.

5 A rich loam soil which retains the moisture but which allows good drainage.

6 Fairly level, or gently undulating land to permit mechanised farming.

The cultivation of wheat

Wheat is cultivated extensively and intensively. When grown extensively, the wheat farms cover very large areas and it is cultivated with the maximum use of machinery and the minimum use of labour. In some extensively farmed regions, wheat is still the only crop grown, and because of this, extensive farming is sometimes called *monoculture*. However, the rotation of wheat with other crops, such as barley, oats, sugar beet and vegetables, is becoming increasingly common. The yield per unit area of wheat grown extensively is much lower than that grown intensively. The most important regions of extensive wheat farming are in Canada and in the U.S.A., in the U.S.S.R., and in China, Australia and Argentina.

Intensive wheat farming involves the cultivation of wheat on small areas of land. Very little machinery is used and the emphasis is on the use of labour. Wheat grown intensively is usually grown in rotation with other crops which means that a given area of land will produce a wheat crop once every three or four years. Intensive wheat farming gives a high yield per unit area, and it is the main type of wheat farming in European countries, and in India, Pakistan and in parts of China.

Wheat-growing regions

Nearly 70 per cent of the world's wheat production is in North America (the U.S.A. and Canada) and Europe (including the U.S.S.R.).

Canada

Most of Canada's wheat is grown in the southern part of the Prairies. The land here is either flat, or gently undulating, which makes possible extensive wheat farming by using highly mechanised techniques. Along the northern side of the wheat belt, the length of the growing season falls below three months and this limits the cultivation of wheat. Along the western side of the belt, wheat cultivation is limited by insufficient rainfall although here dry farming methods are often used, especially when the world demand for wheat is high. Many farmers on the Canadian Prairies rotate wheat with other crops. Some also rear livestock.

Wheat farms vary in size from about 100 hectares in the eastern parts of the belt to over 200 hectares in the western parts. This is because rainfall decreases from east to west across the belt. Wheat farmers plough the land in the autumn and leave the soil bare throughout the winter. Heavy winter frosts break up the clods of earth and kill off the insects and pests in the soil. The land is ploughed again in the following spring, after which it is harrowed and the wheat seed is sown. About 3 to 4 months later the wheat crop is ready for harvesting by the combine harvesters. The grain is taken to the wheat elevators where it is stored until it is required for transporting eastwards by rail to the Great Lakes ports of Fort William and Port Arthur. The wheat is taken by large wheat vessels or by rail to the ports of Buffalo, Halifax, Montreal and New York. Wheat is also transported by rail westward to the ice-free port of Vancouver, and northwards to Churchill on Hudson Bay. This latter route is only open for a few weeks because the Hudson Bay freezes during the winter. The St. Lawrence route is also closed for a part of winter for the same reason.

The U.S.S.R.
Production once centred on the black soil region of the Ukraine, in southern Russia, but wheat cultivation has spread eastwards in recent years.

The main area of production is now in northern Kazakhstan and south-west Siberia where it forms an east to west belt astride the Trans-Siberian Railway. This eastward extension lies between the cold lands and the dry lands. The wheat region occupies the steppes which are temperate grasslands similar to the prairies. The Ukraine is now a region of mixed crop and livestock farming.

Australia
Most of the wheat is grown on the downs of south east Australia. Farms are large and methods of farming are similar to those of North America. In parts of the 'Wheat Belt', rainfall is below 380 mm (15 in) per year and wheat can only be cultivated in alternate years. This is sometimes achieved by using 'dry farming' methods. The low rainfall of the Riverina, the region between the rivers Lachlan and Murrumbidgee has been partly responsible for this region developing combined wheat and livestock farming based on farms of between 600 and 2000 hectares. A network of railways taking the wheat to the coast was very important in the development of inland areas for wheat.

Argentina
The pampas of Argentina, like the downs of Australia, are cultivated extensively for wheat. The main wheat-growing region extends from Bahia Blanca, in the south, to Rosario in the north. The region is crescent-shaped. To the south and west rainfall decreases while to the east it increases. Both act as limiting factors to wheat cultivation. Like Australia, the area has a dense network of railways.

Harvesting wheat by combine harvesters on the Canadian Wheatlands.

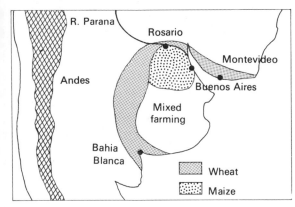

Wheat and Maize in the Pampas (Argentina)

Europe
Wheat is grown in most European countries, but its cultivation is especially important in France and Italy. It is grown intensively and often in association with other crops. Wheat farming is often coupled with animal farming. The yield per acre is one of the highest in the world, often exceeding 100 bushels per hectare as compared with 50 bushels or less per hectare for regions practising extensive farming.

China
The main region of cultivation lies in northern China where, each year, almost twice as much wheat is grown as in the whole of Canada. Heilungkiang (Manchuria) which is a part of China, is becoming increasingly important as a grower of wheat.

India and Pakistan
The main wheat regions are in the upper and middle valleys of the Indus (Pakistan) and the Ganges (India), with the Punjab being especially important. It is grown as a winter crop, mainly on irrigated land.

Wheat production and world trade
The U.S.S.R. is the largest grower of wheat, producing more than 25% of the world's total annual production. However, the U.S.S.R. is not self-sufficient in wheat every year which means that in some years it has to import wheat from other countries, especially from the U.S.A. and Canada. The U.S.A. is the largest exporter of wheat, followed by Canada and Australia. Most of the exported wheat goes to Japan, China, India, the U.K. and West Germany. Wheat moves out:

(i) across the Great Lakes, whose rapids and waterfalls have now been by-passed by canals which permit ocean-going vessels to reach Lake Superior,

(ii) by rail via Vancouver, New York and Halifax. Most of the wheat exported by the U.S.A. and Canada goes to Europe, although these two countries also supply wheat to China and the U.S.S.R. Nearly all wheat in North America is stored in huge towers, called *elevators*, which are located near to the railways. Both Australia and Argentina export wheat, again mainly to Europe, although the export of wheat to Asian countries from Australia has increased appreciably in recent years. Their southern hemisphere location means that they can export at a time when wheat stocks are low in Europe.

Loading wheat grain from elevators at Duluth, Minnesota (U.S.A.)

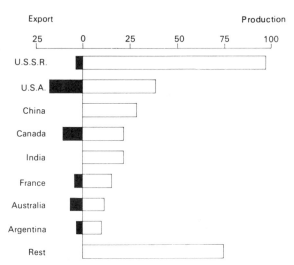

Annual World Production and Export of Wheat (average for the last five years in millions of metric tons). Total approximate annual production 320 000 000 metric tons: Approximate annual export 50 000 000 metric tons.

MAIZE

Maize, which is also known as *corn,* is a cereal which grows to a height of about 1.5 metres (about 5 feet). It has a very long stalk and broad, long leaves. When the plant is mature, the fertilised flowers, which arise in the axils of leaves, produce cobs at the ends of the flower stalks (see diagram).

Growing conditions for maize

There are several varieties of maize, each of which is adapted to different climatic and topographic conditions. Some varieties mature in 90 days while others require as long as 200 days to reach maturity. Some varieties can withstand drought while others require abundant moisture. Some grow well on lowlands, e.g. maize plants grown on the lowlands of eastern U.S.A., while others are adapted to higher altitudes such as those that grow in Peru. As a high-yielding, commercially cultivated crop, maize requires the following conditions:

1 A growing season of between 150 and 210 days with daily temperatures between 20°C (68°F) and 25°C (77°F). Night temperatures should not fall below 15°C (59°F).
2 An annual rainfall of between 620 mm and 1000 mm (25 to 40 in) with most of it falling in the growing season.
3 A dry, sunny season to enable the cobs to ripen.
4 Well-drained deep soils, rich in phosphates, nitrates and potash.

The main maize-growing areas

Most of the world's maize is grown in the warm temperate latitudes, especially in the eastern states of the U.S.A. and in eastern Argentina. Most of the maize grown in these two countries is fed to animals whereas most of that grown in other parts of South America, and in Africa and Asia is grown as a food for man. The following diagram shows that the U.S.A. produces almost 45% of the total world's production.

Important Maize and Oats Regions

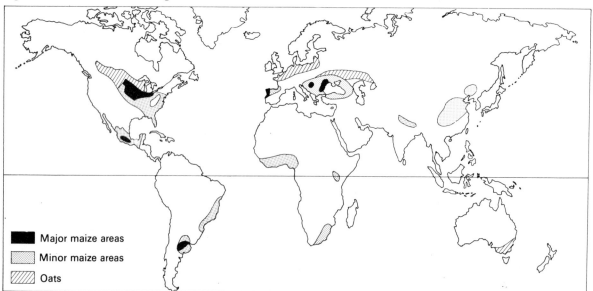

Major maize areas
Minor maize areas
Oats

The Corn Belt of the U.S.A.

The principal maize region of the U.S.A. occupies large areas of the States of Nebraska, Missouri, Indiana, Illinois, Iowa and Ohio. This region is known as the *Corn Belt*. Maize in this region is usually grown in rotation with oats, soya bean, barley, clover and vegetables, partly to maintain soil fertility, and partly to protect the farmers from heavy financial losses in those years when the maize crop fails. The Corn Belt grows more maize annually than any other region in the world. In addition, it is one of the largest regions of mixed farming. Almost 75% of the maize grown is fed to cattle which are moved in, from the high plains to the west, for fattening. In addition, large numbers of pigs are fattened on maize in the autumn for slaughter in the following spring. The mixed maize farms of the Corn Belt are highly mechanized, and machinery is used for almost every farming operation. The principal products of this Belt are meat and dairy produce.

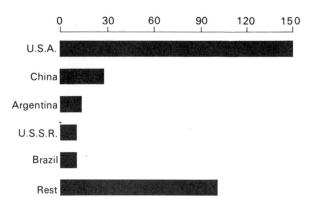

Annual World Production of Maize (average for the last five years in millions of metric tons). Approximate annual production 300 000 000 metric tons.

Other maize regions

Argentina

The eastern part of the Pampas in Argentina is an important producer of maize and the amount of maize exported annually by Argentina makes her the second most important maize exporter. Most of the maize is exported to countries of North-Western Europe where it is used as an animal feed. Maize is also used as fodder for Argentina's beef cattle. Cattle are reared on estancias to the west of the Pampas and when they are old enough they are moved to the maize-growing estancias, to the east, for fattening.

Europe

In Europe, important maize-producing regions are the Hungarian Basin, the Walachian Plain of Rumania, the Basarabia Plain in the U.S.S.R. and the Po Basin in Italy. In South-west Spain, Portugal and France, maize is produced in moderate quantities.

The rest of the world

Other maize-growing countries are China, Brazil, India, Indonesia and Egypt, but in all these countries, most of the maize is consumed locally. The Republic of South Africa and Angola are minor exporters of maize.

BARLEY

Barely is used mainly for making malt and as an animal feed, although some of it is used in the preparation of soups and other food preparations. Barley requires a growing season of only two months. It can be grown under a great variety of climatic conditions and it is also tolerant of many types of soil. For these reasons, barley is more adaptable than wheat and is, therefore, more widely distributed. At the one extreme it can be grown in regions with a sub-polar climate, and at the other it can be grown in oases of the hot deserts of northern Africa. However, it is sensitive to both high temperature and high humidity, and therefore it cannot thrive in the humid tropics. Because of its hardy nature, barley becomes a principal grain in arid, mountainous and sub-polar regions which are unfavourable for other more lucrative crops such as wheat and maize.

The main barley regions

The largest barley producers are the U.S.A., the U.S.S.R., Canada, Australia, China and India. It is grown in practically all countries of Europe, especially in Mediterranean countries where it has been cultivated for several thousand years. In addition, it is a subsistence crop of the tropical highlands of Bolivia, Tibet and Peru.

International trade in barley

Much of the world's barley is used for local consumption and as a feed for animals raised in the areas of barley production. However, western Europe is a significant importer of barley as a feed for dairy cattle and for making malt which is used in the making of beer. Good quality barley, packed in tins, is also exported to countries throughout the world for human consumption.

Annual World Production of Barley by Countries (average for the last five years 120 000 000 metric tons)

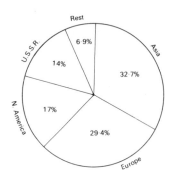

Annual World Production of Barley by Continents

OATS

The ears of oats do not arise from the end of the stem as do those of wheat and barley. The ears hang in clusters of twos or threes which arise from off-shoots of the stem Oats are, therefore, quite distinct in appearance from the other cereals. Like barley, oats are a very hardy cereal and are also used extensively as food for human beings as well as feed for poultry and animals, particularly cattle. Even the straw makes good fodder for cattle.

Oats prefer a cool damp climate, but they cannot

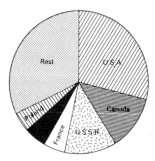

Annual World Production of Oats by Countries (average for the last five years 48 000 000 metric tons)

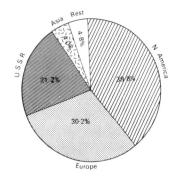

Annual World Production of Oats by Continents (U.S.S.R. shown separately)

withstand cold as well as wheat and barley. Quality of soil is not as important as it is for wheat or barley. Oats thrive best in the humid regions of the cool temperate belt. However, they do not grow well in areas which have temperatures above 15°C (59°F), and are therefore not grown extensively in tropical and sub-tropical lowland regions.

The main oats regions

Important oats producing regions are the northern parts of the U.S.A., southern Canada, north-western Europe (particularly Britain and Scandinavia), Poland and the central and eastern parts of the U.S.S.R. Minor production regions are in northern Japan and Argentina.

International trade in oats

Only a very small percent of the world production in oats enters into international trade partly because oats are not widely eaten as a staple food, and partly because of their weight and bulk which make transport costs high. The tendency is to grow oats in those localities where they may be consumed.

MILLETS

Millets are a member of the cereal family. The small seeds form a head at the end of a tall stalk, 2·5 to 3·5 metres (8 to 11½ feet) high. The seeds, when ground to powder produce a poor type of meal which is used as a staple food by poorer people in countries where it is grown. Millets are essentially a subsistence crop and they hardly enter into international trade. Several varieties of millets are grown, and these are adapted to different climatic conditions, although in general, they tolerate poor soils. The most common type of millets are called *sorghum* which includes Indian millet. Other types are *kaoliang* or Heilungkiang millet and *dhurra* or African millet.

The sorghum and dhurra millets are grown in tropical monsoonal regions where the annual rainfall is between 500 mm and 1000 mm (20 in to 40 in) while kaoliang requires a mild climate and thrives in the summer in Heilungkiang. Too much rain is harmful to millets.

The main millet regions

The chief millet regions are in Asia especially in the north-west Deccan of northern India, Pakistan, the Hwang-ho basin and the Outer Provinces of China, Manchuria, the Dry Zone of Burma, the interior basins of central Asia and the upland farming regions of Japan. Millets are also cultivated extensively by the indigenous subsistence farmers of Africa.

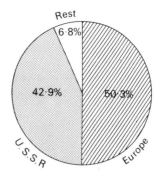

Annual World Production of Rye (average for the last five years 36 000 000 metric tons)

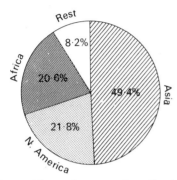

Annual World Production of Millet by Continents (average for the last five years 18 000 000 metric tons)

RYE

Rye is the most hardy of all the cereals. It can be grown on poor soils, and it is able to withstand great cold. It is also able to resist diseases better than any other cereal.

Rye is grown extensively in cool temperate regions, especially in Europe, and because of its hardy nature it can be grown polewards of either barley or oats and on infertile soils which are unsuitable for other cereals.

The chief region growing rye extends across northern Europe from eastern Holland into the U.S.S.R. This is a region of poor, morainic, cold soils.

Rye is made into a dark brown bread, which is very nutritious and which forms an important part of the diet of poor peasant farmers. It is also used in making a type of whisky (mainly in the U.S.A.). Rye is also becoming an important feed for cattle, and pigs. Rye straw is used in the manufacture of a poor quality cardboard. Rye is not an important commodity in world trade, although it enters into the regional trade of several European countries.

EXERCISES

1 Choose *two* of the following: wheat, maize, rice, oats, millet. For *each* one chosen:
 (a) Draw a large sketch-map to show an area which grows and exports the crop.
 (b) For *each* area selected, briefly describe (i) the relief, the soils and climate, (ii) any other factors which affect the cultivation of the crop.

2 Name *two* areas, one in Asia and one in *either* Europe or North America, where rice is grown on a large scale, and for *each* area,
 (a) Draw a map to show its location.
 (b) Describe the conditions under which the crop is grown, and say whether the crop is grown for export or not.
 (c) Say how many crops a year are grown and account for this.

Objective Exercises

1 The successful large scale cultivation of wheat requires all of the following **except**
 A plenty of farming machinery
 B growing season temperatures of about 15°C to 21°C
 C an annual rainfall of about 400 mm to 900 mm
 D a warm, dry and sunny season for ripening the grain
 E good transport and storage facilities

                                            A B C D E

2 The conditions that best favour the cultivation of rice are
  A flat land and high temperatures (about 21°C)
  B flat land and heavy rainfall (at least 2000 mm per year)
  C heavy soils and a hot climate
  D flooded (or irrigated) land and high annual temperatures (at least 21°C)
  E a high density of population

A B C D E

3 Which one of the following statements is correct?
  A China is the world's largest producer and exporter of rice.
  B Burma is one of the smaller rice-producing countries of Asia.
  C Most of the rice grown in Asia is not consumed by the countries that grow it.
  D Asian countries grow sufficient rice to meet their requirements.
  E More rice is grown in Thailand than in Indonesia.

A B C D E

4 Extensive wheat farming in temperate continental latitudes is characterised by
  A a large labour force
  B a high yield per man
  C a high yield per unit area of land
  D crop rotation
  E small farms

A B C D E

5 The main cereal grown in Africa is
  A rice
  B millet
  C oats
  D barley

A B C D

6 Which of the following is **not** a necessity for the cultivation of rice?
  A a large labour force
  B a high rainfall or abundant irrigation water
  C a fast transportation to market
  D flat land or terraced land

A B C D

7 In which way does wheat farming resemble rice farming?
  A it takes place in temperate latitudes
  B it is planted and harvested in a twelve month period
  C wheat enters extensively into world trade
  D most of the wheat is made into flour

A B C D

8 One of the following crops is grown on large farms located on some of the richest soils in the world for the sole purpose of feeding it to animals. The crop is
  A barley
  B oats
  C rye
  D maize

A B C D

The next two questions are to be answered by using the accompanying photograph.

9 This photograph shows the cultivation of
  A wheat
  B maize
  C rice
  D barley

A B C D

10 The crop grown in the area shown in this photograph requires
  A rain throughout the year
  B water-logged soils
  C a dry sunny period for its ripening
  D constant attention

A B C D

20 Vegetables and Fruits

Market Gardening in Holland. Notice the small fields and how these are sub-divided into plots, each growing a different vegetable.

MARKET GARDENING

Market gardening involves intensive cultivation of vegetables. Farms are usually small but the return per acre is high. In the U.S.A. it is called *truck farming*. The types of vegetables grown vary from region to region, partly because of differences in climate and partly because of differences in the eating habits of peoples.

Most of the crops grown on market gardens are perishable and market gardens therefore tend to be near to the regions requiring their produce. They are often sited near to large towns, although there are plenty of examples of market gardens which grow crops for markets, hundreds or even thousands of miles away. This happens when the crops grown are in great demand and can therefore be sold at a higher price, when growing conditions are especially favourable, and when an efficient fast transport system links the growing area with the consuming area.

Market gardening in Europe and North America

The most common vegetables grown in Europe and North America are peas, beans, potatoes, cabbages, celery, carrots, onions, cauliflowers and tomatoes. The truck farms of the U.S.A. normally specialise in one or two crops whose cultivation and handling are highly mechanised as compared with market gardening in Europe.

The mild winters and early springs of southern Europe and north-west Africa, enable vegetables to be grown and harvested several weeks before those of central and northern Europe. This means that the farmers of southern Europe and north-west Africa are able to supply vegetables to the northern markets for about eight weeks in spring and early summer when no locally-produced vegetables are available.

Various methods are used in most of the cooler parts of Europe and North America to encourage vegetables and flowers to mature earlier. Heated greenhouses are often used for early vegetables and flowers. Sometimes the greenhouses are also illuminated throughout the night, thereby enabling plant growth to continue. But whatever method is used, it involves additional capital outlay which means that the grower must have assured markets for his produce. Flowers for festive seasons, which do not coincide with the spring or summer, are often cultivated in this way.

The yield of vegetables per unit area can sometimes

be increased appreciably by intercropping. Some weeks before the first crop is ready for harvesting, a second crop is planted between the rows of the first crop, and so on. The number of crops which can be grown in this way during one year depends upon the type of crop planted (short growing-period plants are often used), the fertility of the soil, the length of the growing season (this depends upon the type of climate), and the techniques used to control soil and air temperatures. The practice of growing several crops in one field during a year is sometimes called *multiple cropping*.

Market gardening in South-East Asia

In South-East Asia, many peasant farmers grow vegetables for their own use. However, in most countries of the region, commercial market gardening is commonly carried on, often by Chinese peasant farmers. Vegetables commonly cultivated are the green leafy types such as "choy sum", Chinese cabbage, spinach, lettuce, vegetable fruits such as the cucumber, brinjal, bitter gourd, snake gourd and tomato, and root vegetables such as potato, sweet potato, "lobak", carrot and various kinds of beans.

Because of their perishable nature, vegetables are not grown for export. They are sold fresh locally to urban populations. Hence, vegetable gardens are usually located on the fringes of towns and cities where the demand for fresh vegetables is high.

There are two types of market garden:
1 The mixed farm where vegetables are grown and where poultry and perhaps pigs are reared.
2 The vegetable farm which specialises on growing vegetables for sale but which sometimes grows flowers and soft fruits.

In South-East Asia, most vegetable growers are mixed farmers. The average size of a market garden is very small, usually under one hectare. The plot is tilled by the farmer and his family whose aim is to get the maximum profits in the shortest possible time. To get good yields, the farmer waters and manures his lands regularly. Animal manure is used, and, sometimes this is supplemented by human manure. In the Cameron Highlands in Peninsular Malaysia, commercial market gardening is of considerable significance. Here, Chinese vegetable farmers specialise in cultivating temperate crops such as radish, turnip, cauliflower, cabbage, celery, asparagus, spinach, etc. The vegetables are taken by lorries to the urban centres of Peninsular Malaysia, and to Singapore.

However, production is insufficient for local needs and vegetables such as Chinese cabbage, bamboo shoots and dried mushrooms are imported from China.

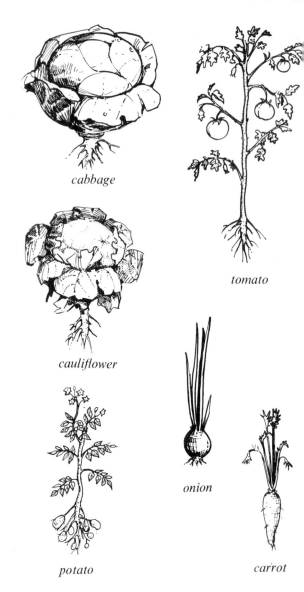

cabbage

tomato

cauliflower

onion

potato

carrot

CASSAVA (MANIOC)

This is a tropical plant which grows to a height of about 3·5 metres (12 feet), producing tubers from which tapioca is made. The plant requires uniformly high temperatures of about 24°C (about 75°F) and a rainfall of over 2000 millimetres (80 inches) a year, and sandy soils which allow the roots to grow freely. It is grown by planting 15-centimetre (6 inch) long pieces of stem which quickly develop roots. Some of the roots swell to form tubers, and after about six months these are ready to harvest. After loosening the soil around the roots, the whole plant is pulled up and the tubers are cut off.

Treatment of the tubers

The tubers must first be boiled and compressed to extract a poisonous substance which they contain. They are then ready for use. Tapioca forms an

216

important part of the diet of many peoples in the under-developed humid tropics. Tapioca flour can also be used for making glue and glucose.

Main cassava regions
Cassava is commercially cultivated in Thailand for making tapioca flour. It is also grown for this purpose in Brazil.

YAMS
Yams may reach a length of 50 centimetres and may weigh as much as 36 kilogrammes. They are used as a starch food, especially in West Africa where large quantities are grown.

TARO
The taro plant produces a large tuber which is perishable and which must be eaten soon after it is dug up. It is grown in many parts of South-East Asia and the Pacific Islands, especially in Hawaii, where a paste called poi is made from the taro tuber.

SWEET POTATOES
The sweet potato plant is very similar to cassava. It requires similar climatic conditions and is grown in hot humid regions between 15°N. and 15°S. although large quantities are grown in the padi regions of China and Japan. The plant is grown from stem cuttings which bear roots (these arise at the stem joints) and as the plant grows it develops tubers which are ready for harvesting after about four months. Like cassava, the sweet potato plant provides an important food for the poorer peoples of the tropics.

FRUITS
Fruits have been an important food for man ever since he was a food gatherer. At one time they were a part of the staple diet of peoples in many regions. Today, fruits are part of the staple diet of peoples in only a few regions. The date and the olive, because of their high food content, are two such fruits. Although fruits are no longer a basic food to most people, vast quantities are grown and consumed annually. Since the beginning of this century, there has been an enormous increase in the demand for fruits which has resulted in a rapid expansion in fruit cultivation, in both temperate and tropical latitudes, and a corresponding expansion in fruit processing and transporting industries. The realisation that fruit was an important source of Vitamin C and sugar, was largely responsible for the increased demand. During this century the use of refrigerated ships has enabled fruits to be moved from one part of the world to another. This means that fruits are available throughout the year because the time of the fruit harvest varies from country to country.

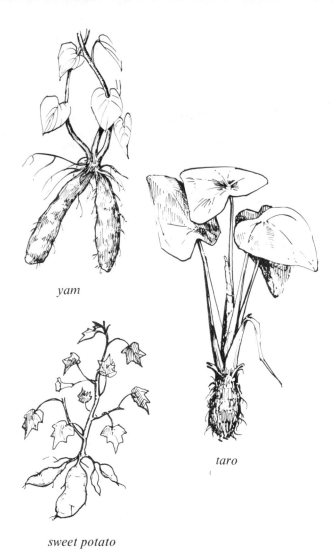

yam

taro

sweet potato

At one time they could be purchased only in the harvest season.

PROCESSING OF FRUITS
Fruits are perishable, and although they can be stored for long periods of time by refrigeration, this is expensive. Large quantities of fruits are now processed so that they can be stored indefinitely at low cost. This can be done by drying, e.g. grapes are dried to produce raisins, currants and sultanas; by canning; by making jams and by turning them into wine (mainly from grapes).

TYPES OF FRUITS
Fruits can be divided into four basic groups:
1 temperate fruits;
2 Mediterranean fruits;
3 citrus fruits;
4 tropical fruits.

217

Citrus Fruit Farm in South Africa

TEMPERATE FRUITS

These are grown mainly in temperate regions such as Europe, North America, northern China and southern Japan and parts of southern Australia and Argentina. The most common temperate fruits are apples, plums, pears and cherries, though there are many shrub fruits such as strawberries, raspberries, black currants and cranberries. Only a very small part of the world's production of each of these fruits enters into international trade, mainly because most of the industrial countries of the Northern Hemisphere produce sufficient quantities to meet their needs. Tasmania and New Zealand have been important suppliers of apples to Asia in recent years.

Harvesting apples in Tasmania

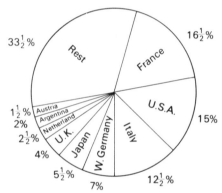

Annual World Production of Apples, by Countries (average for the last five years 27 000 000 metric tons)

MEDITERRANEAN FRUITS

Although these fruits are grown extensively in Mediterranean regions or in regions having a Mediterranean-type climate, some of them, especially grapes, are cultivated in other regions, provided the average summer temperatures are not below 20°C (68°F), and provided there is a long dry period during which the fruit can ripen. The cultivation of grapes, which is called *viticulture*, is particularly important in France and the central Rhine Valley. Over 80% of the world's production of grapes is made into wine. This takes place mainly in Europe. The rest is either eaten fresh or is dried to produce raisins, currants and sultanas. California, South Africa and Australia are also important and these countries now produce both wine and dried fruits for export.

Olives are grown in most Mediterranean countries

and they are, perhaps, the only typical Mediterranean fruit. The oil extracted from this fruit forms an important part of the diet of Mediterranean countries, but the fruit is also eaten fresh. Other fruits of importance are peaches, apricots, figs and certain types of plums. These fruits are eaten as a dessert or are dried, or made into syrup. Some types of plums can be dried without fermenting. When these are dried they are called prunes. All of these fruits are grown on a large scale in parts of California, S.E. Australia and S. Africa and they are exported in four different forms – fresh, canned, dried and as jam.

Trays of prunes drying in the sun in the Californian Valley (U.S.A.)

indigenous to monsoon and warm temperate east margin climates. Oranges are grown, often in irrigated orchards, in California, South Africa and S.E. Australia. Grapefruits are especially important in Florida and California, whilst Sicily and California are the main producers of lemons. Citrus fruits enter into international trade as fresh fruit, canned and in the form of juice and marmalade.

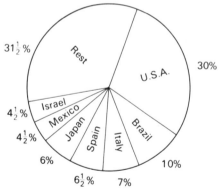

Annual World Production of Citrus Fruits by Countries (average for the last five years 26 000 000 metric tons)

Vineyards in the Douro Valley of Portugal

CITRUS FRUITS

Citrus fruits can be distinguished by their thick waxy skin and acid taste. The orange, lemon, grapefruit and lime are all citrus fruits. Citrus fruits are grown on a commercial basis in most regions having a Mediterranean climate, but they are really

TROPICAL FRUITS

Bananas

Bananas are of many varieties, differing in size, colour, flavour and use. Generally, the mature plant is about 3·5 metres (11½ feet) high and has a trunk with overlapping, tightly packed leaves which are 1 to 2 metres (3 to 6 feet) long and about 0·5 metre (1½ feet) wide and which fan out at the top. Each plant produces one bunch of fruit.

219

Growing conditions

Banana plants require warmth, abundant sunshine and moisture throughout the year. They require:

1. average temperatures of 21°C (about 70°F) to 26°C (about 79°F);
2. an annual rainfall of 2000 mm (80 in), which should be well distributed;
3. fertile, friable and well-aerated soil;
4. lack of strong winds after the fruit has ripened.

All of these conditions occur in parts of the wet tropics.

Planting

The banana plant is grown on a commercial basis in large plantations where production is highly organised. Small pieces of rhizomes (rootstock) are selected from healthy plants and are planted in

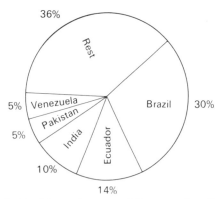

Annual World Production of Bananas by Countries (average for the last five years 26 700 000 metric tons)

shallow holes about 2 metres (6 feet) apart. During growth, weeds and undergrowth must be cut away several times a year. After about 18 months, the plant matures and bears fruits.

Harvesting begins just before the bananas ripen. After the fruit is harvested, the trunk of the plant is cut down. Soon after, several suckers develop from the stump to form separate branches, each of which later bears a bunch of bananas.

Marketing

Since bananas perish quickly, cultivation must be carefully planned and harvesting must be so timed that when the bananas reach the importing countries, they ripen and are ready for sale in shops. During transit, great care must be taken to prevent the bananas from being damaged. Bananas must be handled carefully and plantations are always situated near ports to avoid warehouse storage problems and harmful delays. When the carefully wrapped bananas reach the dockside, they are loaded almost immediately into the waiting ships.

Banana trade and areas of production

While bananas have long been a staple food and grown as a subsistence crop in many under-developed tropical regions, the major commercial producing centres are the Caribbean Islands which are near the U.S.A., the world's largest consumer of bananas. These islands include Jamaica and Cuba and the mainland countries of Honduras, and Costa Rica. Other important exporters are Colombia, Panama, Mexico, Ecuador and Brazil. The Canary Islands and Nigeria export their bananas to European countries, while in Asia, Taiwan is the chief banana supplier to Japan and Korea.

Harvesting bananas in Ecuador

Pineapples

The pineapple is a low plant whose single fruit develops at the end of the short stem. The base of the stem is surrounded by a cluster of long thorn-edged leaves. When the fruit is ripe, it turns orange-yellow and it weighs from one to two kilogrammes. The plant grows best in hot wet climates where average temperatures are about 26°C (about 79°F) and annual rainfall of at least 1900 millimetres (75 inches). However, it can be grown in hot, drier regions provided water is available for irrigation. The plant grows especially well near the sea on land which has good drainage.

Planting

Generally, preparation for planting consists of clearing and levelling the land and then laying down the drainage system. The land is fertilized, after which pineapple slips are planted in rows. While the plants are growing, the land is kept clear of weeds. A year later, a bud develops and the plant begins to flower. After the flowers wither, the bud enlarges into a fruit which ripens six months after flowering.

In Hawaii, dry farming methods are practised, and pineapple cultivation is highly complex and mechanized. To prepare the land for planting, old plants are cut up and ploughed into the soil. Machines then lay down strips of black mulch paper on the soil surface while, at the same time, fertilisers are spread under the mulch paper. The mulch paper serves to conserve moisture in the soil and to discourage weed growth. Then a narrow steel trowel is used to make a hole in the mulch paper for each pineapple slip. During plant growth, fertilization is carried out regularly.

Harvesting

When the fruits are ripe, harvesting begins. On large plantations in Hawaii, fruits are harvested by huge machines, but in less developed countries, labourers use long knives to cut the fruits from the stems. The fruits are taken to the factories by lorries, where they are prepared for canning.

Pineapple canning

Over 80 per cent of the world's pineapple output is canned. In Hawaii, canning is almost entirely done by machines. In the cannery, the fruits are sorted into various sizes by machines. They are then cleaned after which they are skinned and cored, again by machines, and then sliced or cubed ready for canning.

Pineapple-growing countries and trade

The Hawaiian Islands (particularly Lanai Island) produce about 70 per cent of the world's pineapples. Other pineapple-producing countries are Malaysia, Taiwan, Indonesia, the Philippines, the West Indies, Northern Australia and Mexico. Most of these countries produce pineapples for local consumption. Countries which import Hawaiian pineapples are the U.S.A., Canada, New Zealand, Japan and Western Europe. Malaysia's canned pineapples are exported to the U.K. and some Middle-East countries.

EXERCISES

1 (a) Draw a sketch-map of a region you know which contains a market gardening area. Your map should show (i) the location of the market garden area in

Picking pineapples on a pineapple plantation in Johore (Malaysia)

relation to main roads, and settlements, (ii) the location of water supply.

(b) Write explanatory notes beneath your sketch-map on (i) the types of vegetables grown, (ii) the nature of the market for which they are grown, (iii) the distance of the market from the market garden area.

2 (a) Locate the main market garden areas in *either* North America, *or* Europe, *or* your country by drawing a map to show their locations in relation to physical relief and settlements.

(b) Briefly describe market gardening in the region you choose.

3 (a) Name any *four* tropical fruits which are grown on a commercial scale.

(b) For *one* of these:
   (i) locate one area where it is grown;
   (ii) briefly state the conditions under which it is grown;
   (iii) name the countries to which this fruit is exported and explain how it is exported.

4 (a) On an outline map of the World, insert and name:
   (i) *four* major citrus fruit areas;
   (ii) *two* major banana-exporting areas;
   (iii) *two* major pineapple-importing areas;

(b) (i) briefly describe how *either* bananas *or* pineapples are grown;
   (ii) describe how the fruit you choose is transported;
   (iii) name three other fruits which enter the world trade on a large scale.

5 Briefly explain the meaning of 'viticulture'.
   (i) Name *two* regions, *one* in the Northern Hemisphere and *one* in the Southern Hemisphere, where viticulture is an important activity.
   (ii) Describe the climate of *one* of these regions and illustrate your answer by a simple climatic graph.
   (iii) Draw a sketch-map of this region to show the main areas of viticulture and related physical and human features, such as, rivers and settlements.

Objective Exercises

1 Which of the following should **not** be included as preserved products of a Mediterranean fruit growing area?

A wine
B sultanas
C beer
D canned peaches
E frozen orange juice

A B C D E

2 The main pineapple exporting country is

A Hawaii
B Malaysia
C Taiwan
D Indonesia
E Thailand

A B C D E

3 Market gardening areas may produce all of the following **except**

A vegetables
B fruit
C flowers
D poultry
E beef

A B C D E

4 Which one of the following is applicable to market gardening?

A the use of sophisticated equipment on a small unit area of land
B the use of a small labour force on a large unit area of land
C the use of intensive labour on a small unit area of land
D the use of sophisticated equipment on a large unit area of land
E the construction of a close network of irrigation canals

A B C D E

5 Market gardening does **not** need

A fertile soil
B a large labour force
C fast transport to market
D a large land area

A B C D

6 One of the conditions that is necessary for the cultivation of bananas is

A high temperatures and humidity
B a clay soil which retains water

C strong winds which cause little damage
D a cool winter for 4 to 6 months

A B C D

7 Which one of the following countries is **not** an important grower and exporter of pineapples

A Mexico
B Malaysia
C Hawaii
D Brazil

A B C D

8 Which one of the following statements about the cultivation of bananas is true?

A Bananas are grown on fairly small farms in all the major banana-growing countries.
B Bananas are grown on large estates (called plantations) in some of the major banana-growing countries.
C Bananas are grown mainly as a subsistence crop in the Canary Islands.
D Banana trees are grown from seedlings.

A B C D

9 Viticulture is the name given to

A the cultivation of Mediterranean fruits
B the method of extracting juice from grapes
C the cultivation of grape vines
D the manufacture of wine

A B C D

Drinks and Sugar

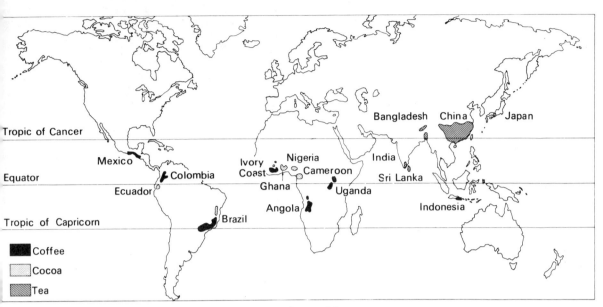

The main coffee, cocoa and tea producing regions of the world

The most common drinks of Man are tea, coffee, cocoa, wine and beer.

Water, which is essential to animal and plant life is obtained from rivers, springs, wells and reservoirs. However, it is neither evenly distributed over the land surfaces of the world, nor is it always fit for human consumption until it is purified. Contaminated water in the past caused numerous epidemics and plagues which brought death to large numbers of people.

During the past one hundred years, the demand for water, for drinking, for agriculture and for industry, has increased enormously. So has the contamination of water, with the result that today many heavily populated industrial regions are faced with the possibility of experiencing acute water shortages before the end of the century. This problem is especially serious in the U.S.A., Japan and West Germany. Industrial development and increased urbanisation often cause atmospheric pollution by poisonous waste gases, which, in turn, results in the pollution of rain water. It also causes pollution of river water because waste products from factories and towns are emptied into rivers. Water, like soil, forests and mineral deposits, is a natural resource which must be conserved.

TEA

The tea plant is an evergreen shrub which by pruning is kept to a height of about 1·5 metres (5 feet). The plant grows best in tropical and sub-tropical monsoon regions which have:

1 between 150 and 750 centimetres (60 to 300 in) of rain a year;
2 average monthly temperatures of at least 18°C (about 64°F) during the plant's nine-month growing season;
3 well drained soil;
4 high humidity producing morning mist or dew which enables the young leaves to grow quickly.

The cultivation of tea

The plant is grown from seed in a nursery. When it is about 0·3 metre (1foot) high, it is transplanted into sloping land. After about 3 years the plant is mature, and its leaves are ready for picking.

Picking and processing

The young leaves at the tips of branches are the first to be picked. Further pickings take place, usually weekly, and in some parts of India and Sri Lanka there may be as many as 20 pickings in one season. In China and Japan there are usually about 4 pickings per season. Picking the leaves calls for skill, and it is usually done by women. Much labour is required during the harvest season and because of this the commercial cultivation of tea is usually located in regions where there is an abundance of cheap labour.

Tea may be processed into two forms – *green tea* and *black tea*. When the leaves are brought to the factory they are dried on trays, either in the sun or in heated rooms. This is called *withering* and it causes the leaves to lose most of their moisture and to become soft so that they do not break when rolled. The tea roller crushes the leaves which are later sifted. If black tea is required, the leaves are allowed to 'ferment' to reduce the tannic acid content. Fermentation is stopped by roasting the leaves which cause them to turn black. They are rolled once more and then graded before being packed for export. If green tea is required, the fermentation stage is omitted.

Tea growing regions and trade

The tea plant flourishes in humid tropical, and sub-tropical regions, especially in the monsoon lands of East and South Asia. It can also be grown as far north as the Black Sea if the winters are mild and there is sufficient rain. Tea is extensively grown on the Assam Hills and the Nilgiri Hills in India; on the hills of Central Sri Lanka around Kandy; and in Sylhet in Bangladesh. Other important tea-growing regions are the Yangtze Basin in China; Skikoku and Southern Honshu in Japan; and West Java in Indonesia. There are small tea-growing regions in Kenya, Brazil, the U.S.S.R. and in Northern Argentina.

The Main Tea-producing Regions of Asia

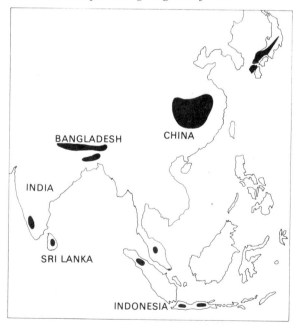

Tea production in India and Sri Lanka

India's annual production of about 400 000 metric tons of tea represents almost 15% of its exports by

Picking tea in Japan

value. Most of India's tea is grown on large tea estates which are concentrated in three main areas which are shown in the following maps. These are:
1 The Darjeeling District on the southern slopes of Himalayan foothills.
2 On the slopes of the Assam Hills along the northern side of the Brahmaputra Valley.
3 On the Nilgiri and Cardamon Hills of Southern India. Sri Lanka's annual production of about 240 000 metric tons comes from the central hilly region around Nuwara Eliya and Kandy.

The Main Tea Regions of Bangladesh and North-East India

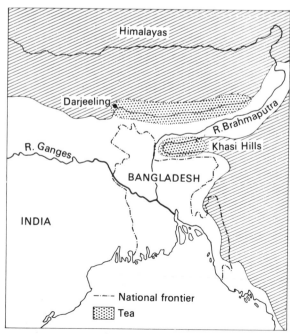

The Main Tea Regions of Southern India and Sri Lanka

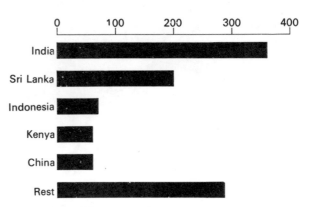

Annual World Export of Tea (average for the last five years in thousands of metric tons). Approximate annual export 1 000 000 metric tons.

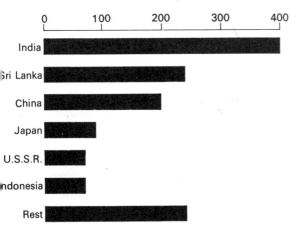

Annual World Production of Tea (average for the last five years in thousands of metric tons). Approximate annual production 1 300 000 metric tons.

The largest tea exporters are India and Sri Lanka as the following graph shows. The U.K. is one of the largest tea importers and most of its tea is imported from India, Sri Lanka and Bangladesh. Australia and West European countries also import their tea from these countries.

COCOA

The fruit of the cacao tree is a pod about 20 centimetres (about 8 in) long. It contains seeds, called beans, which are embedded in a pulpy mass. When processed the beans produce a powder which is called *cocoa*. The pods grow directly on the trunk and branches of the tree. Although the cacao tree is an indigenous tree of Latin America, it is now extensively cultivated on cleared forest lands in West Africa between the equator and 10°N. The tree is grown in regions which have:

1 A mean annual temperature of about 26°C (79°F) with a range of temperature of less than 10°C (50°F).
2 An annual rainfall of at least 2000 mm (about 80 in) which is evenly distributed throughout the year.
3 No strong winds (such winds could easily damage the pods).
4 Deep well-drained soils which are rich in potash.

The cultivation of the cacao tree

Cacao trees are usually grown from seeds which are sown in a nursery. After the seeds germinate, the seedlings are allowed to grow to a height of a few centimetres before these are transplanted into a piece of forest which is cleared of all plants except tall trees which provide shade for the seedlings. Some cacao trees are grown from cuttings which are planted direct into the cleared forest land. After about five years the young trees bear their first pods but they do not come into full production until they are about eight years old. Good trees can remain in commercial production for about thirty years.

Cacao trees, like most other cultivated plants, have to be protected from insect pests and fungus growths. The capsid pest, which feeds on the pods,

225

Splitting cocoa pods to extract the beans in Ghana

Cocoa regions and trade

The chief cocoa regions of the world are West Africa especially Ghana and Nigeria which together produce one half the world's output of cacao beans and parts of Central America, the West Indies, Brazil, Venezuela and Ecuador.

Most of the cacao beans produced are exported to the U.S.A. and western Europe.

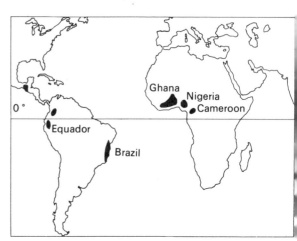

The Main Cocoa-growing Regions

can be controlled by spraying the trees with insecticides. This is not always done on the small holdings owned by peasant farmers in West Africa and because most of West Africa's cacao trees are grown on such holdings, there are substantial crop losses each year.

Picking and processing

The cocoa pods are cut from the trees twice a year after which they are split open so that the pulpy mass of beans can be extracted. The beans are spread on trays in the sun to allow the pulp to drain away and the beans to ferment, after which they are dried and packed in sacks before being exported.

Most cocoa regions export their beans to Europe and North America. On arrival the beans are taken to the factories where they are cleaned, roasted and cracked so that their hard skins, called hulls, can be removed. The beans are then crushed and ground into a paste, after which the fat, which the beans contain, is separated out. This forms a solid fat called cocoa butter. The remainder of the paste is dried to a powder, called cocoa. Cocoa butter is used for making cosmetics and other similar goods. The powder is used, together with cocoa butter milk and sugar for making chocolate.

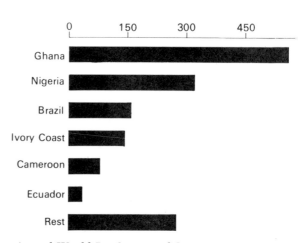

Annual World Production of Cocoa Beans (average for the last five years in thousands of metric tons). Approximate annual production 1 600 000 metric tons.

Cocoa production

Ghana is the world's largest producer of cacao beans. The principal cacao regions are (i) in Ashanti Province around Kumasi, and (ii) in the western part of the country (see the above map). Conditions in these parts of Ghana are ideal for growing cacao trees but insect pests and fungus diseases cause great damage to the crop. This happens mainly because many of the farmers of the small holdings are not able to take adequate action to protect the cacao trees. The cacao farms of Ghana employ almost half a million workers and the export of cacao beans forms more than 50% of Ghana's total exports by value. Most of the exported beans leave the country from the ports of Tema and Accra.

COFFEE

The coffee plant is a perennial shrub which is extensively cultivated in latitudes between 30°S and 28°N. The shrub is pruned to a height of about 2 metres (6½ feet), and after flowering the shrub produces clusters of berries along its branches. The coffee plant grows best in tropical and sub-tropical regions and it is commercially cultivated in regions which have:

1 An average monthly temperature of between 21°C (about 70°F) and 26°C (about 79°F) with a minimum daily temperature of 15°C (59°F), accompanied by high humidity.
2 An annual rainfall of about 1700 mm (about 67 in) with a maximum in the growing season and a minimum when the plant is flowering and when the berries are ripening.
3 Rich deep loamy soils which are well-drained to enable the roots to spread easily.
4 Reasonable shade (although some coffee plants grow well without shade, as in Brazil).
5 A cool dry season to facilitate the picking and drying of the berries.
6 A plentiful supply of cheap labour for weeding, spraying, picking, drying, sorting and bagging the berries.

The cultivation of ooffee

The plant is grown from seeds which are sown in nurseries. About 6 months after germination, the young seedlings are transplanted into the plantations where they grow under tall trees, which provide shade to protect them from the sun. The trees bear fruit in the fifth year. During this period they are pruned and the soil around is manured.

Picking and processing

When the berries are ripe, they are either picked by hand or by shaking the branches over canvas sheets. This takes place in the autumn. The berries do not ripen all at the same time and several pickings take place over a two or three month period. The berries are cured by spreading them out on trays which are left in the open for about three weeks. The trays are covered at night to protect the berries from rain or dew. When the berries are cured, the

Drying coffee beans in the sun in Brazil. These beans are being raked over to ensure that all the beans are properly dried.

dried pulp is removed by machines to release the coffee beans. Sometimes the berries are put into water to remove the pulp after which the beans are dried in the open. The cured beans are packed in bags for export. Before the beans can be used they have to be roasted and ground. This is usually done in the importing country.

Coffee regions and trade

Most of the world's coffee is grown on the tropical highlands of Latin America and Africa at heights from 450 to 750 m (1476 to 2460 feet) above sea level. The largest producer and exporter is Brazil, with almost one half of its coffee exports going to the U.S.A. Other important producers are Colombia, Mexico, El Salvador and Costa Rica (Latin America) and the Ivory Coast, Uganda, Angola, Zambia, Ethiopia, Kenya and Zaire (all in Africa). The only other fairly important producers of coffee are Indonesia, Sri Lanka and India.

Coffee forms a very important part of the economy of some countries, such as Brazil, which means that

The Main Exporters of Coffee Beans on a percentage basis (average for the last five years).

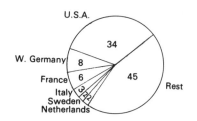

The Main Importers of Coffee Beans on a percentage basis (average for the last five years).

fluctuations in the price of coffee have important effects on their economic stability. The International Coffee Agreement which was set up in 1962 now operates controls to regulate production and prices. This sometimes results in the deliberate destruction of vast numbers of coffee trees, as a means to prevent over-production.

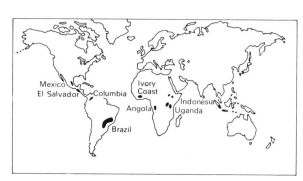

The Main Coffee-producing Regions

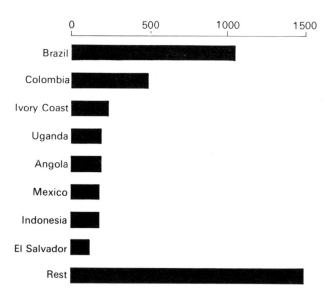

Annual World Production of Coffee Beans (average for the last five years in thousands of metric tons). Approximate annual production 4 000 000 metric tons.

the cultivation of coffee. The coffee tree makes heavy demands on the soil and soil exhaustion soon takes place. This is often followed by soil erosion if preventive precautions are not taken. Another threat which faces coffee growers is frost, which can cause considerable damage to the flowers and young berries. A large amount of Brazil's coffee is grown on large estates called *fazendas* which cover enormous areas and which may have as many as a million trees. But small estates are becoming increasingly important.

The Coffee Region of Brazil

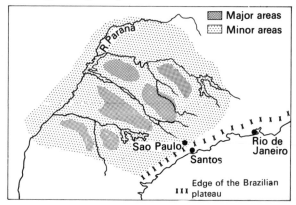

Coffee production in Brazil

Most of Brazil's coffee trees are grown on the Brazilian Plateau, in the State of Sao Paulo, in an area between the edge of the Plateau which parallels the coast, and the Parana River. This region has rich soils called *terra roxa* and an ideal climate for

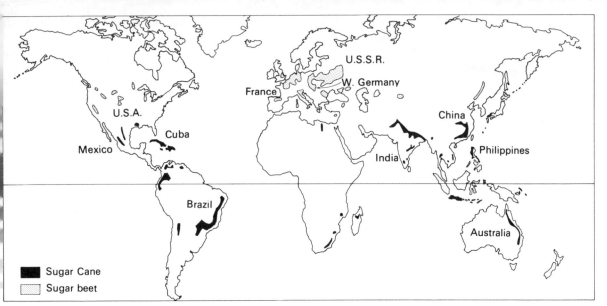

The Main Sugar Cane and Sugar Beet Regions

Sugar Cane
Sugar beet

SUGAR

Although sugar can be obtained from several plants, the bulk of the world's sugar comes from sugar cane and sugar beet, with between one half and two thirds of the total world sugar production coming from sugar cane.

Sugar Cane

Sugar cane is a tropical grass which resembles bamboo. It grows to a height of 2.5 to 3.5 metres (about 8 to 12 feet). The plant can be grown commercially in regions which have the following conditions:
1 High monthly temperatures of between 21°C (70°F) and 30°C (86°F).
2 An annual rainfall of about 2000 mm (about 80 in) with most of it falling in the growing season, or, the equivalent in the form of irrigation water.
3 A dry warm season to enable the canes to ripen so that a high sugar content forms in the pith of the canes.
4 Rich, deep well-drained soils.
5 A large labour force for digging irrigation/drainage canals and for planting, weeding and harvesting the cane, in those areas where mechanised farming is not practised.

The cultivation of sugar cane

Small pieces of the cane, consisting of one of two joints, are planted. These soon begin to grow. The plant reaches full maturity after about one year. Whilst it is growing the cane fields have to be constantly weeded.

Harvesting and processing

The canes are cut off a few centimetres above ground level. The roots are left intact because new canes grow from them for the next crop. This can take place two or three times before new cane stock has to be planted. After the canes are cut, the leaves are removed and these are often used as fodder for cattle and pigs. Sometimes the cane fields are set on fire to burn off the leaves (the cane is not harmed), and to drive out rats and other pests, after which the cane is cut either by hand or by machine. The burning method, though quicker, does not produce any fodder for cattle. The cut canes are transported to the sugar factories by road or rail, or in the poorer areas, by bullock carts. In the factory the cane is cut into small lengths and is passed through a crushing machine which extracts the juice, which is then purified and boiled until the sugar crystallises. The thick syrup which remains is called molasses and when this is removed a crystalline residue of raw sugar remains. It is exported in this form. Refining of the sugar takes place in the importing countries. Molasses is used for making alcohol, the spirit rum and treacle. It is also mixed with other substances to make an excellent animal food. The residue of canes is known as *bagasse* and this is sometimes manufactured into paper, or fibreboard. Sometimes it is ploughed into the cane fields to enrich the soils, or it is fed to cattle, or it is used as a fuel in the sugar factories.

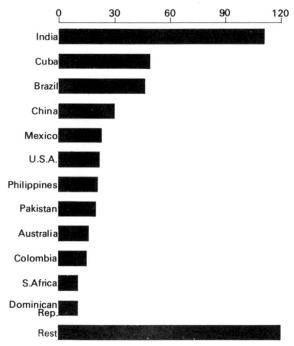

Annual World Production of Sugar Cane (average for the last five years in millions of metric tons). Annual production approximately 500 000 000 *metric tons.*

Sugar cane regions and trade

The commercial production of sugar cane takes place in four main areas:

1 The Caribbean region (Cuba, Dominican Republic, Barbados and Jamaica).
2 Latin America (Mexico, Peru and Brazil).
3 Asia (India, Pakistan, Bangladesh, Taiwan, the Philippines and China).
4 The Australasia Pacific Region (Australia and Hawaii).

The largest producers, such as India and Brazil, are not necessarily the main exporters. The largest exporters are Cuba and other Caribbean producers and Australia.

Sugar Cane production in Cuba

Cuba is one of the world's largest exporters of sugar. Until a few years ago the U.S.A. was Cuba's main buyer of sugar but today most of Cuba's sugar exports are to the U.S.S.R. and China. Cuba has vast stretches of lowland whose soils are so rich that they do not have to be fertilized at any time. The climate is almost ideal for the cultivation of sugar cane. Practically all of the large state-owned estates have excellent railway communications which link with the large sugar factories. More than one half of Cuba's cultivated land is under sugar cane and about one third of the country's labour force is engaged in growing or processing sugar cane.

Sugar Cane areas

Loading sugar cane into cane crusher in Cuba

Sugar beet

The sugar beet plant grows best in cool temperate regions and it is grown mainly in Europe and North America. The plant develops a large, bulbous root in which the sugar forms. The cultivation of sugar beet in Europe became important at the end of the 19th Century to avoid the dependence of European countries for sugar on tropical countries in time of war.

The cultivation of sugar beet

The plant is grown from seeds which are sown in the spring. It is harvested in the autumn. For it to be commercially cultivated, a region should have:
A growing season of about five months with summer temperatures of between 18°C (64°F) and 22°C (72°F), with clear sunny weather when the beet is ripening.
A rainfall of about 600 mm (about 24 in) which should be evenly distributed throughout the growing season, or the equivalent in irrigation water.
Rich, deep well-drained soils.
A fairly large labour force for hoeing, weeding and harvesting the crop.

Sugar beet makes heavy demands on the soil which has to be fertilized both before and after cultivation. Most farmers grow sugar beet in rotation with other crops in order to maintain soil fertility.

Harvesting and processing

The beet is lifted either by hand or by machine in the autumn. After the leaves are cut off the beet is taken to the sugar factories where it is washed and pulped and then soaked in water in large tanks. The liquid is then boiled and the sugar crystallizes out. The pulpy residue is made into cattle food. The leaves are usually ploughed into the soil to enrich

Annual World Production of Sugar Beet (average for the last five years in millions of metric tons). Annual production approximately 210 000 000 metric tons.

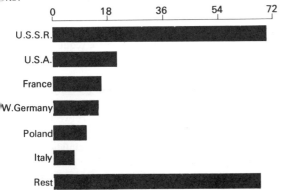

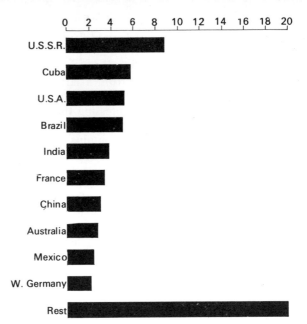

Annual World Production of Raw Sugar processed from Cane and Beet (average for the last five years in millions of metric tons). Total approximately 63 000 000 metric tons.

SPICES

Spices are extensively used in flavouring foods in most parts of the tropics. They are also used in the manufacture of sauces and some medicines.

Pepper is made by grinding the dried, unripe fruit of a climbing plant which grows extensively on hills in southern India, Indonesia and Sarawak (East Malaysia). White pepper is produced by grinding the seeds of the berries. Black pepper is made by grinding both the seeds and the outer fleshy skin.

Cloves are the dried, unopened flowers of an evergreen tree which grows in the Moluccas (eastern Indonesia), and Zanzibar and Pemba islands, off the east coast of Africa.

Ginger is made from the rhizome of a grass-like plant which grows very well in China, India, parts of South-East Asia, Nigeria, Sierra Leone and Jamaica. Ginger is either exported dried or in sugar syrup.

Nutmeg and mace both come from the fruit of a tree which is cultivated in the Banda Isles (near to the Moluccas), and Grenada in the Windward Islands.
Mace is made from the outer fleshy part of the seed. The kernel (inner part of the seed) is dried and forms the nutmeg).

Cinnamon is made from the bark of a tree which grows well in Sri Lanka, Brazil and Egypt.

WINE

Although wine can be made from a large variety of flowers, fruits and vegetables, most wine is made from grapes This is done by pressing the fruit to extract the juice which is then fermented. The bloom on the skin of the fruit contains yeast, and this causes the juice to ferment as soon as it comes into contact with the yeast.

During fermentation, the sugar in the juice is changed into alcohol and carbon dioxide. After fermentation, the juice is left for several weeks to allow the sediment to sink to the bottom. The clear wine is then run off into open vats where fermentation begins again; at the end of which it is bottled.

The type of wine produced depends upon:
1 the actual process used in fermentation
2 the nature of the soils in which the vines are grown.

Wine-producing regions

Almost all Mediterranean countries produce large quantities of wine. France, Italy, Portugal and Spain are especially important. However, flourishing wine industries have been developed in several countries in the southern continents which have a Medi-terranean-type climate. Chile, South Africa an Australia are the most important of these.

Wine production and trade

Approximately 6 million gallons of wine are pro duced annually but of this, only about 75 per cen enters international trade.

BEER

Although beer can be made from most cereals, is most commonly made from barley. The seeds this plant are used for making malt by allowin them to germinate after which they are dried an made into powder. The malt is boiled with wate and the whole is subsequently fermented with yea and then flavoured with hops. Hops are the flowe of a vine-like plant which have to be roasted befo they can be used.

Main beer-producing countries

These are in temperate latitudes (the home o barley and hops), and the U.K. and Germany ar especially important producers.

Beer production and trade

The world produces about 12 million gallons beer annually, most of which does not enter int world trade.

EXERCISES

1 Choose *two* of the following: coffee, cocoa, tea, wine, beer. For *each* one chosen:
   (i) Draw a sketch-map to show an important area which produces the commodity.
   (ii) For the area chosen, briefly describe the physical features, including soils, and the climate, and show how these favour the production of the commodity
   (iii) Carefully explain how the commodity is produced after the crop has been harvested.
   (iv) State whether the area chosen exports the commodity, and if so, name *two* countries which import it.
2 (a) Name *two* beverage crops which are grown in tropical latitudes, and *two* which are grown in temperate latitudes, and for *each* crop chosen:
   (i) Name *one* area where it is grown.
   (ii) State whether it is grown on plantations or small-holdings.
   (b) For *two* of the areas chosen:
   (i) Draw sketch-maps to show the main physical features and settlements.
   (ii) Write a short description of the methods used for growing the crops.

3 (a) On an outline map of the world:
   (i) Mark *five* areas, in each of which, one of th following crops is grown on a large scale: coco coffee, grapes, sugar-cane, tea, hops, sugar-bee Write the names of the crops in the areas chosen
   (ii) For each selected area, mark and name *one* tow which handles either the crop or a product of i
   (b) Name the product obtained from each of any *tw* of the named crops, and for *each*, briefly state ho it is obtained.

Objective Exercises

1 The successful cultivation of coffee requires all of th following conditions **except**
   A rich, deep, well-drained soils
   B annual rainfall of about 1800 mm evenly distribute throughout the year
   C average monthly temperature of between 21°C an 26°C with a daily minimum temperature of 15°C
   D early morning mists followed by cloudy skies
   E a large labour force

   A B C D ‖

Tea and coffee are grown on hill slopes mainly because
A the slopes are terraced
B the temperatures over the slopes are lower than over the lowlands
C it is difficult to grow other crops on the slopes
D the slopes are well-drained
E it is easier to grow these crops on slopes

A B C D E

Which one of the following countries is the largest producer of cacao beans in Africa?
A Ghana
B Ivory Coast
C Nigeria
D Cameroon
E Zaire

A B C D E

The largest exporter of cane sugar is
A India
B Brazil
C Cuba
D China
E U.S.A.

A B C D E

Sugar cane producing regions include
A the West Indies
B Northern China
C Western Europe
D South Africa
E Western Australia

A B C D E

Which one of the following statements, all of which are about sugar cane cultivation, is **not** an essential condition?
A It requires uniformly high temperatures (21°C to 25°C) and heavy rainfall (about 2000 mm per year).
B It requires uniformly high temperatures (21°C to 25°C) and fertile moist soils.
C It requires level land which can be irrigated.
D The growing season should be hot and wet, and this should be followed by a dry sunny season to enable the cane to ripen.

A B C D

Sugar beet is **not** an important crop in the farming economy of
A U.S.S.R.
B West Germany
C Cuba
D Poland

A B C D

8 Conditions favouring the growth of coffee trees include
A a clay soil
B frost in the winter only
C rain only in the summer
D a well-drained soil

A B C D

9 The making of cocoa involves all of the following **except**
A the pods are laid on trays in the sun
B the beans are fermented
C the beans are dried
D the beans are exported in sacks

A B C D

10 Which of the following pairs of countries is important for the cultivation of coffee and sugar cane respectively?
A Brazil and Australia
B Ghana and Poland
C China and Brazil
D Australia and Ghana

A B C D

11 Which one of the following reasons is the most important in respect of the cultivation of sugar beet by temperate latitude countries?
A Sugar beet is a good rotation crop.
B The cultivation of sugar beet by a country reduces its dependency on overseas sugar cane producers.
C After the extraction of sugar from sugar beet, the beet pulp can be used as a cattle food.
D Sugar beet is a more profitable crop to grow than some other root crops.

A B C D

22   Vegetable Oils

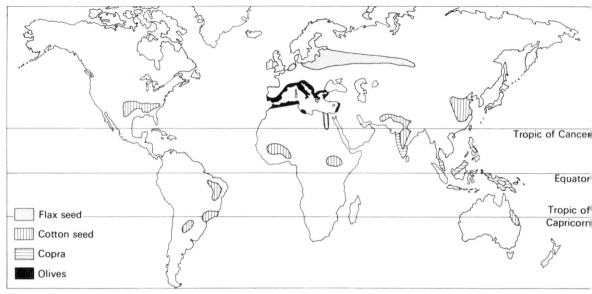

The Main Flax seed, Cotton seed, Olive and Copra Regions of the World

Oils and fats are essential for good health. At one time the only oils and fats used by Man were those obtained from animals. Experiments to find suitable alternatives to animal oils and fats were made in the 19th Century when the population of Europe was expanding so rapidly that there were insufficient animal oils and fats to meet the growing demand. Eventually a substitute for butter using vegetable oils, especially coconut and groundnut oil, was successfully manufactured.

THE OIL PALM

The oil palm can be grown commercially in humid tropical regions which have:

1 Uniformly high temperatures of 26°C (about 79°F) throughout the year.
2 About 2000 mm (about 80 in) of rain per year, evenly distributed.
3 Fairly fertile, well-drained soils.
4 Plenty of labour to weed the young seedlings, harvest the fruit and transport it to the oil mills.

Cultivation of oil palms

The oil palm is germinated from seeds which are sown in wet sand in wooden trays. About one month after the seeds have germinated, the young seedlings are transplanted into a nursery whe they remain for about six months, after which th are transplanted in the estate or plantation. A cov crop is usually grown to prevent the loss of moistu from the soil and to provide the seedlings with shad The oil palm begins to bear its first fruit when is about four years old. During this time, consta weeding, manuring and pruning take place. Aft the bunches of fruit have been cut, they are carri to the light railway or road for collection and delive to the oil mill.

Processing the fruit

The fruit is made up of two parts: an outside flesh layer, called the *pericarp*, which is rich in oil, an an inner seed which, in structure, is similar to coconut. The seed has a hard shell inside which the hard kernel. This is also rich in oil.

The fruit gives two types of oil. The first, which called palm oil, is extracted from the pericarp b boiling and pulping it. This is usually done on th plantations. The oil is either exported in its crud form or it is purified before export. Palm oil is use for making soap and edible fats and oils. The secon type of oil is extracted from the kernels. Thes must first be crushed to release the kernels whic

234

Oil palm fruit in Malaysia

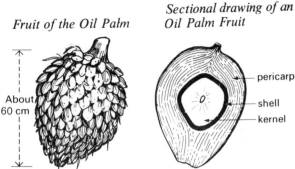

Fruit of the Oil Palm

Sectional drawing of an Oil Palm Fruit

About 60 cm

pericarp
shell
kernel

Zaire, Angola, Central African Republic, Gabon, Cameroons and Dahomey. In South-East Asia, Peninsular Malaysia and Indonesia, especially the island of Sumatra, are important palm oil-producers. Most of the oil is imported by the industrial countries of western Europe and the U.S.A.

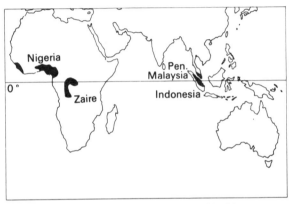

The Main Oil Palm Regions of the World

Loading bunches of oil palm fruit onto an estate railway in Malaysia. This photograph indicates the size of the bunches.

are then pulped. The extraction of kernel oil is usually done in the importing country. This oil is also used for making soap and edible oils.

Palm oil-producing countries and trade
Most of the world's supply of palm oil comes from countries in West Africa, for example, Nigeria,

Annual World Production of Palm Oil (average for the last five years in thousands of metric tons). Approximate annual production 1 800 000 metric tons.

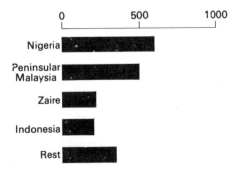

THE COCONUT PALM

This tree grows in most wet tropical coastal lowlands between latitudes 15°N. and 15°S. The tree often grows to a height of 25 metres (80 feet) with all its leaves and fruit arising from the end of the trunk. The tree lives for about 60 years and each tree bears several bunches of coconuts which may number as many as fifty. The coconut consists of an outer fibrous cover which encloses a hard shell lined with white meat, called *kernel*. The kernel is of commercial value because of its high content of oil.

Uses of the coconut palm

The palm is easily cultivated because it requires little investment and little knowledge of specialised agriculture. Under favourable conditions, the palm propagates by itself. This fact, coupled with its many uses, makes it indispensable to people who live in regions favourable to its growth. Almost every part of the palm is used: the leaves and the trunk provide materials for man's shelter; the fruits provide drink and food; the hard shells can be made into utensils. Thus, the palm provides the peasant with his basic needs. Since the beginning of the 20th Century, the palm has become commercially important because its fruits are an important source of vegetable oil which is increasingly demanded by industrialised countries of the temperate zone for the manufacture of many products, particularly margarine and soap. In recent years, the fibre, which is called *coir*, has also become commercially important in the manufacture of cordage and strong matting.

Growing conditions

The coconut palm is commercially cultivated in several tropical regions which have:
1 Daily temperatures of about 26°C (about 79°F), with plenty of sunshine.
2 An evenly distributed annual rainfall of at least 1875 mm (about 75 in).
3 Sandy, well-drained soils (it can tolerate fairly salty soils).

The plant is grown on a commercial basis, frequently on plantations in several wet tropical regions. The plant is grown from seed (which is the coconut), first in a nursery where shade is provided. After about 6 months the seedlings are ready for transplanting in the plantation. They are set in rows, and after about seven years the young trees bear their first fruit.

Processing the fruit

After the coconuts are collected, the outer husks are removed. The nuts are then split open, and are either dried in the sun (on peasant holdings), or

Drying coconut kernels in the Philippines

in smoke kilns (on estates). When dried, the kernels shrink and are dug out from their hard shells by the use of a curved knife. The dried kernel is called *copra* and about 60 per cent of its weight is oil. It is bagged and sent to the oil mills.

Splitting coconuts in Malaysia to extract the 'nut' from the husk.

236

Extracting the oil

Machines cut the copra into small pieces which are then passed through heavy rollers to extract the oil. The residue is pressed into round slabs known as copra cakes, and these are sold as poultry feed. Although most producing countries export a large part of their production as copra, the extraction of the oil in the producing countries is becoming increasingly important.

Main regions

Most of the world's copra and coconut oil output comes from the countries of insular South-East Asia and certain tropical islands of the West Pacific. In insular South-East Asia, the eastern islands of the Philippines, Malaysia and Sumatra, Sulawesi and Kalimantan, are important commercial producers. In tropical Asia, Sri Lanka and India are important centres of commercial production. Producers of lesser importance are New Guinea, the Seychelles, Fiji and the Solomon Islands.

Copra and coconut oil trade

Malaysia, Sri Lanka and the Philippines are three important exporters of coconut oil and copra. India and Indonesia, because of their large populations, consume most of the oil they produce and are less important exporters. Exports from the Philippines and the Pacific Islands go almost entirely to the U.S.A. whereas exports from the other countries go mainly to European countries.

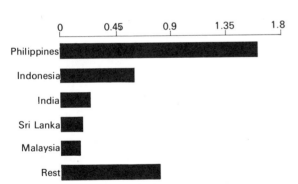

Annual World Production of Copra (average for the last five years in millions of metric tons). Annual production approximately 3 590 000 metric tons.

THE OLIVE TREE

The tree probably originated in the eastern Mediterranean countries from which its cultivation has spread throughout the Mediterranean area and to

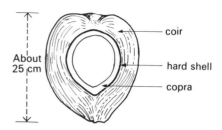

Sectional drawing of a Coconut Fruit

other regions having a similar climate. The tree fruits once a year and it reaches a maximum production of about 12 kg (about 7 lb) approximately 15 years after planting.

The fruit is crushed to extract the oil and unlike most other vegetable oils requires no further treatment.

Main producing regions

Spain and Italy together produce about one half of the world's production of olive oil, and most of the other half comes from the remaining Mediterranean countries. In recent years cultivation of olive trees has become important in California.

OTHER OIL-PRODUCING PLANTS
Groundnuts

This plant is indigenous to South America but is now extensively grown in many tropical and subtropical countries, especially in China, the southern states of the U.S.A., India and parts of west Africa. The plant bears small yellow flowers, which bend towards the ground after fertilisation. The flower stalks then grow rapidly to push the flower head below the soil. The fruit, which is a pod, usually containing 2 seeds (the nuts), develops a few inches below the ground and reaches maturity after about 3 months.

The plant, which is an annual, grows best in regions which have an average temperature of about 26°C (about 79°F), and rainfall of about 500 mm (about 20 in), or its equivalent in irrigation water, during the 4 to 5 months when the plant is growing. The plant grows best in well-drained soils. After the fruit is harvested, the nuts are extracted from the shells and they are crushed to extract the oil. The oil is used for making margarine, edible oils, soap and plastics. The nut residue is rich in protein and it makes a high quality cattle food. In recent years this residue has also been used for making a synthetic fibre which is used for making cloth.

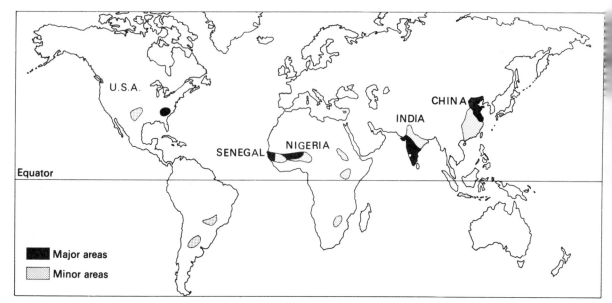

The Main Groundnut-producing Regions of the World

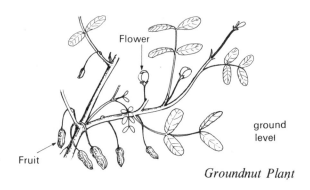

Groundnut Plant

Soya beans

The plant is indigenous to the warm temperate parts of Asia and it is grown extensively in Northern China. It is also grown in the U.S.A., especially in

Annual World Production of Groundnuts (average for the last five years in thousands of metric-tons). Annual production approximately **18 000 000** *metric tons.*

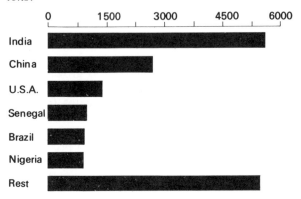

the States of Illinois and Indiana, in Japan, Korea and in the U.S.S.R.

Soya beans grow best in regions which have:

1 Summer temperatures of between 21°C (70°F) and 25°C (77°F).

2 About 700 mm (about 28 in) of rain most of which should fail in the growing season.

3 Deep, well-drained soils, rich in potash.

The soya bean plant, like other legumes such as peas, clover and groundnuts, enriches the soil by extracting nitrogen from the air and turning it into nitrates in the soil. For this reason it is often used as a rotation crop, with crops such as maize (especially in the U.S.A.)

Processing the bean

The beans are crushed to extract their oil, but this forms only 15 per cent of the weight of the bean. The oil is both an edible oil and an industrial oil which is used for making paint, soap and plastic, etc. The bean residue makes a good cattle food. The bean is rich in protein and has formed an important part of the diet of the Chinese in northern China. In recent years the plant was introduced into the U.S.A. and this country and China are now the world's greatest producers. The leaves and stalks of the plant, like those of groundnuts, can be dried to form hay which is used as a cattle feed.

Linseed

Linseed is the seed of the flax plant and it is extensively grown in Russia, North and South America and India. The oil is obtained from the seeds and, because of its drying property on exposure to air, is used in making paints, printing inks and oil cloth.

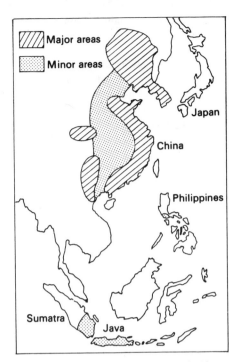

Soya Bean-producing Areas of China and Indonesia

Cotton seeds

Cotton seeds contain about 25 per cent of their weight in oil. When extracted, this is edible and is used for making margarine. It is also used for making metal polish. Cotton seed oil does not enter very greatly into international trade because many cotton-growing countries need the oil for home use. After the oil is extracted the seed residue is used as a cattle feed.

Castor Seeds

The plant, which grows in many tropical regions, produces seeds like those of the bean plant. They contain oil to about 40 per cent of their weight, and this is extracted by crushing the seeds. The oil is used as a high temperature lubricant, in the manufacture of enamels and varnishes, and as an ingredient of plastics. The seed residue cannot be used as a cattle feed because of the poison it contains. However, it makes an excellent soil fertilizer.

Soya Bean-producing Areas of the U.S.A.

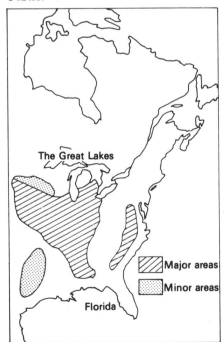

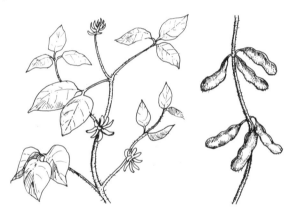

Soya Bean Plant   *Soya Bean Fruit*

Annual World Production of Soya Beans (average for the last five years in millions of metric tons). Annual production approximately 45 000 000 metric tons.

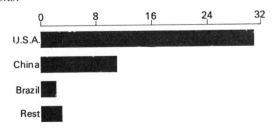

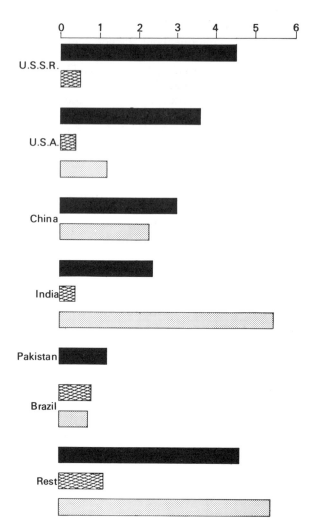

Annual World Production of Cotton Seed, Linseed and Groundnuts (average for the last five years in millions of metric tons). Annual production approximately
Cotton Seed – 20 000 000 metric tons;
Linseed   – 2 400 000 metric tons;
Groundnuts – 15 000 000 metric tons.

EXERCISES

1 (a) On an outline map of the world:
  (i) Mark *five* areas, in each of which, one of the following crops is grown on a large scale: oil palm, groundnuts, soya beans, olives, coconuts. Write the names of the crops in the areas chosen.
  (ii) For each of *three* of the selected areas, mark and name one town which handles either the crop or a product of it.

(b) Name the product obtained from each of any *tw* of the named crops, and for *each*, briefly state ho it is obtained.

2 Choose *two* of the following: oil palm, coconut, groun nut, olive, soya bean and linseed. For *each* one chose:
  (a) Draw a large sketch-map to show an area whi grows and exports the crop.
  (b) For *each* area selected, briefly describe (i) the reli the soils and the climate, (ii) any other factors whi affect the cultivation of the crop.
  (c) Describe the conditions under which each crop cultivated.

3 Vegetable oils are used in the daily diet of a large sectic of the world's population, and they are also us extensively in industry.
  (a) Name any one vegetable oil and briefly describe:
    (i) the plant from which it is made;
    (ii) the conditions under which the plant is grow
    (iii) *one* area where the plant is grown extensively
  (b) Name one other vegetable oil and describe the wa in which this is used in industry.

Objective Exercises

1 Which one of the following plants is grown specificall for the oil that is contained in its fruits/seeds?
  A soya bean plant
  B oil palm
  C coconut palm
  D olive tree
  E maize plant

  A B C D B

2 Vegetable oils are in great demand in industrial countrie because they are mainly used
  A in the manufacture of synthetic fibres
  B as a substitute of petroleum oil
  C for making soap and margarine
  D for making paint
  E by diesel engines

  A B C D F

3 Conditions favouring the growth of groundnuts do no include
  A sandy soils
  B tropical conditions
  C continuous high temperatures
  D moderate rainfall
  E shady slopes facing away from the sun

  A B C D F

4 From which of the following is copra obtained?
   A coconut coir
   B coconut kernel
   C oil palm kernel
   D oil palm pericarp
   E groundnuts

   A B C D E

The following three questions are to be answered by using the accompanying map which shows seven areas in which the cultivation of crops, rich in oil, is important.

5 Which one of the following numbered areas produces a crop from which palm-oil is made?
   A 1
   B 2
   C 3
   D 4

   A B C D

6 Which of the following pairs represents coconut oil and cotton seed oil?
   A 3 *and* 4
   B 4 *and* 1
   C 2 *and* 3
   D 4 *and* 2

   A B C D

7 The areas lettered *A* and *C* are respectively associated with the crops
   A flax *and* olives
   B groundnuts *and* flax
   C olives *and* linseed
   D olives *and* groundnuts

   A B C D

8 Which one of the following plays an important part in the manufacture of soap?
   A petroleum products
   B animal fats
   C soft water
   D vegetable oils

   A B C D

9 Where in South-East Asia are the greatest concentrations of areas growing oil palm, and on what agricultural practices do they depend?
   A on coastal plains based on small holdings
   B on undulating lowlands receiving rain throughout the year, and based on plantations
   C on coastal plains based on plantations
   D on cleared foothill slopes based on small holdings

   A B C D

10 Coconut trees are important to the rural populations of many tropical regions because
   A they are easy to grow
   B coconuts are often grown as a cash crop
   C the fruit is an important food and practically every part of the tree can be utilised
   D coconut trees can often be grown where other trees cannot be grown

   A B C D

11 Which one of the following does **not** use palm oil in its manufacture?
   A soap
   B margarine
   C cooking oil
   D lubricant

   A B C D

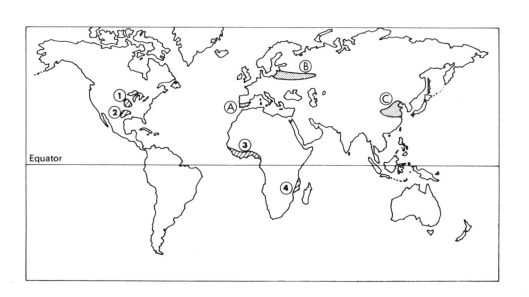

Equator

23   Animal Farming

At one time all animals were wild but between 4000 and 2000 B.C. Man learnt to domesticate some animals which had proved most useful to him. Before this happened Man had learnt to eat the flesh of some animals, to use their skins as clothing and their bones for making useful tools and weapons. After domestication he soon discovered that they had many uses other than just providing meat, skins and bones. He used them as draught animals for pulling his ploughs and for carrying his goods. He also learnt to use the milk of animals for making butter and cheese which provided him with another food.

Over the last two hundred years enormous advances have been made in the extensive development of commercial animal farming, especially cattle and sheep farming. These changes were in part associated with the agrarian and industrial revolutions that affected most European countries in the 19th Century. Industrial development led to the migration of peoples to the towns from the farm lands of countries like the U.K. and Germany. The populations of countries becoming industrialised increased dramatically at the same time and this led to increased unemployment and the emigration of large numbers of people overseas to areas such as Australia, New Zealand and the Americas. Many of the people who emigrated went from the farmlands of the U.K. and Europe and they started rearing sheep and cattle and growing cereals on the extensive grasslands of these 'new' lands. These lands later became important markets for industrial goods in return for farm produce such as meat, hides, wool and grain. The development and expansion of communications, especially railways and refrigerated ships, in the second half of the 19th Century, encouraged rapid expansion in the farming activities of the 'new' lands. The grasslands and the temperate continental interiors of Australia, and North and South America today, are among the most important grain and livestock producing regions of the world.

CATTLE FARMING

Grass is the natural food of cattle and because of this they are reared in almost all temperate and tropical grasslands. It is estimated that there are about 1000 million cattle in the world of which over 150 million are in India. Most of these cattle are bred for beef, the rest are bred for hides and dairy produce, though in large parts of Asia cattle are still bred as draught animals for pulling ploughs and carts.

Two main types of cattle are bred. These are:

1 *European cattle* such as Aberdeen Angus, Shorthorn, and Hereford. These are reared extensively in the cool and warm temperate regions of Western Europe, South America, Australia and the U.S.A., for beef and for dairy produce.

2 *Zebu* or *hump-backed cattle*. These are reared mainly in the tropical grasslands of Asia. They are reared for dairy produce and for draught. In recent years Zebu cattle have been successfully crossed with British breeds and breeds indigenous to parts of South America and Africa, and the resulting cross-breeds have proved to be very resistant to diseases caused by insect pests, and to the summer heat of the sub-tropical regions. These cross-breeds are now widely reared in Northern Australia, parts of Africa, Brazil and Venezuela and in the Southern States of the U.S.A.

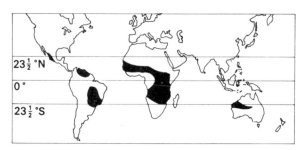

Tropical Grasslands of the World

COMMERCIAL CATTLE FARMING

Australia and Argentina are the most important beef-exporting countries of the world. Most of their beef is exported to Europe and North America because these continents are unable to produce sufficient beef to meet their requirements. Beef production in these two countries, and in other countries in the Southern Continents, expanded enormously because of several factors arising out of the Industrial Revolution. The main factors were:

1 The migration of people from Europe to the new

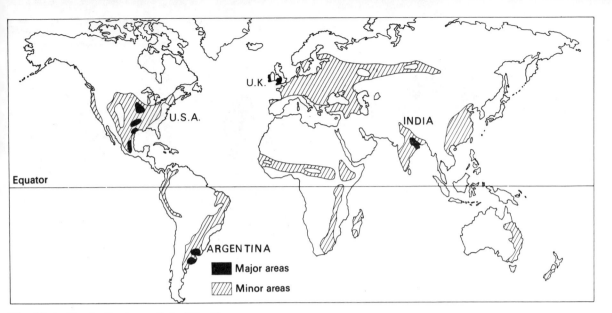

The Main Cattle Regions of the World

lands.

2 The increase in population of the industrialised countries which gave rise to an increase in demand for beef.

3 A rise in the standard of living of people in the industrialised countries which again resulted in more demand for beef.

4 The introduction of fast and cheap transportation both by land and by sea, and the introduction of refrigerated ships and canning.

5 The introduction of special crops, both inside and outside the grasslands, for feeding to the cattle.

Commercial cattle farming is of two main types – the breeding and rearing of young cattle, and the fattening of cattle for slaughter. Beef cattle can tolerate relatively poor grasses and because of this, the general practice is to rear them on poor pastures, leaving them to graze in the open throughout the year. If necessary, supplementary food such as hay and root crops are fed to them in the winter. During this period they require the minimum of attention, and hence only a small labour force is needed. When the cattle are about two years old they are taken to the fattening areas where they are fed on such crops as alfalfa. These areas are near to the beef factories where the cattle are slaughtered. The meat is then chilled or frozen or processed in some other way, such as, canning, after which it is transported to markets in both the producing countries and in overseas countries.

The main beef cattle regions

The U.S.A.

Most of the cattle are reared in the foothill country of the Western Cordilleras after which they are moved to the central plains and prairies to the east for fattening. Here they are fed on maize and fodder crops before being slaughtered in the cattle centres such as Kansas City, Chicago and Omaha. The fattening areas of the U.S.A. have been moving steadily westwards over the past twenty years, a process which has been made possible by the irrigation and fertilization of the soils of large dry areas in the west. This movement has also been accompanied by a shift in the slaughter houses and meat-packing factories. Today there are over 100 million cattle in the U.S.A., most of which are reared for beef, the annual production of which exceeds 10 million metric tons.

Argentina

Most of Argentina's cattle are reared and fattened on large cattle ranches, called *estancias*, on the Pampas. Alfalfa grass, maize and other fodder crops are grown on the ranches for fattening the cattle which are then taken to the slaughter houses near to the coastal cities such Buenos Aires. The meat is chilled or frozen or canned, and most of it is exported. About 3 million metric tons of meat and meat products are produced annually by Argentina.

Australia

Beef cattle farming is expanding in the savana

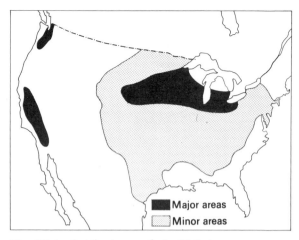

The Main Cattle Areas of the U.S.A.

Inside the freezer of a meat-packing plant in Buenos Aires (Argentina)

regions of Northern Australia and the production and export of beef is steadily increasing as transport facilities are developed. The great distances of the cattle farms from the eastern coastal towns and markets plus the unreliable nature of the rainfall are two factors which make it difficult for the cattle industry to develop rapidly. Better cross-breeds are being introduced and these have already resulted in an appreciable increase in beef production.

Brazil

The grasslands of the inland plateau of the Mato Grosso in Brazil could become a very important beef-producing region.

Africa

The vast grasslands of Africa are of tremendous potential value as beef-producing regions but because of diseases caused by insect pests and the extent to which nomadic pastoralism takes place, there is as yet only limited development. In many parts of Africa, ownership of cattle is a sign of wealth and very often the number of cattle owned is more important than the quality of cattle.

India

Most of India's cattle are undernourished and are in a very poor condition. Cattle are reared either for their milk or as draught animals.

Problems facing cattle farming in tropical regions

Most of the potential cattle-rearing regions of tropical latitudes are affected by three basic problems. These are:

1 *Disease* The most common disease is sleeping sickness which is fatal to cattle and which is spread by the *tsetse fly*. This fly is difficult to destroy and the only really effective control is to hand spray trees and bushes where it breeds,

and if necessary, destroy them by burning. The tsetse fly breeds over very large areas in the tropical grasslands of Africa. Other diseases are spread by ticks especially in the Llanos of Venezuela and the Campos of Brazil.

Annual World Beef Production (average for the last five years in millions of metric tons). Approximate annual production 40 000 000 metric tons.

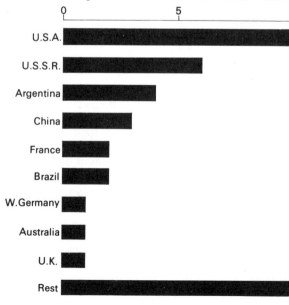

244

2 *Remoteness of the savana lands* The larger part of the savana lands are at considerable distances from the coastal ports, and road and rail communications between the two are still not well-developed. Because of this, it is uneconomical to fatten cattle in the rearing areas because by the time they reach the coast, they are usually in a poor condition.

3 *Climate* The hot wet summers and warm dry winters of the savana lands, present real problems. Summer flooding and winter drought cause very real problems. During the winter, the grass dies and this sometimes causes death to many cattle.

Possible solutions

One of the most effective ways of combatting diseases is to develop disease-resistant breeds. This is being done fairly successfully by crossing Zebu with other cattle. In conjunction with this, measures must be taken to control insect pests which spread the diseases. Shortage of water, especially in the winter season, can be partially overcome by tapping artesian water, e.g., in Australia, and by building temporary dams to hold back the summer rains. It is also necessary to improve, and extend, road and rail communications to enable fattened cattle to be transported quickly to the slaughter houses and markets. All of these measures require very large amounts of capital and an efficient administrative system. At the same time it is necessary to overcome traditional and religious practices, which at present prevent commercial cattle breeding. This can only be achieved by radical and enlightened improvements in educational facilities. This applies particularly to India where over 160 million cattle are regarded as sacred by Hindus.

Dairy cattle farming

For dairy cattle to give a high yield of milk throughout the year they must be reared on rich pastures and fed with high quality fodder crops during the winter. Moist, mild climates favour the growth of good pastures and these occur mainly in temperate climatic zones in such regions as North-West Europe, North-East U.S.A., South-East Australia and New Zealand. The breed of cattle is also important, and some of the best breeds are Jersey, Aryshire, and Guernsey although Friesian and Dairy Shorthorn are becoming increasingly popular because they can be reared for both milk and meat.

Dairy farming is concerned with producing milk for sale in the liquid form, or for making it into butter, cheese, condensed and powdered milk. Most dairy farms are small and they are highly mechanized. The cattle are usually grazed in the open. They are able to get sufficient fresh grass in the summer, but

Undernourished tropical cattle

in the winter they have to be fed with hay, root crops, vegetables and silage which is a cattle food made from hay or maize usually mixed with molasses. The supplementary winter food is sometimes produced on the dairy farms but more and more farms are buying it from cattle food manufacturers. Some dairy farms specialize in producing milk for sale in that form. This type of farm is common in the north-eastern coastal regions of the U.S.A. Other farms produce milk which is used for making butter and cheese. Most of the dairy farms in the mid-western states of Iowa, Minnesota and Wisconsin, are of this type because of the great distances of the farms from the cities.

Europe

Dairy farming is very important in North-Western Europe where it has become a main agricultural activity. In some countries, e.g., the U.K., most of the milk produced is consumed fresh by the large urban populations and only small quantities, as in in Yorkshire and Cheshire, are turned into butter and cheese. In Denmark, the Netherlands and Switzerland, more milk is produced than these countries require and the surplus is made into butter and cheese which form important exports.

245

Annual World Production of Milk (average for the last five years in millions of metric tons). Approximate annual production 350 000 000 metric tons.

Dairy cattle coming in for milking near to Rotorua (New Zealand)

North America

Dairy farming is concentrated in the mid-western states. Most of the milk from these areas is turned into butter and cheese. But there are important dairy farm belts around large cities, such as, New York, Detroit and San Francisco which produce milk for sale in liquid form to these cities. In the Los Angeles Metropolitan Area, for example, there are 200 000 dairy cattle living in sheds, and which are fed on fodder produced in other parts of the U.S.A. This type of dairy farming is known as *dry lot farming*.

New Zealand

North Island has excellent conditions for dairy farming and like Denmark and the Netherlands, most of its produce is for overseas markets.

SHEEP FARMING

Sheep can live under a great many types of climatic conditions, but they cannot tolerate very cold climates or very hot, humid climates. Sheep are reared therefore in almost every part of the world except the polar and equatorial regions. In all other regions, they thrive on land which is unsuitable for cultivation or for rearing cattle and other livestock excluding goats. Vast numbers of sheep are reared in the semi-arid, marginal region around the Australian Desert and around some of the deserts in the Americas, Africa and Asia.

In some regions sheep are reared for their meat, and in other regions for their wool. Cross-bred sheep are reared for both meat and wool. Sheep which are bred for their meat have to be reared on good pastures which are often sown with special grasses. This type of sheep rearing takes place in mild, moist climates such as those of New Zealand,

South-east and Central England. Sheep which are bred for their wool are reared on poorer pastures. This type of sheep farming takes place on large farms on the dry grasslands of Australia, parts of North and South America, parts of central Asia and the U.S.S.R. and the Spanish Plateau.

The pastures for all sheep, whether they are bred for meat or wool, must be well drained because sheep are very susceptible to foot-rot disease.

The main sheep rearing regions

The rearing of sheep on a commercial basis is concentrated in the temperate grasslands which have an annual rainfall of from 350 mm (about 14 in) to 750 mm (about 29 in). The number of sheep per unit area depends on the amount of rain, that is, the number in areas of low rainfall is smaller than the number in areas with a higher rainfall.

Australia

This country has approximately 150 million sheep which makes it one of the most important sheep-rearing countries in the world. Most of the sheep are reared for their wool and each year Australia produces about 10 million metric tons of wool, most of which is exported. The most common breed of sheep kept is the merino sheep and these are

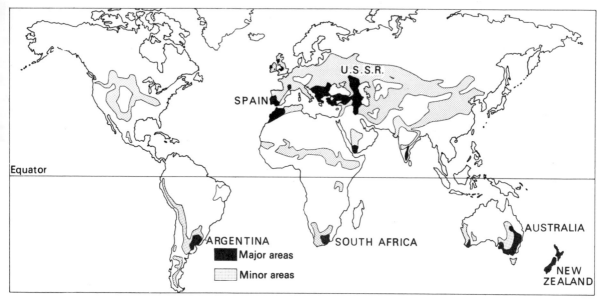

The Main Sheep Regions of the World

reared mainly on the pastures in Western Australia and Murray-Darling Plains. The merino sheep produces a very high quality wool. Australian sheep farms, which concentrate on rearing sheep only, are very large and most of these are in central Australia which has a semi-arid climate where rainfall is unreliable. Because of this, the farmers have to rigidly control the number of sheep per unit area to ensure that any increase in sheep population, which usually occurs in years when the rainfall is heavier than the usual, is eliminated to prevent over sheep population in the following drier years. Farms which rear sheep and which cultivate wheat also, are much smaller. These farms are concentrated in South-East Australia.

Every spring, the sheep on the farms which rear them for wool, are rounded up for shearing, that is, for cutting off the thick coat of wool. This work is done by sheep shearers who travel from farm to farm. After shearing, the fleeces (wool coats) are graded, weighed and baled, and transported to the wool auction centres, such as, Sydney.

New Zealand

The sheep farms of New Zealand produce both meat and wool. The country exports more frozen mutton and lamb than any other country and the total value

Sheep farming in New Zealand

of all sheep products is over 50% of the country's total exports. The moist, mild climate of New Zealand makes it possible for sheep farms to specialise on fattening lambs for export. The lambs are bred on the hill farms around the Canterbury Plains of South Island. They are then moved onto the Plains for fattening.

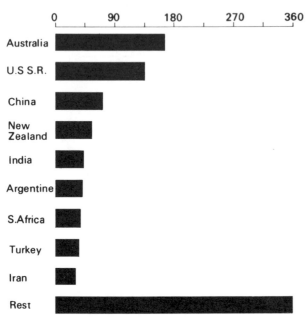

	0	90	180	270	360

World sheep population in millions (average for the last five years)

PIG FARMING

The domestication of the pig first took place in China, but it was not until the early 19th Century that pig farming featured in the agriculture of the of the west. Today, the rearing of pigs is a major activity in Europe, North America and China.

The main pig regions

Pigs can be reared in most climatic regions although the chief rearing regions are located in temperate latitudes.

1 *North America* The main area of rearing and fattening coincides with the Corn Belt, especially in Iowa which is a part of the belt. The pigs are fed on maize and a high percentage of the meat produced each year is canned for export. Meat processing and packing is located in such towns as Cincinnati, Omaha and Kansas.

2 *Europe* Pig-rearing is especially important in northern Europe where the pigs are fed on skimmed milk, which is a by-product of the dairy industry. Skimmed milk is that which is left after the cream has been removed from the milk for making butter. Denmark and Holland are two of the main pig-rearing countries of Europe. They are also two very important dairy farming countries.

3 *New Zealand* Pig-rearing is again closely connected with dairy farming, and, like the U.S.A. Denmark and Holland, it produces pork products for export.

Products of pig farming

The three main products are:
1 pork, which is the non-processed meat of the pig;
2 bacon, which is meat from the body of the pig and which has been treated with salt;
3 ham, which is meat from the pig's thighs which has been cured and smoked

Note Although pigs can be reared in almost all temperate and tropical climates, pig farming is unimportant in many regions because of religious and social prohibitions.

POULTRY FARMING

Poultry farming refers to the rearing of fowls, turkeys, ducks and geese, for meat or for eggs. Poultry can be reared in most temperate and tropical climates but turkeys are susceptible to damp.

Poultry are reared extensively in almost every country, though intensive poultry farming only takes place on a large scale in countries which have a well-developed agriculture. Several countries, e.g. Denmark, Holland, Eire, Australia, have geared poultry farming to overseas markets, and they export annually, large quantities of both eggs and poultry.

EXERCISES

1 (a) Name *three* types of animal farming and for *one of* these:
  (i) Name *one* area where it forms a major activity.
  (ii) Say why it is particularly suited to the area named.
  (iii) Name the main products and say whether they are for export or not.
  (b) Explain why animal farming in the tropics is more hazardous than it is in temperate latitudes.

2 Dairy farming is practised in many regions which have a cool temperate west coastal climate.
  (a) Mark, on an outline map of the world, *three* areas where dairy farming is an important activity.

(b) Choose *one* of these areas and describe how the climate, the soils, and any other factors favour dairy farming.

(c) Name *one* town in the area chosen which either produces dairy products or which exports them.

Objective Exercises

3 Commercial animal farming is mainly practised with all of the following animals **except**
A dairy cows
B pigs
C sheep
D goats
E poultry

A B C D E

Which one of the following statements is **not** true?
A Sheep can be reared on poorer pastures than can cattle.
B Wheat farmers in Australia often rear sheep.
C Goats often cause soil erosion in semi-arid regions.
D It is better to use grain, such as maize, as a foodstuff for animals than as a food for man.
E Most of the cattle in India are under-nourished.

A B C D E

3 There are more sheep in Australia than there are in any other country because
A most of Australia is too hot and too dry for crops
B a large percentage of the Australian grasslands are not suitable for cattle
C one of Australia's principal exports is wool
D most of the sheep are reared for wool
E most Australians eat a large amount of mutton

A B C D E

4 Dairy farming near to large urban settlements is located there mainly because
A of the availability of capital to finance dairy farming
B of special food that dairy cattle need that only urban centres can supply
C there is often surplus labour in urban centres which can be employed on the farms
D of the need for a nearby market because dairy produce is perishable
E agricultural land values are higher near to urban centres

A B C D E

5 Beef and dairy farming differ in that beef farms require
A less labour
B less fertile pasture (except for fattening)
C smaller water supply
D better transport to the centres of population

A B C D

6 Pigs are kept in dairying regions because
A they need labour at different times of the year
B pigs eat skimmed milk
C both need fertile damp areas
D both need fast transport to market

A B C D

7 During the 19th Century tremendous improvements were made to commercialize beef farming in South America, especially in Argentina. Which one of the following pairs of factors was largely responsible for this improvement?
A better quality cattle *and* improved methods of breeding were introduced from Europe
B special crops were grown for feeding to the beef cattle *and* refrigeration ships were introduced
C increased availability of capital *and* the introduction of meat packing factories
D faster cattle ships *and* an increase in immigration into South America

A B C D

8 Although there are more beef cattle in India than in any other country, the annual production of meat is extremely low. All the following factors help to explain why this is so **except**
A most of the cattle suffer from disease
B very few of the cattle are kept on farms which grow special crops to supplement the cattle's diet
C cross breeding to improve the quality of the cattle is rarely undertaken
D many Indians are too poor to buy meat

A B C D

9 Which one of the following statements is *most* applicable regarding the production and export of meat?
A The Southern Continents in general produce meat in excess of their requirements.
B Europe produces considerable quantities of meat but these are insufficient to meet its requirements.
C Great Britain and Europe together form the largest importer of meat while parts of the Southern Continents and the U.S.A. form the largest exporter of meat.
D The U.S.S.R. both exports and imports meat.

A B C D

24   Plant and Animal Fibres

Man's first clothing was probably made from leaves, grasses and bark, and the skins of animals. Today, his clothing is made from natural fibres, animal and plant, and synthetic fibres, which are made in the laboratory. Before natural fibres can be made into clothing they must first be joined together by spinning to make a continuous thread, called *yarn*, which is then woven into cloth. Woven materials are called *textiles*.

MAIN TYPES OF FIBRES
Plant: cotton, jute, manila hemp, sisal hemp
Animal: wool, silk
Man-made: nylon, rayon, terylene

COTTON
Cotton is made from the fibres of the fruit of the cotton plant. This plant, which is an annual, grows to a height of about 1·5 metres (5 feet) before it flowers. The fruit is like a pod and is called a *boll*. It contains black seeds which are embedded in a mass of white fibres, which is often called *cotton lint*.

Cotton is classified according to the length of its fibre. When this is less than 3 centimetres (1·2 inches), it is called *short staple cotton*. When it is over 3 centimetres it is called *long staple cotton*. Long staple cotton produces finer yarn which makes into finer cotton than short staple cotton. The latter is therefore used for making coarse cloth. However, most of the cotton grown is short staple.

Cotton plants in India

Growing conditions
The commercial cultivation of cotton can only be undertaken successfully in regions which have:

1 An annual rainfall of between 500 mm (about 20 in) and 1000 mm (about 40 in), most of which falls in the growing season, or the equivalent in irrigation water.
2 A growing season of about 200 days with temperatures of about 21°C (about 70°F) and the complete absence of frost.
3 Abundant sunshine when the flowers are forming and when the fruit is ripening.
4 A dry harvest season.
5 Rich, deep soils which contain plenty of organic matter.
6 Abundant labour for tending the plants and picking the bolls.

The cultivation of cotton
The plant is grown from cotton seeds which are planted in rows about 1 metre apart. The seed is sown in the spring in sub-tropical regions, but in tropical monsoon regions it is sown just before the wet season begins. After the seeds germinate, and throughout the growing season, the cotton fields must be continuously weeded and the plants sprayed to control insect pests, such as, the boll weevil. This is a small insect which hatches from an egg laid in

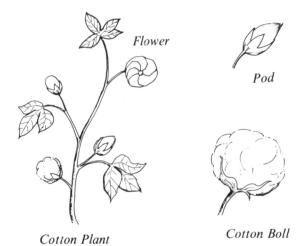
Flower

Pod

Cotton Plant

Cotton Boll

the cotton flower bud. It eats its way out of the bud which causes the bud to fall off. The amount of damage that weevils can do is tremendous because one pair of weevils can produce more than 10 million insects in one season.

Cotton makes great demands on the soil and it quickly removes the organic matter. Soils must therefore be continuously fertilized. If this is not done, the crop will probably fail and if this happens and is followed by heavy rains, soil erosion often takes place, especially on sloping land. Erosion can also be caused by growing only cotton on the same land year after year. This causes the soil structure to deteriorate because of the extraction of some mineral salts by the cotton plant. One way of preventing this type of soil erosion is to interplant cotton with other crops, especially leguminous crops such as groundnuts and soya beans, or to grow other crops in rotation with cotton. The cotton lands at one time experienced very severe soil erosion, but this has largely been prevented by mixed crop farming and the use of fertilizers.

Cotton-picking machine at work

Cotton picking
Cotton bolls are harvested as soon as they burst because any rain which falls after they have opened damages the fibres in the bolls. Although picking is still done by hand in many cotton regions, machine pickers are being increasingly used. But hand picking is more selective than machine picking because the bolls do not all open at the same time. In the U.S.A., a variety of cotton plant has been developed, all the bolls of which are open at the same time. Harvesting the bolls by machine from this type of plant is very efficient.

Treatment of the lint
The fibres in the bolls are called *lint*, and this has

to be separated from the cotton seeds. The process by which this is done is known as *ginning* and it takes place in a machine called a *gin*. This lint is then packed into bales and sent to factories where the lint is spun into a continuous thread called *yarn*. The manufacture of yarn into cotton cloth is done by *weaving*. Although the spinning and weaving is undertaken in some cotton-producing regions, most of the main cotton textile factories are in regions far removed from the cotton-producing regions (Chapter 31).

Cotton-producing regions
The U.S.A. produces more cotton than any other region, and most of this is grown in the south-east

Workers thinning cotton plants in Gezira (Sudan).

part of the country. This cotton region is known as the Cotton Belt and it extends from Georgia and South Carolina to Texas. At one time the main areas were in the eastern part of the Belt but because of soil exhaustion and periodic ravages of the crop by the boll weevil, cultivation steadily moved westwards. Some of the most important areas are now west of the Mississippi. This western part of the Belt is drier and this makes it difficult for the weevil to flourish. The use of irrigation and highly mechanized farming has enabled cotton cultivation to be extended westwards.

Other major cotton producers are China, India and Pakistan. The main cotton region of China is the lower Yangtse Valley, while in India the main region for cotton is the black lava soils of the Western Deccan.

Cotton production and world trade

The U.S.A., the U.S.S.R., China and India are the four largest producers of cotton in the world, but of these only the U.S.A. and the U.S.S.R. are important exporters. China and India require most of their cotton production for their home markets. Egypt, Mexico and Brazil, though relatively small cotton producers, are major cotton exporters.

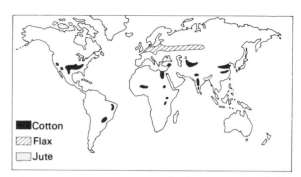

The Main Cotton, Jute and Flax Regions of the World

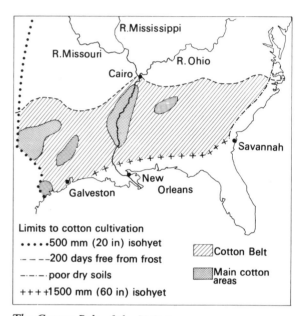

Limits to cotton cultivation
•••• 500 mm (20 in) isohyet
– – – 200 days free from frost
–·–·– poor dry soils
+ + + + 1500 mm (60 in) isohyet

▨ Cotton Belt

▨ Main cotton areas

The Cotton Belt of the U.S.A.

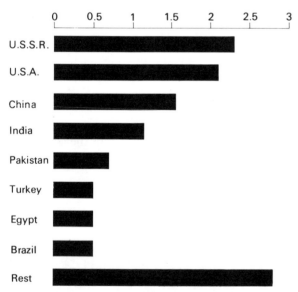

Annual World Production of Cotton (average for the last five years in millions of metric tons). Approximate annual production 12 000 000 metric tons.

1 Uniformly high temperatures of at least 24°C (75°F)
2 At least 2000 mm (about 80 in) of rain, most of which should fall towards the end of the growing season.
3 Moist, alluvial soils.
4 Abundant cheap labour, because machinery cannot be used.

The cultivation of jute

Most of the world's jute is grown in the Ganges – Brahmaputra Delta which is in India and Bangladesh. The land is ploughed in the autumn and the jute seeds are sown in the fields in the following spring before the South-West Monsoon rains begin. After the seeds germinate, the young seedlings grow very fast and by late June the plants are almost mature. It is at this time that the plants require

JUTE

Jute fibres come from the stems of a tropical plant which grows in hot humid regions. Its commercial cultivation requires the following conditions:

252

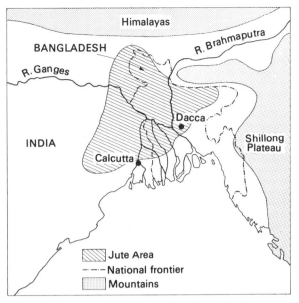

The Main Jute Areas of Bangladesh and India

abundant water and this is brought by the South-West Monsoon winds. The fields are usually deep in water when the plants are harvested in late July. The plants are cut by hand and the cut stems are tied into bundles and left for a few days before being dumped into ponds or streams where they remain until the bark of the stem rots. This process is called *retting*. When retting is completed, the fibres are easily extracted. The fibres are then hung on poles to dry in the sun.

The processing of jute fibres
The fibres are woven into a coarse cloth which is used mainly for making gunny sacks. These are used for packing sugar, rice, flour and other similar products. Jute cloth is also used as a base for carpets.

Annual World Production of Jute (average for the last five years in millions of metric tons). Approximate annual production 3 500 000 metric tons.

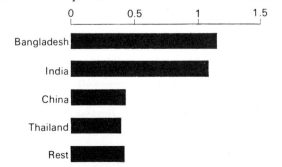

Extracting the fibres from the stems of jute plants in the Ganges Delta. The fibres are removed by pulling the stems backwards and forwards through the water.

In recent years the use of jute cloth, especially for making gunny sacks, has declined because of the use of plastic sacks. Plastic which is manufactured from the by-products of petroleum, is increasing in price and this may result in a return to jute. The two largest jute ports in the world are Chittagong in Bangladesh and Calcutta in India.

FLAX
The fibre of flax is obtained from the stem of the plant, and is used for making a cloth called *linen*. The fibres are about 1·5 metres (5 feet) long. The plant produces good fibres when it is grown:
1 in cool, damp climatic regions which have temperatures between 4·5°C (40°F) and 18°C (about 64°F), and evenly distributed annual rainfall of about 760 mm (30 in);
2 on rich, well-drained soils.

Harvest and processing
The plant is sown from seeds in spring and it grows to a height of about one metre (3 feet) in four months,

253

when it reaches maturity. The plants are then pulled up by hand and after this, the leaves are stripped off and put into tanks of soft water for several weeks so that the resinous matter in the stems rots away. This is called *retting*. The stems are then dried, and beaten in machines to remove the woody cores. After the fibres are separated they are sorted, and the longer ones are spun for weaving into linen. The shorter ones are spun for making rope.

Flax regions and trade

At one time, flax and wool were the principal fibres used for the manufacture of clothing in temperate regions, but with the introduction of cotton, flax fibre has lost much of its importance. Most of the world's output of flax comes from the North European Plain, especially the part which is in Belgium, Holland and N.E. France, Northern Ireland and central Russia. International trade in flax fibre is relatively unimportant as much of the flax produced is used by the textile industries located in the regions of flax production. Important centres of linen manufacture are in Belgium, N.E. France and Ireland.

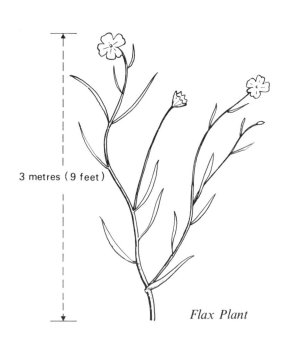

3 metres (9 feet)

Flax Plant

ABACA (MANILA HEMP)

Abaca is a tropical plant which is a member of the banana family. The plant, which is grown from rhizomes, grows best in regions having:
1 average monthly temperatures of over 24°C (about 75°F);

2 an annual rainfall of at least 2000 mm (80 in) which must be evenly distributed;
3 fertile loamy or lava soils.
For commercial cultivation it requires abundant cheap labour.

The fibres, which are in the stem of the plant, are used for making high quality ropes which are especially favoured by seamen because sea water does not rot them easily. Like jute it has suffered from competition with nylon.

Cutting and processing

The plant takes about two years to mature, after which it is cut down and stripped of its leaves. The stem is then crushed between rollers in a factory and the flattened stem is laid out as a sheet. Its fibres are then scraped off and dried.

Abaca regions and trade

Although abaca is grown in several parts of the humid tropics, the most important region, without doubt, is located in the Philippines, especially on the islands of Leyte, Mindanao and Samar. The plant is grown on plantations.

Countries of production and trade

Many parts of the humid tropics meet the plant's growth requirements, but the Philippines monopolise manila hemp production. This is because besides having suitable climate and soils and abundant, cheap labour, the Philippine hemp cultivators possess the necessary skills involved in the production of the fibres. Some abaca fibre is cultivated in eastern Sabah (East Malaysia). Most of the fibre from the Philippines is exported to the U.S.A. and the countries of Western Europe.

SISAL HEMP

This is another tropical plant whose fibres are used for making string, rope and a coarse cloth used for matting. The plant is a low shrub from whose short thick stem arise a mass of long, sword-shaped leaves whose thorned edges are very sharp. The plant is grown from suckers in a nursery until it is about 30 centimetres (12 in) high. It is then transplanted into the plantation.
The plant grows well in regions where:
1 the average monthly temperatures are over 24°C (about 75°F);
2 the annual rainfall is about 890 mm (35 in). Like abaca, it requires an abundant supply of cheap labour if it is to be grown commercially.

Cutting and processing

Unlike flax and abaca, the fibres of sisal hemp are

ontained in the leaves. This means that the plant, which is a perennial, does not have to be cut down. The plant is evergreen, and a few leaves can be cut from it, two or three times a year, for as long as twenty years. The first leaves are cut about four years after the sucker has been planted. They are scraped and crushed by machines so that their fibres can be extracted. These are then washed, dried and graded.

Sisal regions and trade

The most extensive sisal hemp regions are in east Africa where there are sisal plantations in Tanzania, Kenya, Uganda, Angola and Mozambique.

Sisal hemp originated in Yucatan, which is a part of Mexico. It is still grown there commercially as it is in parts of Central America and the West Indies. The bulk of sisal hemp fibres is exported to Great Britain, Western Europe and the U.S.A.

WOOL

For a long time wool has been used in the making of woollen garments for the peoples of most temperate regions. However, wool represents only about one-tenth of all the fibres used in industry.

The nature of wool, that is its strength, its curliness, its thickness and the length of the fibres, largely depends on the type of sheep from which it comes. Sheep wool can be divided into three types.

1 *Merino wool* This has a long, fine strong fibre and it is extensively used for making cloth for manufacture into clothing. This type of wool comes from Merino sheep which are reared in many countries, but especially in Spain and the sheep regions of the southern continents.

2 *Crossbred wool* This type of wool has a shorter fibre and it is not as good as Merino wool. It comes from crossbred sheep which are reared for both meat and wool.

3 *Carpet wool* This wool is coarse and it has a short fibre. It is used mainly in the manufacture of carpets and it comes as a by-product from sheep reared for meat.

Wool-producing regions

Over one-half of the world's production of wool comes from the southern continents and Australia is by far the largest supplier. The main producing countries of these continents are as follows:

(i) *Australia* Almost one-half of Australia's sheep population is located in the dry parts of the Murray-Darling Basin. Most of the sheep are Merinos, and sheep farms are extremely extensive. Altogether, Australia produces about one-third of the world's wool.

(ii) *New Zealand* Most of the sheep in New Zealand are reared for meat. This is because the moist climate favours crossbred and other sheep which make good meat. However, wool is an important product of the country and it amounts to about one-tenth of the world's annual production.

(iii) *South Africa* Most of the sheep are Merinos and they are reared on the dry grasslands of the High Veldt in Transvaal and the Orange Free State. The sheep farms are often large and at times the soil of the grasslands suffers from extensive erosion when conditions are exceptionally dry.

(iv) *South America* Argentina and Uruguay are the two main wool-producing countries of this continent. Most of the sheep are reared on large farms and most of them are crossbreds. The sheep region of these two countries extends from northern Uruguay southwards through eastern Argentina into Tierra Del Fuego.

2 Most of the wool produced by countries of the northern hemisphere comes from sheep which are reared for meat. The main wool-producing countries are as follows:

(i) *The U.S.A.* The bulk of the U.S.A.'s sheep are reared in the dry, semi-arid regions of the western states. Very little mutton is eaten in the U.S.A. and most of the sheep are reared for wool. However, insufficient amounts are produced for home use and wool must be imported from the southern continents.

(ii) *The U.S.S.R.* As in the U.S.A. the chief wool-producing regions coincide with the drier regions, especially those parts of the steppes. The U.S.S.R. comes next after Australia as a world producer of wool.

(iii) *The U.K.* The U.K. is still important as a world-producer of wool, though its importance is now very much reduced. There are important sheep regions on the limestone soils of the Pennines and south-east England, in the hills of Wales and the Lake District, and in the Southern Uplands of Scotland. Sheep are also important on the Wicklow Mountains of eastern Ireland and on the limestone plateaus of Galway in western Ireland.

SILK

Silk is a fibre which is made by a caterpillar often referred to as a silkworm. During the chrysalis stage of its life cycle, the caterpillar extrudes a fine thread in which it encases itself to form a cocoon. The filament may be as long as one mile. It is this filament which is spun into silk thread.

Silkworms are fed on the leaves of the mulberry tree and silkworm farming, which is called *sericulture*, is confined to regions where mulberry trees grow well. This tree requires a Mediterranean or Mediterranean-type climate where winters are short

Feeding mulberry leaves to silkworms

Revolving cocoon holders. After the silkworms mature, they stop eating mulberry leaves, and they are put into special cages where they spin their cocoons

and mild and summers are hot, with an annual rainfall of about 500 millimetres (20 inches)

Sericulture

The moth lays about 500 eggs in autumn. These hatch out into small grubs in the following spring. The grubs are kept on trays on which regular supplies of fresh mulberry leaves are placed. The grubs grow rapidly and after a few months they grow into silkworms about 7.5 centimetres (3 inches) long. They now enter the chrysalis stage and each silkworm spins a cocoon around itself. The chrysalises, except for a few which are allowed to turn into moths for egg-laying in the next autumn, are put into heated ovens to kill the pupae inside them.

During the few months of a silkworm's life, it eats large quantities of mulberry leaves, and it is essential that the silkworm farmer has easy access to un-limited quantities of this foodstuff.

The preparation of silk

The first task is to unwind each cocoon. This is sometimes done by hand and sometimes by machine in factories. As the cocoons are unwound, the threads of perhaps ten or more are twisted together

to give a strong single silk thread. Each cocoon yields about 450 metres (1450 feet) of good quality thread which comes off first, and from 450 to 1350 metres (1450 to 4400 feet) of poorer quality thread which comes off last. The latter is spun into a weaker thread which is called *spun silk*.

Silk regions and trade

China and Japan are the main silk-producing countries of the world. In China, silk production is concentrated in the densely populated regions of the Si-kiang delta near Canton, the Shangtung Peninsula, the North China Plain near Shanghai and parts of the Szechuan Basin. In Japan, sericulture is located in the southern and central islands particularly on the mountainous slopes of Honshu Island. Sericulture is also important in southern Europe, especially northern Spain and regions which border the eastern shores of the Mediterranean Sea. Other centres of silk production are the mountainous slopes of N.E. Thailand and in some Indian states bordering the Himalayas. The chief silk exporters are Japan, Italy, China and Thailand. The United States and countries of north-west Europe are leading importers.

Cotton, wool and silk are natural fibres, all of which are extensively used in the manufacture of cloth.

(a) For each fibre, locate *one* area where it is produced on a large scale.

(b) Briefly describe the best conditions for the production of one of these fibres.

For *each* of these fibres, cotton, jute, Manila hemp, sisal hemp and flax:

(a) Name *one* important area of production.

(b) Name the type of climate in which it grows best.

(c) State whether it enters into international trade on a large scale.

Cotton and wool are still the most extensively used fibres. For each:

(a) Name *two* areas, one in the Northern Hemisphere and one in the Southern Hemisphere where the fibres are produced in large quantities.

(b) Briefly state how it is treated before it can be made into cloth.

(c) List the advantages and disadvantages when its production dominates agriculture in a particular area.

jective Exercises

What is the process called which separates the fibres of the jute stems?

A ginning
B combing
C retting
D carding
E spinning

A B C D E

Which of these is **not** associated with silk production?

A cocoon
B mulberry
C sericulture
D silviculture
E silkworm

A B C D E

The world's largest producer of raw cotton is

A China
B Egypt
C India
D Mexico
E U.S.S.R.

A B C D E

Which one of the following natural conditions would be harmful to the cultivation of cotton?

A an annual rainfall of 500 mm to 1000 mm with most of it falling in the growing season

B a well-drained sandy soil
C temperatures of about 21°C during the growing season
D clear sunny skies after the plant is fully mature

A B C D

5 One of the processes used in the manufacture of cotton yarn is known as *ginning*. This process refers to

A the joining of the cotton fibres to form a continuous thread
B the separation of the seeds from the cotton fibres
C the washing of the cotton fibres
D the gathering of cotton bolls from the plant

A B C D

6 During the processing of flax, an operation called *retting* is used.

This refers to

A the steeping of flax stems in water to rot away resinous matter
B the extraction of the fibres from the stems
C the drying of the flax fibres
D the spinning of the flax fibres into a continuous thread

A B C D

7 Which one of the following plants is cultivated on tropical deltaic lowlands?

A sisal hemp
B jute
C abaca
D flax

A B C D

8 *"The plant is cultivated in certain tropical regions, especially on the islands of Mindanao and Leyte. The plant is grown for its fibres which are contained in the stem, and they are used mainly for making rope."*

The name of the plant is

A jute
B abaca
C flax
D sisal

A B C D

25　Rubber

RUBBER

Practically all of the world's rubber comes from an evergreen tree, called *Hevea Brasiliensis,* which is native to Brazil. This tree produces a regular supply of a white milky fluid, called *latex,* which is obtained from the inner bark of the trunk. Rubber is made by processing the latex.

The world's first supplies of rubber came from wild rubber trees scattered throughout the Amazon forest in Brazil. The invention of the motor car and the manufacture of rubber tyres, and the use of rubber as an insulating material for electric cables and wires, gave rise to an enormous demand for rubber. It soon become obvious that rubber trees had to be cultivated to meet this demand. Rubber seeds were sent from Brazil to Kew Garden, London, where they were germinated in heated glass houses. The young seedlings were then taken to Ceylon (Sri Lanka), and later to Malaysia, Indonesia and Thailand where rubber plantations were developed.

Growing conditions

The rubber tree grows very well in the hot, humid climatic conditions of equatorial lowlands. For its commercial cultivation it requires:

1 Uniform temperatures of above 26°C (79°F).
2 An annual rainfall of over 2000 mm (about 80 in) which should be fairly evenly distributed throughout the year.
3 Well-drained heavy soils.
4 Plenty of shade during its early stages of growth (this is often provided by inter-planting with banana trees).
5 Abundant cheap labour for weeding and tending the young trees and later for tapping the trees to extract the latex.

THE CULTIVATION OF RUBBER

Most of the world's rubber trees are grown either on small holdings or on large farms called *estates* or *plantations,* and the majority of rubber trees are propagated by bud grafting. This is done by sowing rubber seeds in damp, fertile soil in a nursery, where they quickly germinate. After about nine months, the young rubber plants are about 6.3 metres (12 in) high, a bud, from a very high yielding parent tree is grafted onto each of the young rubber trees, which are then transplanted in rows in the plantation. The rows of rubber seedlings are usually interplanted with a cover crop which helps to prevent soil erosion, especially on sloping land, and which enriches the

Tapping a rubber tree in Peninsular Malaysia.

soil, and which provides shade for the young se
lings. The trees are ready to be tapped after ab
seven years. This makes difficult the investment
a large amount capital of which is required for
establishment of a plantation, because for se
years there is no return on the capital investme
Also, it is difficult to forecast what the demand
rubber will be, seven years from sowing the se

Tapping the trees

Tapping usually takes place before day-bre
because at this time the latex flows best. Tappi

1
Bark of rubber
seedling is slit
and the flap
is pulled back

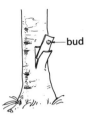

2
A bud taken
from a high
yielding tree
is inserted
beneath the flap
of bark

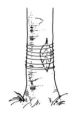

3
The flap of
bark is bound
tightly to the
seedling's trunk

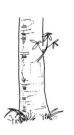

4
The bud begins
to grow

5
The top of the
seedling is cut
off after bud
growth is
established

Bud Grafting

consists of cutting a thin strip of bark from the trunk of the tree. The latex oozes out from the wound and flows along the slanting cut into an earthernware cup. After a few hours the latex ceases to flow. The cups are emptied and the latex is taken to the rubber factory.

Pouring latex from a small cup, which was attached to the tree, into a carrying can.

Preparing rubber sheets
In the factory, the latex is first diluted with water and then poured into aluminium tanks where it is coagulated by adding small quantities of acetic or formic acid. Aluminium separators are placed at 3·5 centimetre (1·4 inches) intervals across the tanks so that after the latex has hardened, rubber strips

of 3·5 centimetres (1·4 inches) thickness are formed. These strips are then passed through rollers and turned into sheets. These are later dried after which they are graded by holding them to the light to see whether they contain holes or bits of foreign matter such as wood. They are then packed into bales to await export.

New rubber processes
Malaysia is developing a new way of preparing rubber for sale. The coagulated latex is crumbled or chopped into small pieces, dried, and pressed into a solid block. Because it is sold with a guarantee of quality, grading each sheet by eye is no longer necessary.

Rubber cultivation in Malaysia
Almost 50% of the value of Malaysia's total exports comes from rubber of which nearly 60% is produced on plantations. The total area of land used for the cultivation of rubber is just over 1.5 million hectares (0.6 million acres) and the greater part of this land is located in the western foothills of the Main Range and in parts of Kelantan State which are in Peninsular Malaysia. Most of the plantation workers are Tamils who originated in southern India, but the majority of the rubber small holdings are owned and worked by Malays and Chinese. The plantations and small holdings mainly occupy land which has been cleared of jungle.

Rubber producing regions and world trade
Malaysia and Indonesia are the largest producers of natural rubber and together they produce each year nearly 70% of the world's production. Countries of secondary importance as producers of rubber are Sri Lanka and Thailand which together produce about 15% of the world's production. The only

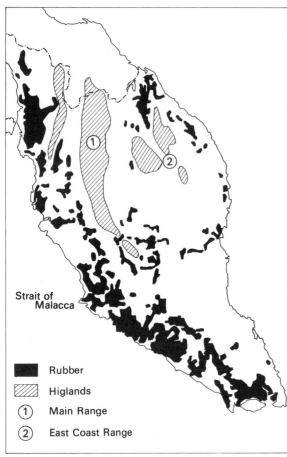

Strait of
Malacca

Rubber

Higlands

(1) Main Range

(2) East Coast Range

The Main Rubber areas of Peninsular Malaysia

countries outside Asia which are significant rubbe[r]
producers are Nigeria and Liberia. The industri[al]
countries of Western Europe, the U.S.A., th[e]
U.S.S.R., China and Japan are the main rubbe[r]
consumers.

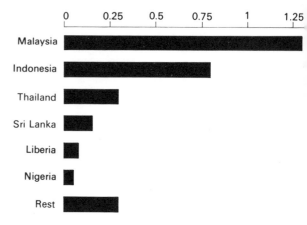

Annual World Production of Rubber (average for
the last five years in millions of metric tons). Ap-
proximate annual production 2 900 000 metric tons.

SYNTHETIC RUBBER

Synthetic rubber was being made, on a small scal[e]
in Germany and the U.S.A. before World War I[I.]
But during the war, when supplies of natural rubbe[r]
were cut off, both these countries rapidly expande[d]
their production of synthetic rubber. Today, th[e]
U.S.A. is the largest producer of synthetic rubbe[r]

Rubber Producers and Consumers

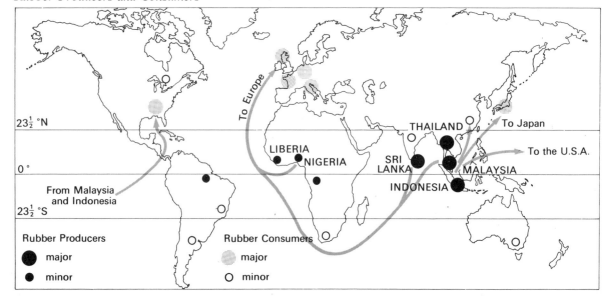

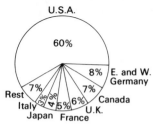

Annual World Production of Synthetic Rubber by Continents and by Countries (average for the last five years 3 200 000 metric tons)

ost of which is made from petroleum. The annual roduction of synthetic rubber is considerable. This oes not mean that it is replacing natural rubber ecause each type of rubber tends to be specially uited to the manufacture of certain types of goods. he price fluctuates greatly with demand and ompetition from synthetic rubber. This fluctuates ith the price of oil which is used to make synthetic ubber. Governments help to keep prices stable but any plantation owners are turning to palm oil roduction for a more reliable income.

TOBACCO

obacco is cultivated in many regions extending om temperate to equatorial latitudes, and it is rown on both large plantations and small holdings. he plant is an annual and grows to a height of om 1 to 1.3 metres (3 to 4 feet), and it bears a umber of large leaves. This plant was indigenous o North America and was first used by the N. merican Indians.

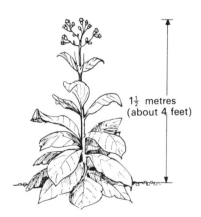

1½ metres (about 4 feet)

Tobacco Plant

Requirements of the plant

The plant grows best in regions which have:
1 a moist, warm climate with at least five months free from frost;

2 rich, heavy soils;
3 skilled workers for weeding and picking the leaves.

Its cultivation

Tobacco plants are grown from small seeds which are usually sown in well-fertilized seed-beds. These are covered with muslin (fine cloth) to protect the young seedlings from heavy rain, insect pests and strong sunlight. After a few weeks the seedlings are strong enough to be transplanted to fields which have been prepared and fertilized.

Harvesting and processing

After about four months the leaves of the plant reach maturity and the flower buds are nipped out to prevent loss of goodness from the leaves. The latter are harvested and hung in drying sheds where they are cured. The leaves are then treated and made ready for manufacture into cigarettes, cigars or pipe tobacco.

Main tobacco-growing regions

The chief regions are:
1 the U.S.A. (Virginia, North and South Carolina, Kentucky and Georgia);
2 Jamaica and Cuba;
3 Rhodesia, Malawi and Zambia;
4 the U.S.S.R., Bulgaria, Greece and Turkey;
5 India, China, the Philippines and Indonesia.

EXERCISES

1 Each year approximately equal quantities of natural rubber and synthetic rubber are produced in the world.
  (a) Name *two* areas where natural rubber is extensively produced.
  (b) List the important conditions which favour the cultivation of rubber trees.
  (c) State the extent to which natural rubber enters international trade and explain why.
  (d) Briefly explain why the production of synthetic rubber has become so important and name *two* raw materials from which it is made.

2 Choose *one* area where tobacco is grown on a commercial scale and for it:
   (a) Briefly describe the methods used in growing the crop.
   (b) Name the chief climatic requirements of the plant.
   (c) Name *one* town associated with processing the crop.

Objective Exercises

I Which of the following countries is **not** a major producer and exporter of natural rubber?
   A Indonesia
   B Malaysia
   C Brazil
   D Sri Lanka
   E Thailand

   A B C D E

2 All of the following conditions are necessary for the commercial cultivation of tobacco **except**
   A fertile, heavy soils
   B plenty of labour skilled in weeding the plants and picking the leaves
   C a market for the leaf tobacco
   D a rapid transport system
   E a moist, warm climate which has at least five months free from frost

   A B C D E

3 How often are rubber trees tapped?
   A every day
   B every 2 to 3 days
   C every week
   D every other week
   E every month

   A B C D E

4 Which of these regions does **not** grow tobacco on a commercial basis?
   A the U.S.A.
   B the West Indies
   C Australia
   D South Africa
   E South-East Asia

   A B C D E

5 Which of these is true of rubber growing?
   A trees produce latex which is tapped throughout their life
   B trees live for 40 to 60 years
   C trees can be tapped for 5 to 7 years
   D trees take 5 to 7 years to mature

   A B C D

6 In which of the following countries is tobacco grown on a large scale for export markets?
   A China
   B U.S.S.R.
   C Greece
   D U.S.A.

   A B C

7 All of the crops sugar cane, oil palm, rubber and coffee are commonly associated with
   A plantation agriculture
   B mechanized farming
   C intensive farming
   D mixed farming

   A B C

8 Rubber is made from latex which is obtained from the following part of the rubber tree
   A leaves
   B wood
   C inner bark
   D seeds

   A B C

9 A young rubber tree begins to produce latex and ready for tapping
   A after seven years
   B after it loses its first leaves
   C when it reaches a height of one metre
   D after three years.

   A B C

he over-exploitation of a natural resource is known
s a *robber economy*. Tree-felling in many regions
became ruthless after man developed metal tools
o enable him to fell trees quickly, and in many
gions, especially in Western Europe (including the
.K.) and in the U.S.A., vast areas of forest have
een removed. Only in a few regions has the removal
f forest been accompanied by *reafforestation* (the
lanting of new trees) and as a result some regions
ave lost the greater part of their original forest
over. Forests have been cut down for several
easons, of which the three most important are:

To enable man to expand his farmlands 2 To
rovide man with constructional timber. 3 To
rovide man with fuel.

hifting cultivation in equatorial and tropical
atitudes, especially in parts of South America,
frica and South-East-Asia, and extensive felling
f trees for fuel and constructional timber is still
ausing *deforestation* on a large scale in many parts
f the world. However, most governments and
nternational organisations are aware of the need
o practise forest *conservation*, and measures are
eing introduced which will slow down and even-
ually stop deforestation.

**he affects of climate on the distribution of
orests**

Trees will grow in almost any climatic region
rovided:

Where there is from 250 to 500 mm (about 10 to
20 in) of rain available for tree growth each year.
2 The average monthly temperature does not fall
below 6°C (about 43°F) for at least three months
of the year.

The nature of the forest cover that a region has
depends partly upon the amount and seasonal
distribution of rain and, temperatures. In humid
equatorial latitudes, heavy, fairly evenly distributed
rainfall and, uniformly high temperature results in
the growth of a dense and luxuriant forest cover.
In tropical monsoon regions, because of the definite
dry season, the forest cover is less dense, whilst
in savana regions, tree cover is broken into small
patches most of which are located near to permanent
sources of water such as rivers.

Types of trees

Rainfall and temperature together affect the rate at
which trees grow. Hot, humid climates enable trees
to grow continuously throughout the year which
means that tree growth is much faster than it is in

Felling coniferous trees in Canada

climatic regions which have either a dry season or
a cold season. Climatic factors also affect the shape
and nature of the leaves of trees. Trees can be
divided into two groups according to the shapes of
their leaves.

1 *Needle-shaped leaf trees* These trees are usually
called coniferous trees. They grow mainly in
cold latitudes, to whose low temperatures they
have become adapted by growing long, needle-
shaped leaves, which reduce the amount of transpiration (see Chapter 15). They do not lose all
their leaves at the same time. It is a gradual
process in which old leaves are replaced by new
ones. Because of this the trees are always in leaf
and they are called evergreens. They have widespread shallow roots and a thick resinous bark—
the former to absorb water from the less-frozen
top soil, and the latter to retain the water in the
trunk. They are conical in shape and have flexible
branches to allow the snow to fall off thus pro-

tecting the trees from damage which would otherwise happen as a result of a heavy covering of snow. Although coniferous trees are adapted to cold areas, they grow to some extent in most climatic regions. Most coniferous trees have soft wood.

2 *Broad leaf trees* These trees grow in temperate and tropical latitudes. Broad leaf trees grow in cool temperate regions where the temperature is over 6°C (43°F) for at least six months. During the winter they lose all of their leaves and trees which do this are called deciduous trees. Examples of cool temperate broad leaf trees are the *oak, elm, ash, poplar, maple* and *chestnut*. In warm temperate and tropical latitudes where there is a pronounced dry season, for example, in Mediterranean and tropical monsoon climatic regions, broad leaf trees are also deciduous. These trees lose their leaves in the dry season. *Teak* and *acacia* are examples of tropical deciduous broad leaf trees.

In humid tropical latitudes, such as lowland equatorial climatic regions, the broad leaf trees are evergreen because there is neither a pronounced dry season nor a cold season, and continuous growth therefore takes places throughout the year. Evergreen broad leaf trees, unlike deciduous broad leaf trees, have thick waxy leaves which reduces the rate of transpiration. Tropical broad leaf evergreens include *ebony* and *ironwood*. The continuous growth of tropical evergreen trees results in some trees being in flower and others in fruit at the same time.

Types of wood
For commercial purposes, trees are classed either as hardwoods or as softwoods. Most broad leaf trees are hardwoods and most coniferous trees are softwoods. The best known hardwoods are *teak, mahogany, ebony, ironwood* and *rosewood*, and all of these grow in tropical broad leaf forests. The best known softwoods are *pine* and *spruce*, and these grow in temperate coniferous forests.

The commercial exploitation of trees
Vast numbers of trees are felled annually. Some of these come from natural forests, both temperate and tropical, and some come from coniferous forests which have been planted by man. Coniferous forests are exploited much more than broad leaf forests, especially tropical forests, because their timber can be put to a much greater number of uses.

Exploitation of coniferous forests
The timber of coniferous trees is the main source of raw material for the manufacture of paper pulp and vast numbers of these trees are felled annually for this purpose. The timber is also used for making crates and boxes for transporting goods, and for making furniture, and matches. In addition, turpentine (used in the manufacture of paints and polishes), and cellulose, are important by-products derived from softwoods.

The natural conditions of coniferous forest regions makes the felling and removal of trees relatively easy. This is because most coniferous trees grow in pure stands, that is, trees of one species only grow together in the same area. This makes commercial felling economical. Also, coniferous forests have very limited undergrowth which permits the felled trees to be dragged out easily.

Coniferous trees are felled during the autumn and the trunks are hauled over the snow-covered grounds, often by tractors, to the frozen rivers.

Log barge carrying spruce and cedar logs down a river in British Columbia (Canada)

or floating downstream to the saw mills in the following spring. This method of getting the timber to the saw mills is being used less in those regions where hydro-electric power stations and irrigation dams have been constructed, because these restrict the use to which the rivers can be used for transporting the timber. Rivers which do not have power stations are, still used, of course.

The planting of coniferous trees either by reafforestation (replaced forest areas), or by afforestation (new forest areas) is becoming increasingly important in many temperate regions. The saw mills in these forests are usually located in central positions to avoid having to transport the timber great distances. A reduction in haulage distance is important because at the saw mill the timber is reduced to about 60% of its original bulk.

Exploitation of tropical broad leaf forests

Although the total area of these forests is very considerable, commercial exploitation is still limited to relatively small areas. This is because many different tree species grow close together in small areas and this increases the cost of commercial felling. Also, many tropical hardwoods have buttress roots which means that the trunks have to be cut through several metres above ground level. This adds to the cost of felling. The dense undergrowth of some tropical forests, especially monsoon forests, hampers the dragging of timber out of the forest which again adds to the cost. Whenever possible, the timber from tropical forests is transported out of the forests by floating it down the rivers, either to the saw mills or to collecting points from where it is transported by road or rail to the saw mills or to ports which export it. Although considerable quantities of tropical hardwood timber are exported to temperate countries for processing, an increasing amount is being processed in the countries of origin. Saw mills and plywood factories are being set up in countries such as Malaysia, the Philippines and in some African countries.

The distribution of the forests
Coniferous forests (softwoods)

The greatest extent of coniferous forests is in the Northern Hemisphere between latitudes 50°N and 65°N, in North America, and in Europe, and Asia. The coniferous forest belt of Eurasia is known as the *taiga*. The most common trees of these forests are *pine, spruce, fir* and *larch,* and the timber from these trees is used for making paper pulp, matches, furniture, packing cases and crates. The principal timber-exporting regions of these forest belts are, the U.S.S.R., Sweden, Finland, Eastern Canada, British Columbia and Western U.S.A.

Tropical forests (hardwoods)

The greatest extent of hardwood forests is between the latitudes of 20°S and 20°N, in the Amazon Lowlands of South America, the Congo Basin of Africa, and the monsoon lands of South-East Asia. The most common trees of these forests are *teak* (mainly in South-East Asia) *mahogany* (mainly in Central America) and *ebony* (mainly in Africa and South-East Asia). The timber of these trees is used for making plywood, furniture and for various types of constructional work.

Forests of mixed temperate hardwoods and softwoods occur in latitudes between 25°N and 45°N and 25°S and 45°S. The largest area of this type of forest is in the Appalachian Mountain area of Eastern U.S.A.

Wood pulp and paper

About 90% of the world's paper is made from wood pulp, and most of this is made from softwoods. Newsprint is made from wood pulp produced by grinding the logs with water to a sticky fibrous mass. Better quality paper is produced by using chemicals. The wood is turned into a fine quality pulp by using acids. It is then bleached, rolled and heated. Clay is often added to give a good surface to the paper made from this type of pulp. A good supply of water and cheap electricity are required in the manufacture of wood pulp which means that pulp mills are usually located near to a plentiful supply of water and to a source of electric power, usually hydro-electric power. Integrated pulp and paper mills are located in such regions as the St. Lawrence Valley in Eastern Canada.

International trade in timber

Since the early 20th Century, the demand for timber, for manufacture into paper pulp to feed the expanding paper industry, has increased steadily. Softwoods provide most of the timber for making paper pulp, and most of this timber comes from the forest belts of the U.S.S.R., and Finland, Norway and Sweden (all in Europe), and from Quebec, Ontario and British Columbia (all in Canada) and from parts of the U.S.A., especially California and the south-eastern states. Most of these regions manufacture large quantities of paper pulp and paper, in addition to timber, which is mainly consumed by the industrial countries of the world. Japan relies heavily on the U.S.S.R. and Canada for obtaining its paper pulp requirements. In the Southern Hemisphere, New Zealand and Australia are becoming important producers and exporters of timber, paper pulp and paper.

The amount and value of tropical hardwood timber in international trade is much smaller than that of softwood timber. Teak and mahogany are two of

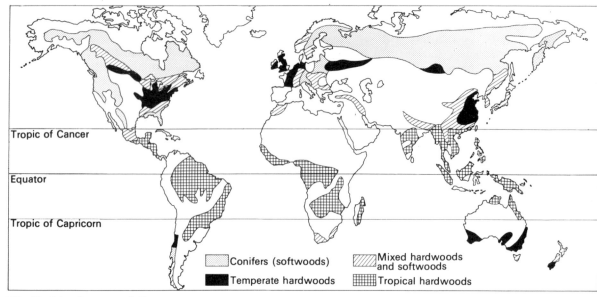

World Distribution of Forests

Conifers (softwoods) Mixed hardwoods and softwoods

Temperate hardwoods Tropical hardwoods

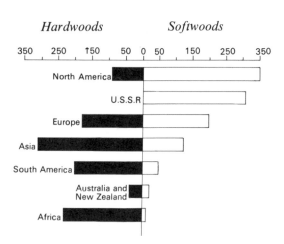

Annual World Production of Timber (average for the last five years in millions of cubic metres). Approximate annual production: Softwoods – 1100 million cubic metres; Hardwoods – 1000 million cubic metres.

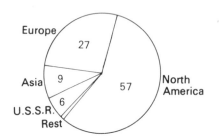

Annual World Production of Paper Pulp (average for the last five years on a percentage basis).

the most important tropical hardwoods, and Burma and Thailand are the main producers and exporters of teak. The Ivory Coast of West Africa is the main producer of mahogany.

Inside a printing factory in Great Britain. This machine is used for printing newspapers. The size of this machine can be appreciated by comparing its height with the height of the operator.

266

Finland, Sweden and Norway are major producers of timber.

(a) Explain how the conditions in these countries are so suitable for forest growth.

(b) Name *two* common types of trees of the forests in these countries.

(c) Briefly describe how the trees are felled and how they are transported to the pulp factories.

(d) Name *one* of the main uses of the timber of these forests.

Trees are an important part of the natural vegetation of several parts of the world.

(a) Make a simple division of the world's forests on a climatic basis.

(b) Name *one* area where each type of forest, named in (a), is important.

(c) State the main advantages and the main disadvantages in the commercial exploitation of tropical forests.

(d) Briefly state the extent to which timber enters into international trade.

bjective Exercises

All of the following trees are deciduous **except**

A teak
B acacia
C beech
D larch
E spruce

              A B C D E

Which of the following produces aromatic (sweet smelling) wood?

A camphor
B mahogany
C teak
D bamboo
E pine

              A B C D E

Which type of forest contains trees which are used for making paper?

A coniferous
B deciduous
C equatorial
D monsoon
E Mediterranean

              A B C D E

Why is it difficult to exploit the trees of an equatorial forest?

A the trees are surrounded by thick thorny undergrowth

B the trees are mainly hardwoods
C the stands are always pure
D the stands are very mixed

              A B C D

5 In which one of the following continents/countries is the production of hardwood timber more important than the production of softwood timber?

A U.S.S.R.
B North America
C Africa
D Norway

              A B C D

6 Lumbering in the coniferous forests of Canada is made easy by all of the following factors **except**

A the forests have very little undergrowth
B large numbers of trees of the same type grow close together
C the trees are softwoods and are easy to fell
D transporting the felled trees to the saw mills is cheap

              A B C D

7 Which one of the following countries is the largest exporter of softwood timber and wood pulp?

A the U.K.
B Canada
C Norway
D the U.S.A.

              A B C D

8 Wood pulp mills are usually located near to

A a supply of cheap electric power
B paper mills
C a plentiful supply of water
D a source of cheap electric power and a plentiful supply of water

              A B C D

27   Produce from the Sea

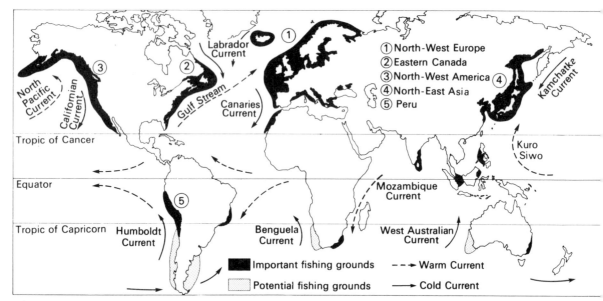

The Main Fishing Grounds of the World

The total weight of fish caught each year in the world is equal to about 70% of the world's annual production of meat. Fishing is a major activity in some countries and sometimes it forms a very important part of a country's economy, e.g. as in Finland. Whilst most of the fish caught in the world comes from the oceans and seas, fresh water fishing is of considerable significance in some regions, for example, the Tonle Sap, in the Khmer Republic, yields an enormous quantity of fish each year and this forms an important part of the diet of the people in that part of Khmer.

Conditions for fish breeding

Fish live on very small plant and animal organisms, collectively called *plankton*. Oceans and seas, which are rich in plankton, usually have an abundance of fish. Plankton grows best in seas and oceans where:

1 The water is shallow, (as in the seas on continental shelves), thus allowing plenty of light to penetrate to the sea bed.
2 The water contains a variety of mineral salts, (as in seas into which large rivers drain).
3 Cold and warm currents meet.
4 Cold water upwells at the surface.

Some or all of these conditions occur in the waters around North-West Europe, the waters along the North-East and North-West coasts of North America, the waters along the coast of Peru, and the waters off the coast of China and arou Japan. These regions form the major fishing groun of the world because:

1 Their waters are rich in plankton.
2 Their indented coastlines provide natural ha bours, many of which have become major fishin ports.
3 The lands bordering some of the coastal wate are not able to produce sufficient food eith because the soils are poor, or the land is to mountainous, or the land has too large a popul tion, all of which have resulted in the peop turning to the sea for a livelihood.
4 Many of the regions near to these waters hav large populations which provide good market

Types of fishing

Sea fishing is of two types – inshore fishing in th waters up to 70 kilometres (about 44 miles) fro the coast, and off-shore fishing in the waters beyon this limit. Small vessels are used for inshore fishin and these usually stay at sea for one or two day only. Much larger vessels, usually equipped wit refrigerated holds, carry out off-shore fishing, an these vessels may remain at sea for several week at a time. Various methods are used for catchin fish but the most common methods used by com mercial fish vessels are by *nets,* by *lines,* and b *traps.*

Fishing by nets

Fish which breed near to the surface are called *pelagic* fish – examples are herring, mackeral, pilchard and tuna. These fish are best caught either by drift nets or by seine nets.

By drift nets These nets hang vertically in the water because they are weighted along the bottom edge, and supported along the top edge by floats. Fish are caught by their gills becoming entangled in the mesh of the nets.

By seine nets These nets are similar to drift nets but instead of being left hanging in the water, the nets are pulled by their ends to surround a shoal of fish. The nets are sometimes stretched between two fishing boats, or, if used near to the coast, they are pulled by fishermen with perhaps one end being attached to, and pulled by, a small boat.

Drift Net

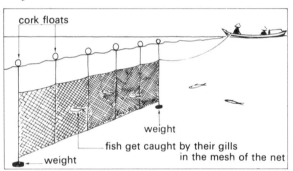

Fish which feed on the sea bed of continental seas are called *demersal* fish – examples are sole, cod, haddock and halibut (all inhabitants of temperate seas), and snapper and garoup (inhabitants of some tropical seas). These fish are usually caught by trawl nets. This type of net has a conical shape which is open at the base. It is weighted and is kept open by a rigid structure. The fish are caught by dragging the net along the sea bed by fishing vessels called *trawlers*.

Trawl Net

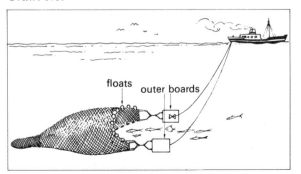

Fishing by lines

Several types of lines are used. Some lines may be as long as 1 or 2 kilometres and these lines carry hundreds of baited hooks. Such lines are trailed by fishing vessels. Other lines are much shorter and are operated individually by hand. Line fishing is still important in Japan, and in Newfoundland, where it is used mainly for catching cod.

Fishing by traps

Shell fish such as lobster, crab and crayfish are usually caught in wicker baskets containing bait. The baskets are lowered into shallow coastal waters and left for one or two days, before they are hauled up. Salmon which are returning to breed in the rivers of British Columbia (W. Canada) are sometimes caught by traps which are set in the mouths of rivers. A more elaborate type of fish trap is used in the coastal waters around Malaysia and parts of Indonesia. This consists of a line of wooden stakes, driven into the sea bed, at the end of which is the trap. This is an area surrounded by close-set poles with a narrow entrance where the line of stakes joins it. A platform is usually built over the trap to enable the fishermen to haul up the catch.

Fish Trap or Kelong in Malaysia

Whaling

The whale is a marine mammal which lives in the colder waters of high latitudes, and because of its great size it cannot be caught by nets. The most common method used is the power-fired harpoon which often carries an explosive charge. The hunting and catching of whales takes place mainly in the Antarctic Ocean by large vessels which not only catch the whales but which also cut them up, process the blubber (fat) into oil, and can the whale meat. The oil is subsequently used in the manufacture of margarine, soap and various lubricants. Most of the ships engaged in whaling belong to the U.S.S.R., Japan, Norway, and the U.K.

Important fishing regions

The world's four major fishing regions are all in the Northern Hemisphere. The only important fishing region in the Southern Hemisphere lies off the coast of Peru. This is rapidly becoming a major fishing region.

North-West Europe

The North-East Atlantic is one of the major fishing regions of the world. It contains the fishing grounds of the Barents Sea, Iceland, the North Sea, and the Bay of Biscay. These seas cover the continental shelf off Europe and their waters rarely exceed a depth of 185 metres (about 600 feet), and because of this, and the large number of rivers draining into them, the seas are rich in plankton. Fishing in this region is undertaken by modern vessels using modern equipment which operate from the large fishing parts of Bergen (Norway), Boulogne (France) and Hull (the U.K.). Trawlers from these and other ports, often travel as far as 2400 kilometres (about 1500 miles) and they stay at sea for up to three weeks.

Norway is the leading fishing country of Europe. The harsh climate makes farming difficult, and this, plus the absence of extensive mineral and forest resources, has caused Norway to turn to the sea. *Cod* and *herring* fishing dominate Norway's fishing industry, although whaling in the Antarctic is of great importance. Although the fishing industry of the U.K. is not as important to the country's economy as it once was, it is highly organised. The industry is centred in the ports of Grimsby, Hull and Aberdeen, and vessels from these ports fish in the North Sea, the Barents Sea and in the waters off Iceland and Greenland. *Cod, haddock* and *herring* are the main catch.

Eastern Canada

The fishing grounds of this region extend for about 1600 kilometres (almost 1000 miles) from Cape Cod to Newfoundland, off the east coast of Canada. There are three reasons why this region became one of the world's major fishing grounds. These are:
1 It has an indented coast with good natural harbours.
2 It has poor soils and a harsh climate which caused people to turn to the sea for a livelihood.
3 It is located near to the meeting of the North Atlantic Drift and the cold Labrador Current which results in the rapid growth of plankton, and waters rich in fish.

Fishing is both inshore and off-shore, with the main fish caught being *cod, haddock* and *sardines*. The

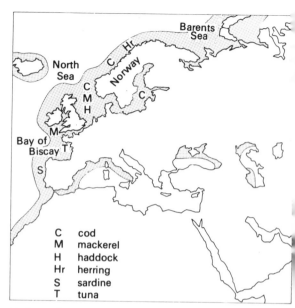

C	cod
M	mackerel
H	haddock
Hr	herring
S	sardine
T	tuna

Fishing Grounds of Western Europe

Fishing by seine net off the coast of British Columbia (Canada)

inshore fishing is undertaken by small boats, called dories, whilst the off-shore fishing, which centres on Halifax, St John's and Lunenburg, is undertaken by large vessels which stay at sea for up to three weeks.

North-West America

The fishing grounds extend from California to

270

Alaska. Although *halibut, cod* and *herring* are extensively caught, perhaps the most important fish is *salmon*. Salmon fishing and salmon canning extend from the Bering Strait to Oregon. The main characteristics of salmon fishing are:

Young salmon hatch from eggs laid in mountain streams and lakes.

When they are about one year old the young salmon swim downstream to the sea.

When they are about four years old the salmon return to the rivers to lay their eggs (this is when they are caught).

Salmon are caught by nets and traps across the rivers, and by trawls and seine nets in coastal waters.

Dams across rivers prevent salmon from swimming up-river unless special 'ladders' are built to by-pass the dams.

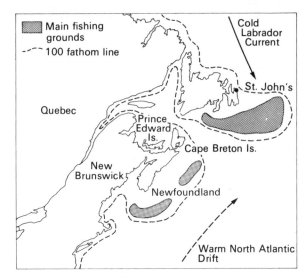

Fishing Grounds of Eastern Canada

plankton and fish, and the indented coastline of Japan has many natural harbours.

The cold waters of the northern area yield *cod, halibut, herring* and *salmon*, while the warmer waters of the south yield *sardine, tuna* and *mackeral*. Shell fish such as *prawn, lobster, crab* and *cuttlefish* are

Salmon jumping 'ladders' as they swim up-river in British Columbia.

North-East Asia

The fishing industry of this region is dominated by Japan which is one of the three most important fishing countries in the world. The per capita consumption of fish in Japan is the highest in the world. This is partly the result of the Japanese turning their attention to the sea because of the great pressure of population on agricultural resources, and partly because of the abundance of fish in the waters around Japan. Natural conditions in these waters are very similar to those in the waters off eastern Canada. The meeting of the warm Kuro Siwo and cold Oya Siwo currents in the broad epi-continental shelf seas gives rise to waters rich in

Fishing Grounds of Japan

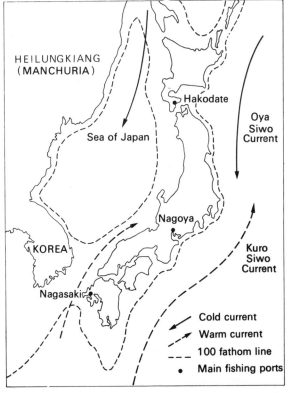

caught in the shallow coastal waters in large quantities. As in the waters off Eastern Canada, small vessels look after inshore fishing, while larger vessels which stay away for several weeks, look after off-shore fishing. Japan also has an important fresh water fishing industry which centres on fish farms.

China and the U.S.S.R. also operate large fishing industries in this region. The principal fish caught in the waters off China are *garoup*, *carp* and *mullet*. Like Japan, China operates a fresh water fishing industry based on fish farms, and as in Japan, fish is a major constituent of the diet of the Chinese.

The fishing grounds of Peru
Only since the end of World War II has the fishing industry of Peru assumed world importance. In fact, it is now one of the most important in the world. The Humboldt, or Peruvian Current, which flows along the coast of Peru, is a cold current and because this operates in a tropical region, the coastal waters are rich in plankton and also in fish. The most common fish caught is *anchovy* and enormous quantities of this fish, in processed form, are exported annually. Chimbote is the main fishing port.

Problems facing the fishing industry
Fishing is still a *robber economy* in many parts of the world, that is, the catching of fish is not balanced by replenishing the fish stock. With the introduction of faster and more powerful fishing vessels, and modern fishing equipment to meet the ever-increasing demand for more food for the world's growing population, over-fishing has become a serious problem. Fish are being caught faster than they are being replaced by reproduction, and unless effective control, through international agreement, is practised, some fishing grounds will become seriously depleted in fish. Some effective control must set a limit to the quantity of fish any nation can catch in a given period, and it must protect breeding, and enforce vessels to use nets whose mesh prevents young fish from being caught.

Pollution of the sea
The coastal waters of many industrial countries are becoming heavily polluted by industrial waste which is discharged into the rivers, or directly into the sea. Some industrial waste contains poisonous chemicals which become stored in the body. The cumulative storage effect of chemicals such as compounds of mercury can cause serious illness and eventually death. This has happened in parts of Japan. Oil pollution, particularly oil slicks on the surface of water, can deprive fish of oxygen and cause large numbers of them to die.

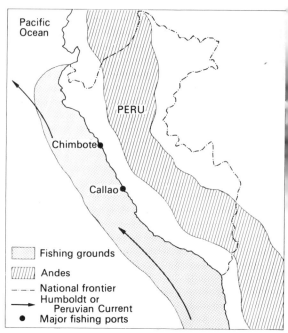

Fishing Grounds of Peru

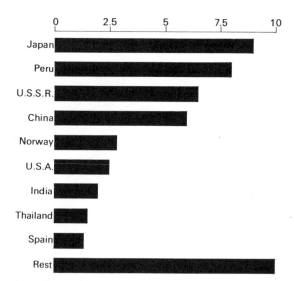

Annual World Catch of Fish (average for the last five years in millions of metric tons). Annual catch approximately 51 000 000 metric tons.

Possible solutions to the problem
1 The prevention of over-fishing by international agreement and control.
2 The removal of all poisonous and harmful chemicals from industrial waste before it is discharged into the rivers and seas.
3 The introduction of fish farming to breed fish to replenish the fish caught. This is already being done by some countries.

Fish and fish product form an important part of the diet of a large part of the world's population.

(a) Name *three* important fishing grounds in the Northern Hemisphere.

(b) Draw a large sketch-map of *one* of the grounds named, and on it, mark and name:

   (i) any important ocean currents.

   (ii) one important fishing port.

   (iii) two lines of longitude.

(c) Briefly describe any one method by which sea fish are caught on a commercial scale.

Write a short essay on the importance of the sea to man.

bjective Exercises

Over-fishing means catching

A more large fish than small fish

B more fish than the number replaced by natural means

C fish larger than standard size

D more fish than in previous years

E fish caught outside the international fishing limits

           A B C D E

Pollution affects fish by causing them to

A die

B store posions in their body fat

C taste of pollutants

D become a health hazard

E accumulate chemicals such as mercury and chromium in their respiratory channels

           A B C D E

Improvements in efficiency have been brought about by the use of all of the following **except**

A nylon nets

B radio weather forecasting

C refrigeration

D motorised boats

E over-fishing

           A B C D E

Which one of the following conditions is more closely related to the location of the major fishing grounds than any of the others?

A indented coasts

B large coastal populations

C meeting of cold and warm currents

D shallow water

E temperate latitude location

           A B C D E

5 An important disadvantage to the fishing industries of Europe is caused by

A a wide continental shelf

B an indented coastline

C lack of an adequate labour force

D a concentration of urban populations in the coastal areas

           A B C D

6 The importance of Japan's fishing industry is partly explained by all of the following statements **except**

A Japan has a large number of sheltered harbour

B Japan is not able to grow all the food its population requires.

C Warm and cool currents meet in the waters around Japan.

D There are not many manufacturing industries to provide alternate employment.

           A B C D

7 All of the following countries are major exporters of fish **except**

A Iceland

B Japan

C Norway

D India

           A B C D

8 Which one pair of the following factors *most* helps to account for the large number of fish in the waters around North-Western Europe?

A the meeting of warm *and* cool currents near the shallow epi-continental waters

B the high latitude *and* the abundance of plankton

C low salinity of the water *and* low water temperature

D strong westerly winds *and* often turbulent seas

           A B C D

9 All of the following may favour the development of a fishing industry **except**

A the meeting of cold and warm currents

B an indented coastline providing safe harbours

C a fertile coastal plain providing a plentiful food supply

D a shallow continental shelf encouraging plankton growth

           A B C D

28   Sources of Power

Wind, water, wood and charcoal, were once the most important sources of power and they played an important part in the early development of agricultural techniques such as irrigation and drainage, and later, in some of the techniques used in the early development of industrial activities. Today, the main sources of the world's power are water, coal (including lignite), petroleum, natural gas, and certain minerals such as uranium (from which nuclear power is obtained). The most important of these sources, in respect of the amount of power derived from them, are coal, petroleum, natural gas and water, from which is obtained over 95% of the world's total power production. Of these four sources, petroleum has become increasingly important over the last ten years. Over the same period, hydro-electric power (derived from water), and nuclear power, have both increased, though not at the same rate as power developed from petroleum. However, the deposits of coal, petroleum, and natural gas, will eventually be exhausted because of their continual use. This will not happen to hydro-electric power and nuclear power, which will probably become two of the most important sources of power in the 21st. Century.

POWER FROM WATER
Most of the industries of the early 19th Century, in Europe and North America, were driven by water power, and they were located near to either large, or fast-flowing rivers. With the invention of the steam engine, which used coal for turning water into steam, there was a migration of industry to the coalfields, but during the early 20th. Century when water power was used for driving generators to produce electricity, it no longer become necessary for all industry to be confined to the coalfields. Electricity made from water power is called *hydro electricity,* while electricity made from water which is turned into steam by burning coal, petroleum or natural gas is called *thermal electricity.* Electricity made in nuclear reactors is called *geothermal electricity.*

The production of hydro-electric power
Moving water possesses energy which can, under certain conditions, be converted into electrical energy. The larger the volume of water and the faster it flows from a higher to a lower level, the greater the amount of power it contains. The most favourable conditions for the production of hydro-electric power are:

1 A regular supply of water derived from (i) heavy annual rainfall evenly distributed throughout the year, or, (ii) meltwaters from mountain glaciers, or (iii) reservoirs (man-made lakes), or natural lakes.
2 A good vertical descent of water, which is called the *head* derived from (i) a waterfall, or (ii) a dam. However, if the volume of water is very large, only a small head of water is needed.
3 A large market for the hydro-electric power. This should be within 500 kilometres (about 315 miles) of the power station because of the difficulty of transmitting electric power across distances greater than this.

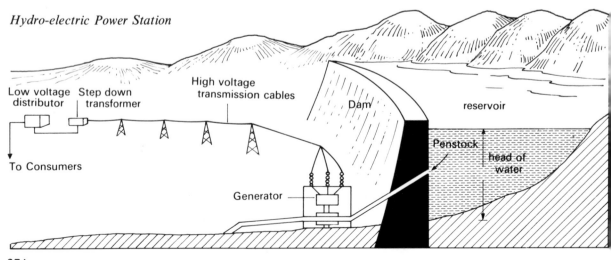

Hydro-electric Power Station

Low voltage distributor   Step down transformer   High voltage transmission cables   Dam   reservoir   Penstock   head of water   To Consumers   Generator

The Grand Rapids Power Project on the River Nelson in Manitoba (U.S.A.). The electric power grid is in the foreground.

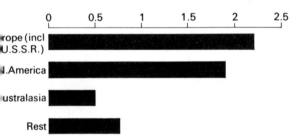

The Approximate Annual Production of Electric Energy (hydro-electric, thermal and nuclear) in millions of kilowatts

The generation of hydro-electric power

The water passes through pipes, called *penstocks*, from the lake behind the dam, under the force of gravity. The penstocks lead the water to the blades on the wheels of the turbine generators which causes them to rotate.

The mechanical energy of the rotating wheels is converted into electrical energy by the generator dynamos, and this is transmitted through heavy cables to the areas where it is required.

The construction of dams and hydro-electric power stations, requires heavy capital outlay, some, or, all, of which is often provided by government bodies.

Because of the high cost of construction, many hydro-electric schemes have multi-purpose functions such as, providing irrigation water, flood control, and improved water communications, in addition to providing electric power.

Some important hydro-electric power schemes

Some of the most important H.E.P. schemes are located in mountainous countries, such as Norway, Switzerland, Western Europe, Japan, and in the mountainous regions of the U.S.A., Canada, and Australia, all of which have a regular supply of water and numerous waterfalls. The construction of H.E.P. stations on the lowlands involves high costs and it is for this reason that mountainous regions are favoured for the production of hydro-electric power.

The Kitimat Scheme of British Columbia

The Kitimat Scheme is located in the coastal ranges of British Columbia, about 600 kilometres, (375 miles) north of Vancouver. The River Nechako, which rises in the coastal ranges and which once flowed eastwards, was chosen for providing the water for the power station. A dam was built across this river so that a large lake developed behind the dam. The water of the lake was then led westwards, via tunnels which were cut through the coastal ranges, to the power station at Kemano.

275

Most of the electricity generated is transmitted via cables to Kitimat, which is located at the head of a deep sheltered fiord, about 75 kilometres (about 47 miles) north of Kemano, where it is used in one of the world's most important aluminium smelting plants (see Chapter 29).

The Snowy Mountain Scheme in Australia

This scheme is similar to that of Kitimat in that the waters of a river are dammed, and made to flow in the opposite direction. The scheme is located in the Snowy Mountains of South-East Australia and it involves (i) diverting the waters of the south-flowing Snowy River into the west-flowing Murray River, (ii) diverting the waters of the Snowy River tributaries and the Upper Murrumbidgee River into the Tumut River, which is a tributary of the Murrumbidgee. Several large dams have been built to hold back the waters of these rivers to create reservoirs, the waters of which are led away, via tunnels, cut through the Great Dividing Range. Several power stations are located on the western side of the Dividing Range, and these produce electricity for supply to New South Wales and Victoria. In addition to this, the water passing through the stations is collected in new reservoirs constructed on the Murrumbidgee for the purpose of providing irrigation water for nearly 500 000 hectares of land in the Murrumbidgee and Murray basins.

Other hydro-electric power schemes

North America

(a) Grand Coulee Dam is located on the Columbia River, and much of the electricity generated is used to power irrigation works.

The Grand Coulee Irrigation Project

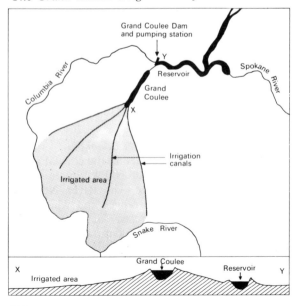

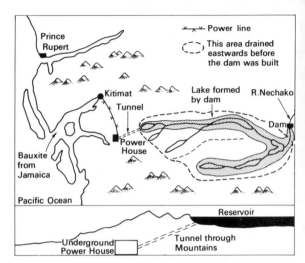

The Kitimat Project. The power station is 776 metres (about 2580 feet) below the level of the lake.

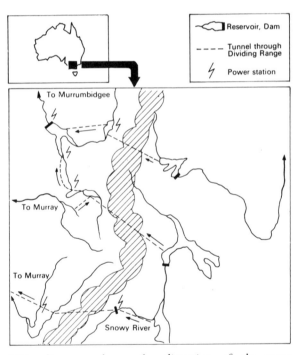

This diagram shows the diversion of the upper Snowy River system and the upper Murrumbidgee into the Tumut River.

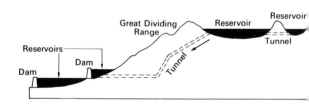

(b) Hoover (Boulder) Dam on the Colorado River generates electricity for Los Angeles.

(c) The St. Lawrence Seaway is a dual purpose scheme which allows ocean vessels to reach the Great Lakes, and which develops hydro-electric power.

Europe and the U.S.S.R.

Most of the hydro-electric power schemes of Europe are smaller than those of North America, but those of the U.S.S.R. are very large. Most of Europe's hydro-electric power is produced in Norway, Switzerland, Sweden and Italy. The construction of hydro-electric power schemes in the U.S.S.R. increased enormously after World War II. Dams were built on such rivers as the Yenisey, Angara and Vakhsh, and the hydro-electric power stations of these have enormous outputs, e.g. that at Bratsk, on the Angara, produces as much electricity as did the whole of the U.S.S.R. in 1950. The station at Krasnoyarsk, on the Yenisey, was completed in 1971 and it produces 5 million kw as compared with 2.1 million kw for Grand Coulee hydro-electric power station (the largest in the U.S.A.). Although more and more hydro-electric power stations are being built, the older ones of western Russia are still very important. These include the stations of Kuybyshev, Dneprodzerhinsk and Volgograd (Stalingrad). The main hydro-electric power-producing areas in the U.S.S.R. are in the valleys of the:

(a) Dnieper-Volga Rivers

(b) Yenisey River

(c) Angara River

(d) Vakhsh River

Japan

The development of hydro-electric power, in Japan, is very similar to that of Norway and Sweden. Japan is one of the world's greatest producers of hydro-electricity.

Malaysia

The hydro-electric power station at the Chenderoh Dam, on the Sungai Perak, supplies power to the surrounding tin and rubber industries. Another station at Abu Bakar Dam, in the Cameron Highlands, supplies power to central Peninsular Malaysia.

Africa

Important hydro-electric power plants are at Kariba, on the Zambesi River, and at Owens Falls, on the Nile. Central Africa, because of its equatorial rainfall, large rivers, and steep slopes, has a great hydro-electric potential.

New Zealand

The super-heated waters and steam of the geysers, around Rotorua, in North Island of New Zealand, are used for generating electricity.

The Rotorua geysers in New Zealand

The Main Hydro-electric Power Stations of the U.S.S.R.

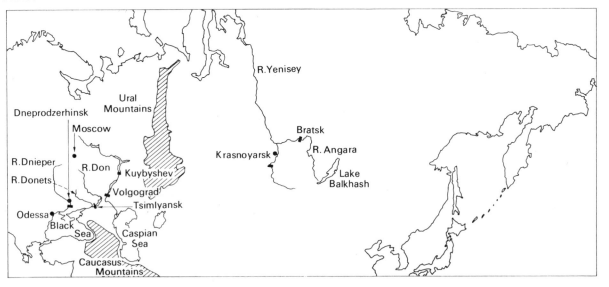

POWER FROM COAL

Coal is formed from the vegetable matter of swamp forests, not unlike those of the Ganges and Mississippi deltas. It occurs in layers, called *seams,* which often alternate with layers of clay and other rocks. The first stage in the formation of coal is *peat,* which is formed of partially decomposed vegetable matter compressed by over-lying layers of silt. As the deposition of vegetable matter and silt continues, the peat is compressed into *lignite,* which later may be compressed into coal. Most of the world's coal deposits were formed in the Carboniferous Period which occurred about 300 million years ago.

Types of coal

There are three basic types of coal:

1 *Lignite* or brown coal which has a carbon content of less than 4%, and which gives out only moderate amounts of heat. It is used mainly for the production of thermal electricity.
2 *Bituminous Coal* which has a carbon content of between 45% and 80%. Because of this, it gives off more heat than lignite. Some bituminous coal has a high gas content. This is called *gas coal* and it is used for making gas. Another type of bituminous coal is used for making coke which is used in blast furnaces. This type is called *coking coal.*
3 *Anthracite* is a very hard coal which has at least a 90% carbon content. This enables it to burn with great heat, and little smoke.

The mining of coal

The methods used for mining coal depend upon the distance of the coal seams below the surface, and whether the seams are horizontal or tilted. Three basic methods are used.

1 *Shaft mining* This method is used when the coal seams lie several hundreds of metres below the surface, and it involves the sinking of vertical shafts down to the level of the seams. Horizontal tunnels, called *galleries,* are formed from the shaft as the coal seams are dug out. This is the most common type of mining.
2 *Opencast mining* When coal seams lie near to the surface, say, at depths not exceeding 50 metres, the rock layers lying on top of the coal seams are removed, and the coal is then quarried by mechanical excavators. This type of mining is used in Fushun in Northern China, and in the provinces of Shansi and Shensi.
3 *Adit* or *drift mining* This method is used to extract coal from horizontal, or gently sloping seams that outcrop along the sides of valleys. It is used in the valleys of the Appalachian Coalfield, in the U.S.A., and in North-East England.

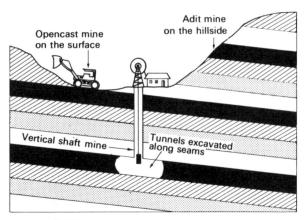

Types of Mines

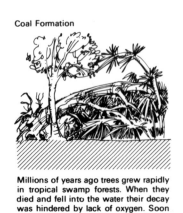

Coal Formation

Millions of years ago trees grew rapidly in tropical swamp forests. When they died and fell into the water their decay was hindered by lack of oxygen. Soon they were covered with mud, and more trees died until a thick layer was formed.

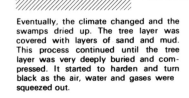

Eventually, the climate changed and the swamps dried up. The tree layer was covered with layers of sand and mud. This process continued until the tree layer was very deeply buried and compressed. It started to harden and turn black as the air, water and gases were squeezed out.

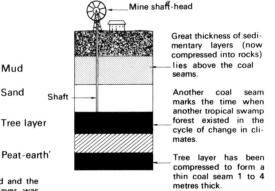

Great thickness of sedimentary layers (now compressed into rocks) lies above the coal seams.

Another coal seam marks the time when another tropical swamp forest existed in the cycle of change in climates.

Tree layer has been compressed to form a thin coal seam 1 to 4 metres thick.

Open cast coal mining in the U.S.A. This photograph shows the almost solid layers of coal which are being mined at different levels. Such mining results in the formation of huge open 'basins'.

Mining by adit as in the Northern Appalachian Coalfield (U.S.A.)

Mining by shaft

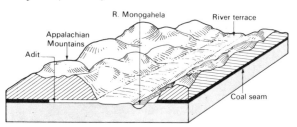

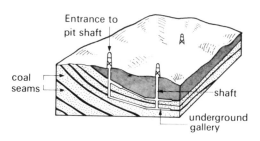

Important coal-producing regions
Coalfields of the U.S.A.

The main coal mining region of the U.S.A. lies in the eastern part of the country, and it is known as the Appalachian Coalfield. There are four parts to this, as shown in the map.

1 *The Pennsylvania Coalfield* This occupies the northern part and it centres on Pittsburg. The coal seams are thick and almost horizontal, and

they extend over an area of almost 5000 squares kilometres (about 20 000 square miles). Because of this, mining is easy and cheap, and this is especially the case in the numerous valleys where the coal seams are exposed on the valley sides. The coal of this field makes good coke and much of it is used in the iron and steel industries of this part of the U.S.A.

2 *The North-East Appalachian Coalfield* The coal

279

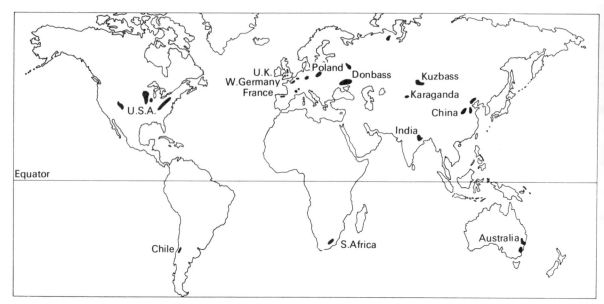

The Main Coalfields of the World

seams of this field are heavily faulted and mining is more difficult. However, the coal is an excellent anthracite.

3 *The West Virginia Coalfield* This lies in the central part of the Appalachian Coalfield. Its coal is of very high quality, and in recent years it has produced more coal than any other part of the Appalachian Coalfield. Its coal is used in the industrial centres of the U.S.A., in Canada, and in some South American countries.

4 *The Alabama Coalfield* This field centres on Birmingham, which is one of the largest iron and steel centres of the southern states. The coal

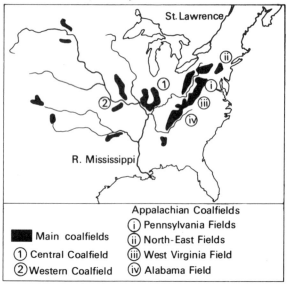

The Main Coalfields of the U.S.A.

Appalachian Coalfields

■ Main coalfields	ⓘ Pennsylvania Fields
① Central Coalfield	ⓘⓘ North-East Fields
② Western Coalfield	ⓘⓘⓘ West Virginia Field
	ⓘⓥ Alabama Field

deposits lie near to extensive iron ore and limestone deposits.

Other important coalfields

These are the Central and Western Interior fields but the coal of these is inferior to that of the Appalachian Coalfield. Coal from the Central Coalfield is mainly used for developing power for Chicago.

Coalfields of the U.S.S.R.

The U.S.S.R. is second to the U.S.A. as a producer of coal, and most of its coal comes from three coalfields.

1 *The Donbass Coalfield* This is located between the River Donetz and the Sea of Azov, in the area known as the western steppe. It produces about one third of the U.S.S.R.'s annual coal production, and most of it is high quality bituminous coal and anthracite. The former is of good coking quality, and is used in the iron and steel industries at Krivoy Rog, which is about 400 kilometres (about 250 miles) to the west.

2 *The Interior Coalfields* These are located in Siberia. The Kuznetsk, or Kuzbass Coalfield, which centres on Novo-Kuznetsk, is the most important of the interior coalfields. It produces over 100 million metric tons of coal each year.

Other coalfields of the U.S.S.R.

The region around Tula produces lignite which is used mainly for generating thermal electricity, as does the Vorkuta coalfield, which is located in the tundra belt of the U.S.S.R.

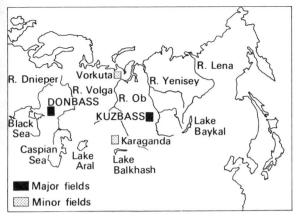

Coalfields of the U.S.S.R.

Coalfields of the European Economic Community (E.E.C.)

The E.E.C. is comprised of the U.K., Holland, Luxembourg, Belgium, West Germany and Italy. There are three main coalfield regions in the E.E.C.

1 *Coalfields of the U.K.* There are several coalfields in the U.K. but of these the most important are in Yorkshire, Nottinghamshire and Derbyshire, in Northumberland and Durham, and in South Wales. The latter is an important producer of high quality coal much of which is made into coke for feeding the iron and steel works at Neath, Port Talbot, Newport, Cardiff and Ebbw Vale. The rest of the coal is used for generating thermal electricity. Coal mining in South Wales is mainly by deep shafts. The heavily faulted coal seams make mining both difficult and expensive to undertake.

Most of the coal mined in the Yorkshire, Nottinghamshire and Derbyshire Coalfield, is used for providing thermal electricity for the northern textile industries, especially of Yorkshire, in the centres of Bradford, Leeds, Halifax and Huddersfield, and for providing fuel, in the form of coke, for the iron and steel industries in towns such as Sheffield and Doncaster.

The coal of the Northumberland and Durham Coalfield makes into good coke which is used in the iron and steel works of Middlesborough and other centres, which produce steel plates and girders for the shipbuilding industry of Tyneside, Wearside and Teeside, all of which are in the north-eastern part of the U.K.

2 *Coalfields of West Germany* West Germany is a major world producer of coal with most of its production coming from the Ruhr Coalfield. The coal seams of this coalfield dip from south to north. Most of the important mining is now in the Lippe Valley area where the coal seams are at considerable depths. Mining is made difficult

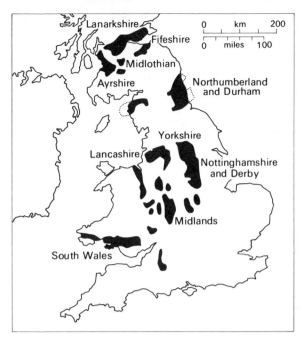

The Main Coalfields of Britain

by folding and faulting. Several types of coal are mined, but the most important is a high quality bituminous coal used for making coke, which is used in the integrated iron steel works at Essen, Dortmund and Duisburg. Much of the iron ore used in these works is imported from Lorraine in France, and from Sweden. The only other important coalfield in West Germany is the Saar coalfield.

3 *Other coalfields* The France-Belgium coalfield is still an important one although its output has

The Main Coalfields of West Germany and the Rhine Basin

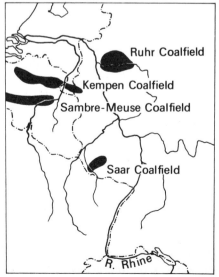

281

declined in recent years. The coal seams are heavily folded and faulted which often makes mining of the coal difficult and dangerous. The main mining centres are Mons, Charleroi and Namur. These towns also have important iron and steel works.

The coalfields of Eastern Europe
East Germany has many extensive deposits of lignite which make it one of the world's major producers of this type of coal. Most of the country's production comes from the Saxony Coalfield, and most of the lignite is used mainly for generating thermal electricity. Another important coalfield is the Upper Silesian Coalfield, in Southern Poland. The coal from this field is used in the iron and steel centres of Katowice and Chorzow.

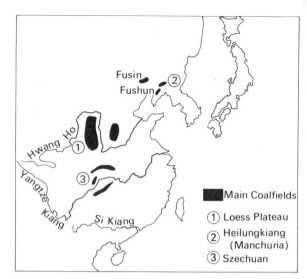

The Main Coalfield of China

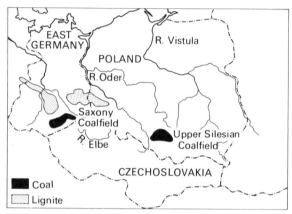

The Main Coalfields of Eastern Europe

The coalfields of Asia
1. *China* There are huge deposits of coal under the loess plateau of North-West China, and in parts of Heilungkiang (Manchuria). Other large known coal deposits are in Sinkiang and Szechuan. China produces about 400 million metric tons of coal each year, which is about the same as West Germany's production. About one third of this output comes from the coalfields of Heilungkiang, especially from the centres of Fushun and Fusin. The coal seams in these coalfields are thick and sufficiently near to the surface to permit opencast mining.
2. *Japan* The coal deposits of Japan are of poor quality and of limited extent. Most of the coal is used for making coke, and the country's chief coalfields are in Hokkaido, at Ishikari; in Kyushu, at Chikuho, and in Honshu, at Joban.
3. *India* Most of India's coal lies in the Damodar Valley region in North-East India. The main coalfield centres on Jharia. The coal is used in

The Main Coalfields of Japan

iron and steel centres such as Rourkela, Durgapur Jamshedpur. Most of the iron ore for these centres is mined near Singhbhum, in Bihar.

Coalfields of Australia
Australia has three coalfields, the most important of which is in New South Wales, extending from Newcastle to Port Kembla. The coal from this field makes excellent coke and this is used in the blast furnaces at Newcastle and Port Kembla. The coalfield in Queensland produces good quality coal. The only other coalfield of any importance is in Victoria. This field produces lignite which is used for generating thermal electricity. It lies near to the surface and mining is by opencast methods.

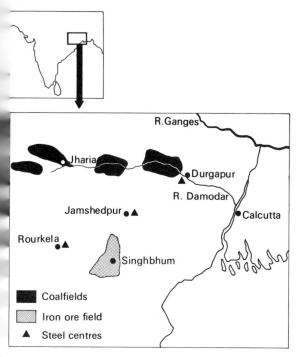

The Main Coalfields of India

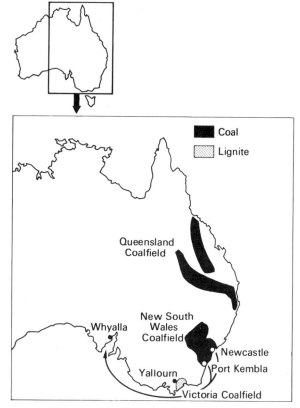

The Main Coalfields of Australia

How coal is used

The principal uses of coal have changed considerably since the end of World War I. During the 19th, and early 20th Centuries, coal was the main fuel used for developing steam power for driving factory machines, locomotives and ships. Coal was also used extensively for generating heat for buildings. Today, very few factory machines, or locomotives are driven by steam, using coal as a fuel. There are three main ways in which coal is now used: (i) as a source of energy; (ii) for making coke for extracting metals from their metallic ores; and (iii) as a raw material from which other chemicals are made.

Coal as a source of energy

Most of the coal produced is used to provide energy either by burning it directly to produce heat and

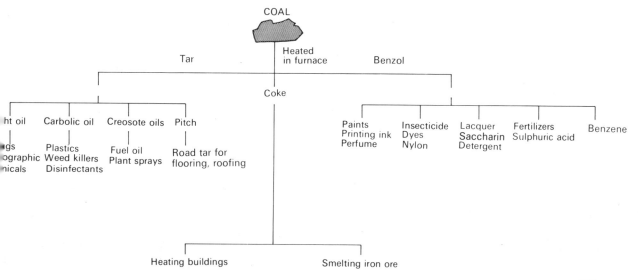

then steam for generating thermal electricity, or by heating it in the absence of air to convert it into gas, coke and smokeless fuels. Thermal electricity is produced in a power station which may consume as much as 20 000 metric tons of coal each week.

Coal for making coke

When coal is burnt in the presence of air, the volatile gases in it are driven off and they burn. Most of the carbon content of the coal also burns. But if coal is heated in a closed furnace, the gases driven off do not burn and they can be led away by pipes and stored in special tanks. After the gases have been driven off there remains a hard mass of carbon, called *coke*. Several liquids such as benzol and tar are also driven off with the gases. The main use of coke is for smelting iron ores in blast furnaces.

Coal as a raw material

The production of gas and coke from coal results in the formation of several valuable by-products the most important being benzol, tar and ammonium compounds. These all condense from the gas as it is cooled. Benzol is used for making benzene, which, when added to gasoline, makes car engines work better. Detergents and insecticides are also made from benzol. Tar has many uses – for making road surfaces and creosole (for preserving timber). The ammonium compounds are used for making fertilizers and other valuable products.

POWER FROM PETROLEUM
Origin of petroleum

Petroleum, which in its natural state is called *crude oil*, is a compound of hydrogen and carbon. It is thought to have formed from the decomposition of minute marine organisms which collected in the sediments on the floors of some seas. During, and after the formation of oil, earth movements probably forced the oil out of the sediments into porous sedimentary rocks such as sandstone and limestone. Sometimes folding occurred, and sometimes the oil-bearing rock layers were sandwiched between layers of impermeable rocks which prevented the oil from seeping away. Oil trapped in this way, slowly moved along the sedimentary rocks and collected in *pools* in anticlines, or in fault traps. Natural gas and water also usually occur in oil-bearing rocks; water collects below the oil, and gas, which has a lower density, above the oil. Sometimes only gas occurs. When gas deposits are large enough, they are commercially worked, e.g. as in the North Sea area between the U.K. and Western Europe.

Oil drilling

When oil-bearing rocks have been located, a hole

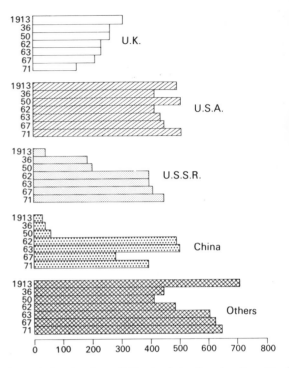

Annual Production of Coal of the five leading Coal Producers in metric tons

Oil in Fault-traps

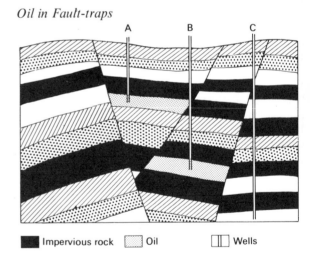

Impervious rock   Oil   Wells

is drilled from the surface to the rocks containing oil. This is done by means of a large metal structure, called a *derrick*, from which a steel pipe, the end of which is fitted with a drill head, called a *bit*, is slowly forced through the surface rocks. As the hole deepens, more steel pipes are added, and this continues until the oil deposits are reached. During the drilling, mud mixed with water is forced down the pipes to lubricate the bit and to flush out the drilled-out rock particles. When the oil deposits are reached, the oil gushes out if it is under natural

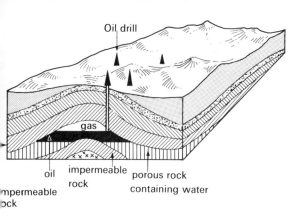

oil impermeable porous rock
rock containing water

impermeable
rock

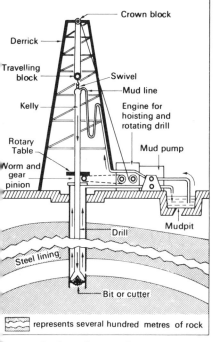
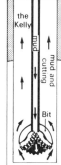

represents several hundred metres of rock

Rotary Drill and Derrick

An off-shore oil rig in the North Sea. An idea of the size of the rig can be obtained by comparing the cage, containing crewmen, with the rig.

pressure. If it is not under pressure then it has to be forced out by pumping.

In recent years the demand for oil has increased enormously, and because of this, oil deposits which lie under the sea bed are now being tapped. This type of oil drilling is called *off-shore drilling,* and the same methods as described above are used. The only difference is that the oil derrick has to be mounted on a platform which is called an *off-shore oil rig.*

Transporting of oil
Most of the oil-consuming countries are considerable distances from the oil-producing countries. The transport of oil from the producers to the consumers involves two basic movements.

1 *By pipeline* The oil at the well head is taken by pipeline either to the oil refinery, or to the nearest port from which it is transported by tankers to the importing countries. Although pipelines are expensive to build, they are cheap to maintain and operate. Pipelines occur both in the producing countries, as just mentioned, and in the consuming countries where they are used to transport the oil from the importing ports to the areas where the oil is used.

2 *By tanker* Oil tankers vary very much in size with largest ones exceeding 400 000 metric tons. Oil tankers have increased in size enormously over the last three decades, partly because the costs of building and operating a large oil tanker are not proportional to the size, and partly because it was economically viable to build large tankers, called *super tankers,* to carry oil from the Middle East oil producers to the Far East (Singapore and Japan), at the time when the Suez Canal was closed. Now that the Canal has re-opened, the demand for such tankers may be less.

Many ports did not have sufficiently deep water to take the super tankers, and special ports were built to handle them. Other parts, such as, Milford Haven in South Wales (U.K.) have deepened their channels.

Other means of transporting oil
Rail and road tankers are used mainly to transport refined oil and petrol to the petrol stations.

Refining crude oil
Crude oil is refined into the various petroleum

A 200 000 ton shell tanker unloading crude oil from the Middle East, at Pulau Bukom (Singapore)

products in a refinery. The refining consists of three basic operations.

1 The breakdown of the hydro-carbon mixture into its different parts. The crude oil is vapourised and the vapours are then allowed to condense into liquids in a tall tower, called a *fractionating column*. During distillation, the separated hydro-carbons, which are called *fractions,* are collected at different levels as they condense into liquids. The fractions are:

   (a) petroleum gases and gasoline which are separated by cooling, when the latter forms a liquid;

   (b) kerosene;

   (c) gas oil;

   (d) residue of heavy oils, which when distilled in a vacuum, produce lubricating oils, wax and bitumen.

2 The conversion of hydro-carbons from one fraction into another. This is done because there is a greater demand for lighter fractions such as petrol, than there is for the heavier fractions. The conversion is achieved by a process called *cracking* which may be either thermal or catalytic. In thermal cracking the heavier oil fractions are heated under pressure until they break down into lighter fractions. In catalytic cracking the same breakdown is effected by using a catalyst, such as, silica, which permits the change to take place at a lower temperature and pressure.

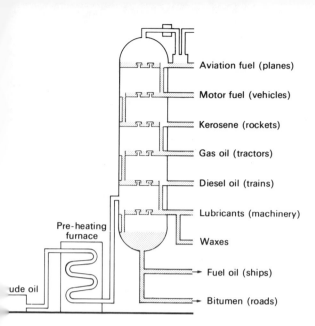

- Aviation fuel (planes)
- Motor fuel (vehicles)
- Kerosene (rockets)
- Gas oil (tractors)
- Diesel oil (trains)
- Lubricants (machinery)
- Waxes
- Fuel oil (ships)
- Bitumen (roads)

Pre-heating furnace

crude oil

Oil refinery on the Gulf Coast of the U.S.A.

3 The purification of the fractions to remove various impurities, especially sulphur compounds.

Location of oil refineries

At one time most of the oil refineries were located in coastal areas near to the oil-producing regions, e.g. Abadan in the Middle East. With the introduction of large oil tankers, it often became more economical to transport the crude oil from the producing countries to the consuming countries, and oil refineries were built at the entry ports of consuming countries. In recent years pipelines have been laid to connect the entry port with the regions where the oil is used, and new refineries have been built in some of those regions.

Oil products and their uses

The main oil products are:

(a) *Natural gas* This occurs in association with crude oil. It is an excellent fuel. Its importance is increasing, especially since the discovery of the large natural gas reservoirs below the North Sea.

(b) *Petroleum gases* These consist of butane, ethane and propane. They are used for making chemicals.

(c) *Gasoline* This is used by the internal combustion engine whose power drives most of the world's land transport.

(d) *Kerosene* This is mainly used as fuel for jet aircraft.

(e) *Gas oil* This is made into diesel oil which is used for powering lorries and some locomotives.

(f) *Bitumen* One of the end products of distillation is a black residue, called *bitumen*. It is used for making roads.

Oil–producing regions
The Middle East

This region produces about 35% of the world's annual production of crude oil and most of this comes from the area around, and underneath, the Persian Gulf. The three leading oil producers of the Middle East are Iran, Saudi Arabia and Kuwait, and the location of the oil areas in these countries, as well as in other countries of the Middle East, is shown in the accompanying map. About one third of the Middle East's oil production is refined in the Middle East at centres such as Abadan and

The Main Oil-Producing Regions of South-West Asia

287

Kirkuk. The rest is exported, in its crude form, mainly to Western Europe, Japan and Canada. Pipelines carry the oil from the oilfields to ports, such as, Tripoli and Saida, on the Mediterranean coast, from where it is shipped to the consuming countries, or it is carried by tankers from the oil-producing regions at the head of the Persian Gulf.

The U.S.A.

There are three main areas of oil-production in the U.S.A. The most important is the Gulf Coast, that is, the coastal areas of Texas and Louisiana. This is followed by the Mid-continental area in the states of Kansas and Oklahoma. The third one is in central California. Oil produced in the Gulf Coast region is sent by pipeline to the Gulf ports of Galveston and Houston, as well as across the continent to large industrial centres such as Chicago, and to conurbations such as New York and Philadelphia.

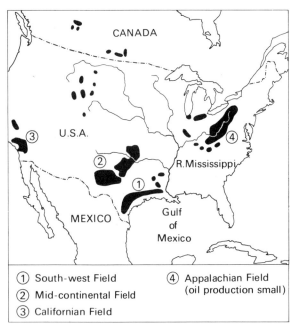

① South-west Field
② Mid-continental Field
③ Californian Field
④ Appalachian Field (oil production small)

The Main Oilfields of the U.S.A.

The U.S.S.R.

Oil deposits occur, and are worked, in several parts of the U.S.S.R. but of these, four stand out as major producers. The most important of these is located between the River Volga and the Ural Mountains, and it produces almost 60% of the country's total production. The two chief oil centres of this region are Ufa and Kuybyshev. Some of this oil is exported to East European countries. The oil is transported by pipeline from Kuybyshev to Bransk and then by branch pipelines to oil

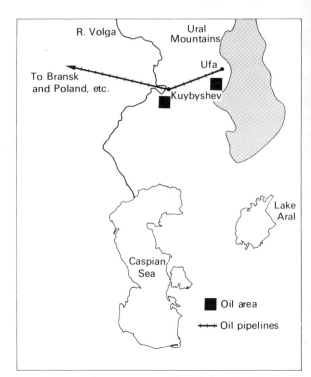

Important Oil Deposits in the Volga Basin

refineries in Poland, East Germany, Hungary and Czechoslovakia. Large oil deposits have been located near the confluences of the Tobol and Irtysh rivers,

Important Oil Deposits in the Ob Basin

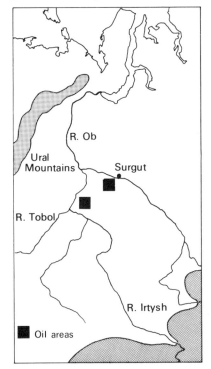

nd the Tobol and Ob rivers, in western Siberia, nd oil production in this region is now increasing apidly. The chief centre is Surgut. The oil region f the Caucasus, which was once the main oil roducer, now produces less than 20% of the U.S.S.R.'s annual production. Baku, on the shore f the Caspian Sea, is still the main oil-refining entre. Another important and expanding oil-roducing region lies to the east of the Caspian ea, where oil deposits are being worked in the Kyzyl-Kum Desert and at Nebit Dag, near to Krasnovodsk, on the south-eastern shore of the Caspian Sea.

Oil Areas of the Caucasus and Caspian Sea Region

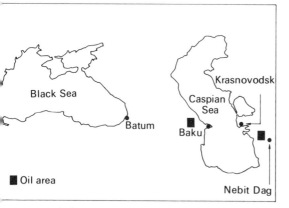

mportant Oil Deposits in North Africa

Africa

ibya and Nigeria are the two main oil producers n Africa, and of these Libya is the more important. ts oil deposits are about 200 kilometres (about 25 miles) inland from the north African coast and hese are worked at Serir, Zelten and Hofra. The il is piped to coastal ports such as Mersa el Brega nd Tobruk. Nigeria's importance as an oil producer increasing. Most of its oil deposits are in the Niger Delta. The oil is refined at Port Harcourt. Off-hore oil deposits have been located and when hese are tapped, Nigeria's output is likely to ncrease appreciably. The only other oil producer Africa is Algeria. Most of its oil is obtained from he oilfield at Hassi Messaoudi, and the oil is sent y pipeline to the port of Arzew.

Venezuela

Most of Venezuela's oil deposits are around the hores of Lake Maracaibo and beneath its bed. At ne time, Venezuela was one of the world's major il producers but with the enormous increase in utput from the Middle East, it now ranks fifth.

Asia

Oil deposits have been located in various parts of Asia, especially in Indonesia. Production, on a world basis, is at present small, but it is possible that it will increase appreciably.

Oil Production and Trade

Although the greatest known oil deposits are in the Middle East, the U.S.A. and the U.S.S.R. are the two largest oil producers as the following diagram shows. Over the last few years, active oil exploration has taken place in several regions, and of these, that of the North Sea appears to be the most promising. If oil reserves here prove unex-pectedly large, then this region could have an important affect on world oil trade. But, at present, the greatest export of oil comes from the Middle East.

World Annual Crude Oil Production (average for the last five years in millions of metric tons). Total approximately 2 360 000 000 metric tons.

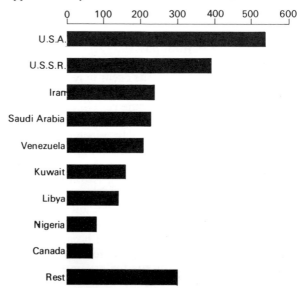

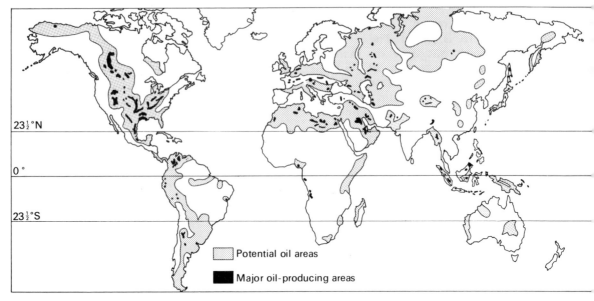

The Main Oil Areas of the World

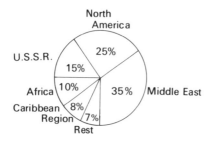

World Annual Production of Crude Oil (average for the last five years on a main area basis). Total approximately 2 360 000 000 metric tons.

Production of natural gas

Natural gas often occurs in association with oil, but sometimes large deposits occur in regions where there are no workable oil deposits. Natural gas, as a source of power, has become very important over the last decade, and it will become increasingly important. The gas is taken by pipeline from the producing to the consuming areas, although sometimes it is liquified and taken by tanker. Gas produced in Algeria (N. Africa), is transported in this way to the gas terminal on Canvey Island in the Thames estuary (U.K.). Like oil and coal, natural gas yields important by-products such as sulphur, hydrogen and helium, and because of this it is an important raw material for the chemical industry. But its main use is for generating power for industry.

The main gas-producing regions
The U.S.A.

At one time the U.S.A. was the largest producer but the combined gas production of Europe and the U.S.S.R. now exceeds that of the U.S.A. The main areas of production are the San Juan basin in New Mexico, along the Gulf Coast, and the central areas of Oklahoma and Kansas, and the Sacramento Valley in California. The gas from these gas fields is taken by over 350 000 kilometres (about 220 000 miles) of pipelines to most of the large industrial and urban centres.

The U.S.S.R.

One of the world's largest reserves of natural gas is located in the Yamal Peninsula of eastern Siberia. The gas deposits here are being extensively developed, and pipelines take the gas to the Leningrad region. There are plans to link this gas field to the pipeline network which is supplying gas, produced in central U.S.S.R. and in the Caucasus, to eastern Europe. Another important gas field lies just north of the confluence of the Ob and Irtysh rivers in western Siberia. Gas is also produced in the Caucasus.

The North Sea Basin

The gas reserves of this region, which extend from eastern England to north-west Germany and Holland, are considerable. The main producing areas are in the North Sea, off the east coast of England, and in northern Holland, at Slochteren

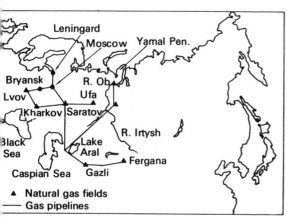

The Main Natural Gas Fields of the U.S.S.R.

Natural Gas Fields of the North Sea Basin

ATOMIC and NUCLEAR POWER

An atom consists of two parts The centre part is called the nucleus and this is composed of protons and neutrons. The outside part consists of one or more electrons. The basis of atomic power is the enormous amount of energy which is released when the nucleus of an atom is split. This is achieved by bombarding the nucleus with a stream of neutrons. When this happens, we say that nuclear fission has taken place. When the nucleus of an atom is split, more neutrons are released which in turn split

nuclear power station in Great Britain.

neighbouring atoms. A chain reaction is then set up which takes place almost instantaneously, resulting in a violent explosion.

Atomic power has been developed only since the end of the last war. Before it could be developed, a way had to be found to control the chain reaction of nuclear fission. This was achieved by slowing down the speed of the neutrons. Atomic power from uranium is now being produced by atomic reactors in several parts of the world. It is interesting to note that one ton of uranium gives as much power as 10 000 metric tons of coal.

Energy released from the reactor is heat energy. This is used for producing steam which drives a steam turbine which in turn powers a dynamo for the making of electricity.

Uranium-producing regions
The chief regions of production are:

1 The U.S.A.
2 Canada
3 South Africa
4 Australia
5 Zaire

Uranium is concentrated into uranium oxide befo it is shipped to the consuming countries. On arri it has to be purified before it can be used the reactors. Nuclear power stations have be developed mainly in the U.S.A., Britain, and Fran Egypt and India are two developing countries whi are building nuclear power stations.

Note Enormous amounts of power can also obtained from atoms by making them join togeth This is called *nuclear fusion*. In time, atomic pow will also be made by nuclear fusion, but this likely to take a great deal of research as it pos very much greater problems for science to harne this power for power stations.

EXERCISES

1 (a) In what ways can the location of a hydro-electric power scheme be influenced by (i) rainfall, (ii) relief, and (iii) the distribution of population? Your answer must refer to specific examples, *one* in a temperate area and *one* in a tropical area.

(b) Name *two* advantages, other than power, that may be provided by hydro-electric power schemes. Name specific examples.

2 There are areas of high coal production in Europe, Asia and North America, and areas of low coal production in Australia, Africa and South America.

(a) For *one* continent of high coal production:
(i) Draw a sketch map of the continent to show the locations of *three* important coal-mining areas.
(ii) Briefly describe *one* method by which coal is mined in one of the areas.

(b) For *one* continent of low coal production:
(i) Describe how that continent supplements its inadequate resources of coal.
(ii) Explain the extent to which the coal resources of the continent chosen have hindered industrial development, if at all.

3 Write a brief description of the coal mining industry of either the U.S.S.R. or the U.S.A. Illustrate your answer with fully-labelled sketch maps and diagrams.

4 (a) Name *two* important oil-producing countries and *two* important oil-exporting countries.
(b) Describe the process by which oil is obtained from the earth's crust.
(c) Briefly explain how crude oil is refined.

5 (a) Draw fully-labelled diagrams to show how oil occu in the earth's crust.
(b) Name *four* areas which contain large oil reserves.
(c) Carefully describe the various uses to which petroleu and its products are put.

6 On a world outline map:
(a) Mark and name *four* areas of active oil productio
(b) Mark and name *three* important oil-consuming area
(c) Mark the routes used for transporting oil from th main producing to the main consuming areas.
(d) Mark and name *two* important towns in oil-producir areas, and *two* important oil-processing towns oil-consuming areas.

7 Write a brief account of the main sources of power an how some of them are utilised.

Objective Exercises

1 Deposits of petroleum (crude oil) which are current being worked, usually occur in
A sedimentary rocks which are adjacent to large bodie of water, such as seas or lakes
B sedimentary rocks lying between layers of imperviou rocks in gently folded structures
C hot deserts
D the bed rocks of epi-continental seas
E low-lying regions whose soils are permanently froze
A B C D I
·· ·· ·· ·· ··

292

Atomic power is derived from the mineral called
A graphite
B felspar
C pyrites
D uranium
E bauxite

A B C D E

Which one of the following conditions is necessary for the generation of a regular supply of hydro-electric power?
A an abundant and constant flow of water
B heavy rainfall
C a fast-flowing river
D rivers which flow down steep slopes
E rivers which flow through lakes

A B C D E

Adit or Drift Mining refers to the mining of coal
A from tunnels extending beneath the sea
B from tunnels extending from a vertical shaft, deep below the surface
C from open pits where coal seams are exposed
D from tunnels excavated into coal seams exposed on the side of valleys or sloping ground
E by opencast methods

A B C D E

One of the advantages of HEP is that
A once established, power continues almost indefinitely
B it is the cheapest form of power known
C it is usually available near to populated areas
D it is cleaner than coal but less clean than oil
E it can be transported cheaply over great distances

A B C D E

In choosing a site for the construction of a hydro-electric power station, which one of the following would most likely provide the greatest natural head (vertical drop) of water?
A a site on a river meandering on a flood plain
B a site beneath a hanging valley in glaciated highlands
C a site on a river delta
D a site on a river flowing from a lake on the floor of a glaciated valley

A B C D

The location of oil refineries depends on several factors, but in general, oil refineries are *usually* located
A on the oil fields
B at the oil-importing ports
C at the oil-exporting ports

D in the industrial regions where the oil is consumed

A B C D

8 Which of these is **not** connected with coal-mining
A shafts
B adits
C galleries
D wells

A B C D

9 Important uses of coal do **not** include the making of
A tar
B chemicals
C paints
D electricity

A B C D

10 Oil traps may occur in
A synclines above impervious rock layers
B anticlines above impervious rock layers
C synclines below impervious rock layers
D anticlines below impervious rock layers

A B C D

11 Atomic power is currently produced by
A fusion
B fission
C radioactivity
D uranium smelting

A B C D

12 Thermal electricity refers to electric power
A contained in batteries
B made from running water
C made from solar radiation
D made from coal or oil or natural gas

A B C D

13 Coalfields are usually located
A in association with igneous rocks
B near to river valleys
C in association with sedimentary rocks
D in industrial regions

A B C D

29 Non-Ferrous Metals and Non-Metallic Minerals

To the geologist, minerals are inorganic substances which have a definite chemical composition. In this book the term is used to include all materials which occur in the earth's crust and which are of economic value.

Definition of terms

A mineral is a chemical compound which occurs in the earth's crust and which forms the basis of rocks.

A metal is a chemical element which can be separated from a mineral by special treatment.

An ore is a rock which has a metallic content sufficiently high to make it worth mining.

Minerals are of three types:
1 Those which can be used to provide power, e.g. coal and petroleum (see Chapter 28).
2 Metallic minerals. These can be put into two groups:
  (a) non-ferrous minerals, that is, minerals containing metals such as, tin, aluminium and copper, etc.
  (b) ferrous minerals, that is, minerals containing iron (see Chapter 30).
3 Non-metallic minerals, e.g. asbestos, sulphur and salt.

Metallic minerals

The cooling of molten magma in the earth's crust, at the time when violent movements were taking place below the crust, resulted in gases and liquids being forced into the joints of the crustal rocks. The magma, the gases, and the liquids derived from it, contained metallic minerals which, on cooling, solidified into large masses in the magma itself, and into thin layers in the joints of the crust. The large masses are called *lodes*, and the thin layers are called *veins*. Sometimes lodes and veins are exposed at the surface. They are then called *reefs*. Most metals occur as compounds, such as oxides or carbonates, or sulphates, but a few, such as gold, occur in the pure state.

Weathering and erosion by rivers, of rocks containing lodes and veins, has resulted in particles of the metallic minerals being transported to the base of slopes where they were deposited along with rock particles. These deposits of metallic mineral particles are called *alluvial* or *placer* deposits. Most

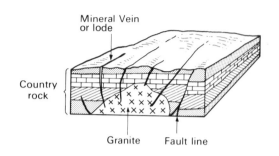

Mineral Vein or lode

Country rock

Granite   Fault line

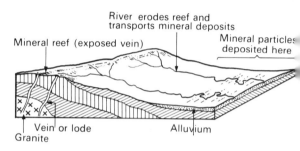

River erodes reef and transports mineral deposits

Mineral reef (exposed vein)

Mineral particles deposited here

Vein or lode

Granite

Alluvium

of the extensive tin deposits of Malaysia are placer deposits.

Metallic minerals are mined in different ways according to the nature of the deposits. Placer deposits are usually mined by opencast methods and sometimes by dredging, whereas vein and lode deposits are mined by shaft (underground) methods. Some metallic minerals contain a high percentage of metal although the more extensive deposits often contain a low percentage. Whether a metallic mineral is mined or not depends on several factors, the most important of which are:
1 The percentage of metal content of the mineral.
2 The percentage of impurities in the mineral.
3 The ease of mining the mineral.
4 The ease of transporting the minerals to the region where it is to be smelted.

NON-FERROUS METALS
Copper

Copper was one of the first metals to be used by man, and at one time, most of his utensils were made from it. Sometimes it was smelted with tin to give a harder metallic substance, called *bronze*, and for a period in man's history, bronze was so

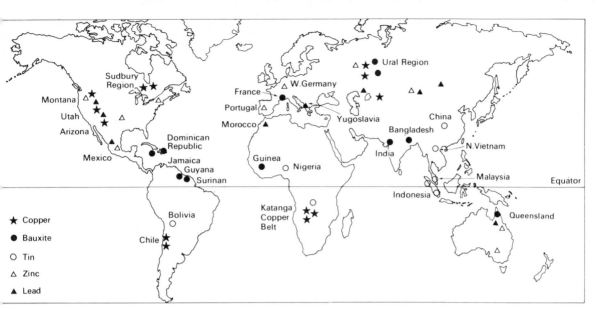

The Main Bauxite, Copper, Tin, Lead and Zinc producing Regions of the World.

mmonly used that this period is known as the
ronze Age. When two or more metals are smelted
gether , the resulting metallic substance is known
an *alloy*.

time man learnt how to make and use iron, and
ter, steel, and from that time the use of copper
d bronze was of less importance. Both iron and
eel are harder than copper, and they can be used
many more ways than either copper or bronze.
owever, many inventions were made during the
dustrial Revolution which gave new uses for
pper, and which resulted in an increased demand
r this metal. The chief of these new uses were:
The manufacture of electrical equipment using
copper, which is a good conductor of electricity.
The manufacture of phosphor-bronze (an alloy
of copper, tin and phosphorus) which is very
hard and which is used for making parts of
engines.
The manufacture of coins because copper tends
to resist corrosion.

oday, just over 50% of the world's annual copper
roduction is used by the electrical industry. The
st is used for making boilers and piping, and in
e manufacture of alloys such as bronze and brass
hich is an alloy of copper and zinc.

he occurrence, mining and processing of
pper

opper occurs in its pure state and also as com-
ounds with sulphur and with oxygen. The most
nportant source of copper is copper sulphide
hich may have a copper content as high as 20%.
ure copper occurs in small particles contained in
rocks. These rocks are worth mining even if the
copper content is as low as 1%.

Most of the world's copper production comes from
opencast workings. After the ore is mined, it is
concentrated, then smelted and finally refined. Since
most of the ores mined have a fairly low copper
content, the ores are concentrated as near to the
mines as possible in order to reduce transportation
costs. Concentrated ores usually have a copper
content of not less than 70%. These copper concentrates are called *matte copper*. Matte copper is
smelted to reduce it to almost pure copper, which
is called *blister copper*. This is done as near to
the concentration plant as possible. In order to
get pure copper, which is essential if the copper
is to be used by the electrical industry, the blister
copper is refined by electrolysis. Refining requires
an abundant supply of electric power and this
process is carried out near to this source of power.

Important copper-producing regions
The U.S.A.

The main areas of production are in Utah, near
to Bingham, in Southern Arizona, near to Bisbec
and Morenci, in Montana, near to Butte, and in
Northern Michigan in the Keeweenaw Peninsula.
Mining in Arizona and Utah is opencast, whilst
that in Montana and Michigan is by shafts.

The U.S.S.R.

The U.S.S.R. is the world's second largest producer
of copper ore and refined copper. The principal
copper deposits are mined in the Ural Mountains
near Magnitogorsk, Chelyabinsk and Sverdlovsk.

Smelting takes place at Krasnouralsk, Karabash and Mednogorsk. Other important deposits occur in the mountains of Central U.S.S.R., near to Namangan, and in the southern part of the Caucasus Mountains at Dilzhan. Other important copper smelting centres are Karsakpay, midway between Lake Aral and Karaganda, near to Ust-Kamenogorsk, on Lake Balkash and at Alaverdi in northern Armenia.

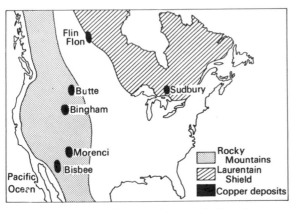

Copper Deposits in the U.S.A. and Canada

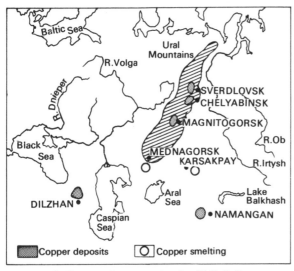

Principal Copper Deposits in the U.S.S.R.

Zambia and Zaire

The belt of country extending from the Katanga District of Southern Zaire to the town of Ndola in Zambia, contains important copper deposits. The copper ores in the Katanga part of the copper belt are mined opencast, but in Zambia they lie deep below the surface and can only be mined by sinking shafts and underground galleries. The copper ores from Zaire and Zambia are smelted in the copper belt at Lumbumbashi and Likasii. Smelting is by electricity which is provided by the power station at the Kariba Dam, on the Zambesi

River in Zambia. The copper area around Ndola in Zambia, produces more copper than the Katanga District. Most of Zambia's copper is exported to the U.K.

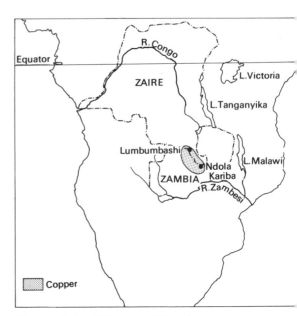

Copper Belt of Zaire and Zambia

Chile

The area lying between the Pacific coastal range and the western edge of the Andean Mountains in southern Peru and northern Chile, is the Atacama Desert which remained uninhabited and undeveloped until the discovery of extensive deposits of copper, nitrates and other minerals. The copper deposits of Chile occur along the eastern edge of this desert belt in the areas around Chuquicamata, Potrerillos and El Salvador, all in northern Chile. The copper ores of these regions are mined opencast. Mining these deposits is made difficult by the arid landscape, the altitude of the mines (about 3000 metres above sea level), and the rugged terrain which makes transportation costs high for taking out the ore to the Pacific ports and for bringing in the mining equipment, food and other essential items to the mining towns. The coastal ports of Antofagasta and Chanaral are merely supply bases, export outlets for the copper, and other minerals, and terminal ports of the transit routes across Chile to Argentina and Bolivia.

Canada

Copper ores occur in many parts of the Canadian Shield, but the two most important mining regions are those near to Sudbury and Flin Flon. These copper ores usually occur in association with other metallic minerals, such as, silver and nickel.

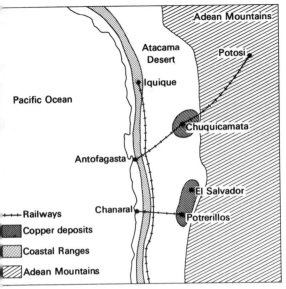

Copper Deposits of Chile

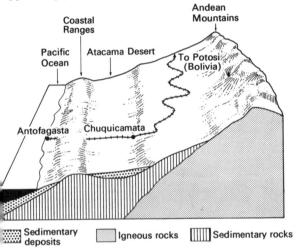

Cross Section of Northern Chile

Annual World Production of Copper Ore (average for the last five years in thousands of metric tons). Approximate annual production 5 700 000 metric tons.

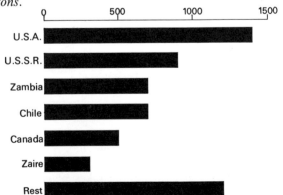

Tin

Tin was being used in the Mediterranean civilisations almost 5000 years ago, mainly for making bronze and pewter (an alloy of lead and tin). Its main use, up to the mid-19th Century, was in the manufacture of these alloys, but from about 1850 it was used in the making of tin plate. Tin plate consists of mild steel sheet, coated with a thin layer of tin to protect the steel from rusting. The demand for tin plate, and hence for tin, grew rapidly after the invention of tin cans as containers for food.

The occurrence, mining and processing of tin

Tin occurs in two types of deposits, lode and vein deposits, and placer deposits, and it usually occurs as tin oxide in a black ore called *cassiterite*. Nearly 75% of the world's annual production of tin comes from placer deposits, the most important of which are in Malaysia, Southern China, Thailand and Indonesia. The rest comes from lode deposits, mainly in Bolivia. Tin ore is mined in placer deposits by tin dredging, by hydraulic mining, and by gravel pump mining.

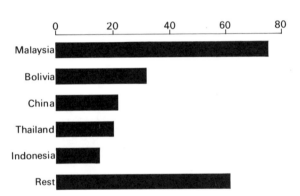

Annual World Production of Tin Ore (average for the last five years in millions of metric tons). Approximate annual production 128 000 000 metric tons.

Tin dredging

A dredge is a huge floating machine which digs out the alluvium containing the tin ore and which separates the cassiterite from the particles of rock. This is done by washing, and the rock particles are thrown out at the back of the dredge as *tailings*. A dredge is very expensive and therefore tin dredging is only operated by large companies. The dredge floats on a pond of its own making, and it digs up the tin-bearing alluvium with an endless chain of steel buckets. Tin dredges are powered by electricity. The following diagram shows how the dredge operates. Tin dredging is extensively used in Malaysia.

Tin dredge in Malaysia

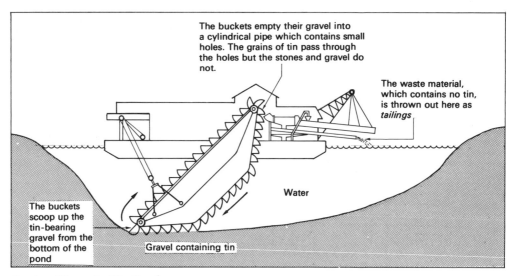

The buckets empty their gravel into a cylindrical pipe which contains small holes. The grains of tin pass through the holes but the stones and gravel do not.

The waste material, which contains no tin, is thrown out here as *tailings*

Water

The buckets scoop up the tin-bearing gravel from the bottom of the pond

Gravel containing tin

Section View of a Tin Dredge

Gravel pump mining

This type of mining involves breaking up the tin-bearing alluvium with powerful jets of water from a monitor (see photograph). The muddy water which collects at the bottom of the slope is pumped to the top of a sloping wooden platform down which it is allowed to run. Wooden boards placed across the platform hold back the cassiterite. The monitors are powered by diesel engines. The sloping platform is called a palong, in Malaysia.

Lode deposits. These are mined in very much the same way as coal which lies far below the earth's surface. Shafts are dug down to the lode deposits and underground galleries are excavated from the shaft into the lode. In Bolivia, where the tin is mined in this way, some of the mines are as deep as 4000 metres (about 13 125 feet).

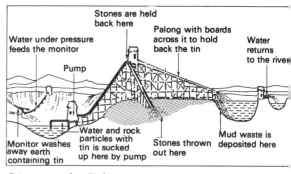

Stones are held back here

Water under pressure feeds the monitor

Palong with boards across it to hold back the tin

Water returns to the river

Pump

Monitor washes away earth containing tin

Water and rock particles with tin is sucked up here by pump

Stones thrown out here

Mud waste is deposited here

Diagram of a Palong

The main tin-producing regions

Malaysia

Most of Malaysia's tin mining takes place in th[e] lowlands lying between the west coast and th[e]

palong at a tin mine in Malaysia

Powerful jets of water break up the tin-bearing rocks at a tin mine in Malaysia

Main Range of Peninsular Malaysia, although there are some important workings in the eastern lowlands of this peninsula, as shown in the map. The chief mining area is the Kinta Valley and it produces about one third of Malaysia's total annual production. The tin ore is mined by both dredging and gravel pump mining, with the latter method accounting for about one half of Malaysia's production. Other methods of mining are by opencast methods and by hydraulic mining. The latter is similar to gravel pump mining except that the water pressure of the monitor is obtained by leading water, through pipes, from dammed rivers on the higher slopes, to the workings on the lower slopes. The cassiterite is smelted in three main centres – Penang, Butterworth and Klang. At these centres, the tin concentrate is smelted into ingots of pure tin each weighing about 220 kilogrammes (about 500 lbs). Most of Malaysia's tin is exported via Penang and Port Swettenham to the U.S.A., Japan and Europe.

Bolivia

The chief tin region in Bolivia extends from Potosi to Oruro, at an altitude of about 3500 metres (about 11 484 feet). Mining is difficult, partly because of the depth of the ores, and partly because of the harsh climatic conditions. At this altitude the air is rarefied which makes breathing difficult. In addition, the region lacks oil, coal and timber, which means that the power required for operating the mines has to be brought in. Most of this power is supplied by hydro-electric power stations located on some of the rivers draining eastward to the Amazon. The power station at Corani supplies hydro-electricity to the tin centre of Catavi. All mining equipment and food, etc., has to be brought into the region by railway from coastal ports such

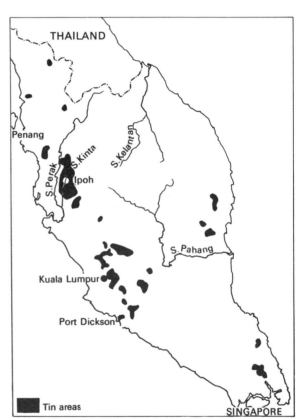

The Tin Areas of Peninsular Malaysia

as Arica and Mollendo, in Chile and Peru respectively. The tin concentrates are exported through these same ports. Because of the altitude of the mining centres, the cost of rail transportation is high, which handicaps the expansion of mining activities.

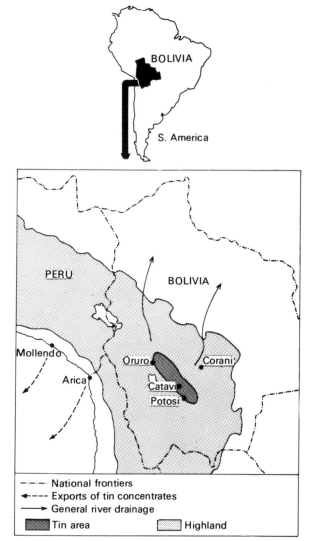

Tin Area of Bolivia

Aluminium

Although aluminium is the most common metal in the earth's crust man did not know how to use it until about 100 years ago, and because of this, it has become a really useful metal to man only over the last 50 years. Today, aluminium is the most important light weight metal used by man. Among its many characteristics, the most important are its light weight combined with strength, its resistance to corrosion, and its good conductivity. In addition, aluminium can be easily rolled into sheets, that is, it is *malleable,* and it can also be easily drawn out into wire and moulded into almost any shape, that is, it is *ductile.* Because of these several properties, aluminium is widely used in the manufacture of aircraft, railway carriages, buses and motor cars, and in the manufacture of electrical goods and domestic utensils such as cooking utensils, refrigerators, washing machines and cooking ovens.

The occurrence, mining and processing of copper

Bauxite and *Cryolite* are the two main aluminium ores. Bauxite is a clay which is rich in aluminium hydroxide. The most important bauxite deposits occur in humid tropical regions, or in regions which once had a tropical climate, where the process of leaching has resulted in the formation of aluminium hydroxide concentrates in the subsoil. Cryolite is an aluminium ore containing aluminium oxide.

The most common method of mining aluminium ores is by opencast mining because most of the ores are near to the surface. After the bauxite is mined it is crushed, washed and then treated chemically to produce aluminium oxide, often called *alumina.* During this process the weight of the bauxite is reduced by about 50%. The alumina is then smelted to give about 99% pure metal, by using electricity. An abundant supply of electricity is required and because of this most alumina smelters are located near to hydro-electric power plants. Sometimes the aluminium mines are near to sources of hydro electricity, but in most cases the ore has to be taken considerable distances to the source of electric power. The Kitimat aluminium plant in British Columbia uses hydro-electricity produced in the nearby coastal ranges, and aluminium ore from as far away as Jamaica. Most of the world's aluminium smelting plants are in regions where there is large scale hydro-electric power development. Usually these regions are isolated from the market areas. However, aluminium, in the form of ingots, is fairly easy to transport to the market areas where it is reheated and made into sheets, wire, machine parts, and other items. Sometimes other metals are smelted with aluminium to produce alloys which have specific characteristics, e.g. a hard alloy is

Annual World Production of Aluminium (average for the last five years in thousands of metric tons) Approximate annual production 8 200 000 metric tons.

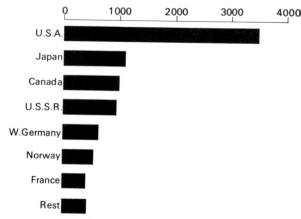

...tained by smelting aluminium with copper, and ...n alloy resistant to corrosion is obtained by using ...agnesium.

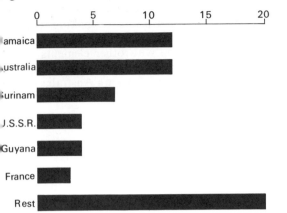

...nnual World Production of Bauxite (average for ...e last five years in millions of metric tons). ...pproximate annual production 60 000 000 metric ...ns.

...luminium-producing regions
...anada
...lmost 25% of the world's annual production of ...luminium comes from the smelters at Kitimat, in ...ritish Columbia, and the smelters in Eastern ...anada at Arvida, Three Rivers and Isle Maligne. ...he latter two places are located on rivers which ...ise on the Canadian shield and flow into the ...t. Lawrence. All of them are near to hydro-electric ...ower stations. The alumina ores used by these ...ast Canadian smelters come mainly from Guyana, ...n South America.

...ustralia
...mportant deposits of bauxite occur along the coast ...n Cape York Peninsula, near to Weipa, in Queens-...and, and near to Gove in the Northern Territories. ...he ore is taken by coastal carriers to the smelters ...t Gladstone. Other bauxite deposits are mined, by ...pencast methods, in the Darling Downs of Western ...ustralia. These are smelted at Kwinana. The ore ...s also exported to Japan and Europe.

...S.S.R.
...arge deposits of bauxite occur near Tikhvin, ...outh-east of Leningrad, and the ore is smelted ...ear to the hydro-electric plant at Volkhovstroy. ...nother large deposit of bauxite is mined near ...erov, in the northern Urals, and the ore is taken ...o smelting plants as far away as Dnepropetrovsk.

...inc
...inc ores occur in many parts of the world but ...specially in parts of Europe and North America. ...he most common ore is *zinc blende* (zinc sulphide)

and this, as well as other zinc ores, often occur in association with lead

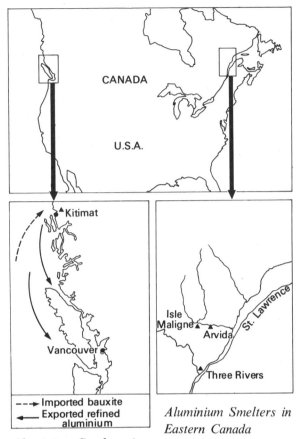

Aluminium Smelters in Western Canada

Aluminium Smelters in Eastern Canada

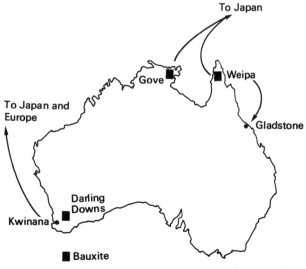

Bauxite Deposits in Australia

Zinc is extracted from its ore either by distillation, or by electrolysis. In the former process, the ore is first roasted, to turn it into zinc oxide. During this process the sulphur is driven off as sulphur dioxide, and this is used for making sulphuric acid. After roasting, the zinc oxide is heated with coke to break it down into zinc vapour which is led away to closed containers where it condenses to liquid form, which later solidifies. In the second process, the zinc blend is first roasted to produce zinc oxide which is then dissolved in sulphuric acid. The solution is subjected to electrolysis which causes the zinc to be precipitated. This process is usually used only in regions which have abundant supplies of hydro-electricity.

The main uses of zinc
Zinc has three main uses.

1 *For making alloys*
Zinc is used in the making of *brass,* which is an alloy of zinc and copper. Its use in this alloy increased with the expansion of the electrical industry at the end of the 19th Century. This alloy is mainly used for making accessories and fittings for the electrical industry. More recently, zinc has been alloyed with aluminium for making car accessories such as hub caps and door handles. These items are usually given a thin veneer of chromium.

2 *For galvanising*
Zinc resists corrosion by air and water, and because of this, a thin coating of zinc applied to iron and steel sheets protects these from rusting. The coating of iron and steel goods in zinc used to be done by electrolysis, and the process was called *galvanising.* Today, such goods are either dipped into molten zinc or they are exposed to zinc vapour.

3 *Miscellaneous uses*
Zinc is used as a roofing material, for making toothpaste tubes, and for making the casing for dry batteries. It is also used in making paints and some antiseptic ointments.

The main zinc-producing regions
Canada is the largest producer of zinc ore, most of which is mined in British Columbia. Other important producing regions are the U.S.A. (Oklahoma, Kansas and Missouri), the U.S.S.R., Australia (Broken Hill), and Mexico.

Lead
Lead is a soft metal which is easy to work, and because of this it has been used by man for a long time. It was an important metal to the Romans who used it for making water pipes, and for lining public baths. The main uses of lead today are for making battery plates, sheaths to protect electric wiring and cables, tanks for storing acids, and solder (which is an alloy of lead and tin). It also used for joining metal seams and for making paints. A more recent use is for making reinforced lead cylinders for transporting nuclear waste. Lead occurs in association with other elements and the most common one is *galena* (lead sulphide). The main lead-producing regions are Australia (Broken Hill,) the U.S.S.R., the U.S.A. (Ozark Plateau in Missouri), Mexico and Canada.

The smelting of lead ores
The ore, which is mainly galena, is first roasted to drive off the sulphur, after which it is heated with coke. The liquid lead collects at the bottom of the furnace and is led away to solidify. This is later refined.

Gold
Gold has been highly valued by man for thousands of years and it has been used, from the earliest civilisations, for making coins and jewellery. Gold, more than any other metal, has had a powerful influence on history. Its main use is as a basis for the monetary system of many countries.

The occurrence, mining and processing of gold
Gold occurs in placer, vein and reef deposits. At one time the main source of gold was placer deposits but today almost 75% of the world's annual production of gold comes from vein and reef deposits. Because of the high value of gold, it is commercially viable to mine gold-bearing rocks which contain as little as 10 grammes of gold per ton of rock in underground mines as deep as 3500 metres (about 11 485 feet). After the gold-bearing rocks have been brought to the surface, they are crushed at the mine head and the gold is extracted by using chemicals.

Annual World Production of Gold (average for the last five years in thousands of kilograms) Approximate annual production 1 400 000 kilograms

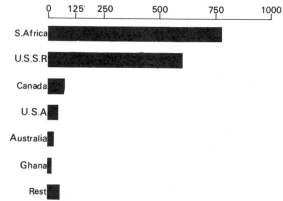

Important gold-producing regions

South Africa

This is the main gold-producing country in the world, supplying almost 60% of the world's annual production. South Africa's gold mining region, is called *Witwatersrand*, or *Rand* for short. The gold is embedded in igneous rocks several thousands of metres below the surface. Some of the gold mines are more than 2500 metres (about 8200 feet) deep. The gold-bearing rocks are mined like coal. The Rand forms a belt about 300 kilometres (about 500 miles) long, and it centres on Johannesburg.

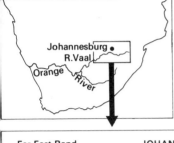

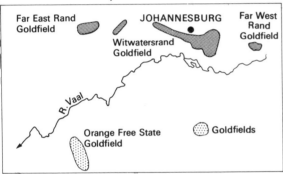

Gold Fields of South Africa

Canada and Australia

Taken together they form the third most important producer of gold. Most of Canada's production takes place in the Porcupine-Kirkland Lake district of Ontario. Australia's main gold-producing region is near to Kalgoorlie, which is situated in the desert region of Western Australia. The gold is obtained from deep mines. All water for this mining centre has to be brought by pipeline from Perth, 512 kilometres (about 320 miles) to the west.

Silver

Silver has been almost as highly-prized as gold throughout history. The metal is both ductile and malleable and it does not corrode except when in contact with sulphur. Silver is extensively used in the manufacture of jewellery, and photographic plates. The main silver-producing regions are in Mexico, the Andean States of Chile and Bolivia, and in the Cordilleran areas of the U.S.A. and Canada.

NON-METALLIC MINERALS

Stone It has been used as an important building material for a long time in many parts of the world. Some stones have a more pleasant appearance and some have more special uses than others. The most extensively used building stones are *limestone, granite* and *marble*. Slate, which splits into thin sheets, is used for making roof tiles and it is extensively used in Great Britain and other industrial countries.

Phosphates These occur in North Africa, the U.S.A. and the U.S.S.R. and are used for making fertilizers

Nitrates The most important deposits occur in the Atacama Desert of northern Chile. Like phosphates, these are used for making fertilizers.
Note Nitrates are now made extensively from atmospheric nitrogen.

Potash The largest deposits occur in East Germany (near Stassfurt). It, too, is used in the manufacture of fertilizers.

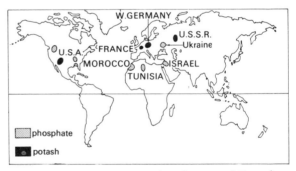

Main Regions producing Phosphates and Potash

Graphite This is a form of carbon. It is soft and is used mainly in the manufacture of lead for pencils, and paints, dry batteries and certain lubricants. The chief deposits occur in Sri Lanka, Malagasy, Korea and New Mexico.

Sulphur It occurs mainly in regions of volcanic activity and hot springs. Sulphur is widely used in the manufacture of a great variety of chemicals and in the vulcanisation (hardening) of rubber and the manufacture of insecticides and medicines. The most important sulphur deposits are in the Gulf Coast region of the U.S.A. Sicily is also an important producer.

Note Sulphur also occurs in iron pyrites which is an ore of iron.

303

Making salt by evaporating sea water in salinas in the Canary Islands.

Asbestos This is a fibrous mineral which occurs in certain types of igneous rocks. The fibres can be spun and woven into textiles. The principal property of asbestos is its great resistance to electricity and heat, and it is therefore used for making fireproof clothing and materials, and insulating materials for preventing the passage of heat and electricity. Eastern Canada is the world's largest producer. South Africa is also an important producer.

Mica It occurs in igneous rocks and it can be easily split into thin sheets. Mica resists great heat and electricity and is semi-transparent. It is used for making windows to furnaces and as insulators in electrical equipment. The two most important producing-regions are North Carolina in the U.S.A., and the Damodar Valley of northern India.

Salt It occurs extensively, in all oceans and seas but especially those of the tropics, and in certain types of rocks. It is essential to all living animals and about one half the world's production is eaten. Its other main use is in the manufacture of many important chemicals. The main producing regions are the U.S.A., East Germany (Stassfurt), the U.K. (Cheshire) and Poland.

EXERCISES

1 (a) Name *two* areas where tin ore occurs on a large scale, and for *one* of these, draw a map of the country to show the principal areas where the ore is worked. Mark and name *two* towns associated with its mining.

(b) Carefully describe *one* method by which tin ore is mined.

(c) Make a list of the more important uses of tin.

2 Some non-metallic minerals play an important part in agriculture.

(a) Name *two* minerals widely used as fertilizers in agriculture.

(b) For each mineral, name *one* area where it is mined on a large scale.

(c) Name some of the main areas where minerals are used in agriculture.

(d) Briefly state any other ways in which the minerals named in (a) are used.

Objective Exercises

1 Which of the following is **not** a non-metallic mineral?
A potash
B graphite
C aluminium
D sulphur
E asbestos

A B C D E

2 In which of the following does a mineral **not** occur?
A vein
B lode
C reef
D placer
E overburden

A B C D E

What is the name of the mineral from which aluminium is made?

A pyrites
B bauxite
C cassiterite
D graphite
E galena

A B C D E
.. ..

Which of the following is **not** relevant to the smelting of aluminium?

A bauxite is brought from the tropics to temperate countries usually for smelting
B large amounts of electricity are needed
C it is usually carried out at HEP stations
D it is usually carried out on coalfields

A B C D
.. ..

Which one of the following statements *most* accurately describes the meaning of *placer deposits*?

A Particles of metallic minerals occuring in a mass of rock.
B Concentrations of water-deposited metallic mineral particles located in sedimentary rocks.
C Mineral particles located in river deposits.
D Deposits of gold particles in river alluviums.

A B C D
.. ..

Two important non-metallic minerals used in the manufacture of materials for protection against fierce heat are

A mica and potash
B asbetos and graphite
C graphite and mica
D asbestos and mica

A B C D
.. ..

Potash is an important source for the manufacture of fertilizers. The largest deposits which are worked in the world are in

A Africa
B the U.S.A.
C East Germany
D the U.S.S.R.

A B C D
.. ..

Tin is used in the manufacture of all of the following. Which one was responsible for the rapid increase in the world production of tin ore?

A pewter
B phosphor-bronze
C tin plate
D bronze

A B C D
.. ..

9 The electrical industry uses several different metals but the most important of these is

A steel
B copper
C tin
D aluminium

A B C D
.. ..

30   The Iron and Steel Industry

Iron and steel are the two most commonly used metals because, in comparison with most other metals, they have many more uses and they are cheaper. Iron and steel are the basis of many types of industrial development and they are extensively used in such industries as engineering, car manufacture, the manufacture of locomotives, railway lines and rolling stock, and in shipbuilding.

The nature and mining of iron ore
Iron ore occurs in several forms in both igneous and sedimentary rocks, and it occurs in almost every part of the world. However, iron ores are usually only mined when the iron content of the ore exceeds 20%. There are four main types of iron ore.

1 *Haematite* This is an oxide of iron and the ore often contains about 70% iron. It commonly occurs in sedimentary rocks. The iron ore deposits to the west of Lake Superior (U.S.A.), and those of the Krivoy Rog region of the Ukraine (U.S.S.R.), and northern Spain, are all of this type.
2 *Limonite* This is also an oxide of iron but the iron content rarely exceeds 60%. It also occurs in sedimentary rocks, and it is the main iron ore of the Lorraine region of north-east France.
3 *Magnetite* This ore is also an iron oxide with an iron content of about 70%. It is a black ore which usually occurs in igneous rocks. It is an important ore of the iron ore regions of the Urals (U.S.S.R.), and the iron ore regions of Kiruna and Gallivare (Sweden).
4 *Siderite* This ore is a carbonate of iron with an iron content of about 30%. It occurs in sedimentary rocks.

The mining and transportation of iron ore
Many of the world's large iron ore deposits are mined opencast. The surface rocks are stripped off and the exposed iron ore is then dug out by diesel-powered shovels. The mining of iron ore from underground galleries is undertaken in some regions, e.g. in the Krivoy Rog region.
Iron ore is transported by the cheapest, not necessarily the quickest, means from the mining regions to the blast furnace centres where it is smelted. Some low grade ores cannot always stand the cost of transportation to the smelting centres, and in some instances blast furnaces have been built near to the ore mines. Coke and limestone, both of which are needed for smelting iron ores, are brought to the ore fields. Iron ores are always transported b water, if navigable waterways occur near to the o fields. Sometimes the distance by water from th ore fields to the iron smelters is much greater tha by land, but it is still cheaper to use water tran portation. In some regions, iron ore is concentrate and made into pellets. This is done by crushir the ore to get rid of a large part of the roc waste. Some of the ore exported from Wester Australia to Japan is sent in pellet form. Anothe way of transporting the ore is to mix it with wat to produce a slurry which is then transported b pipeline or by tanker.

Major iron ore producing-regions
The U.S.S.R.
Vast deposits of iron ore occur at Krivoy Rog, i the Ukraine, at Kursk, in the Ural Mountains, an at Kustanay, to the east of the Urals. In additio there are smaller, but important ore deposits i parts of the tundra region, and in Siberia.

The U.S.A.
Almost 75% of the iron ore mined in the U.S.A comes from the huge deposits to the west and sout of Lake Superior, in the Mesabi Range and i other ranges. The only other region where iron or is mined extensively is near to Birmingham, whic is situated at the southern end of the Appalachia Mountains.

Iron Ore Fields of the U.S.A.

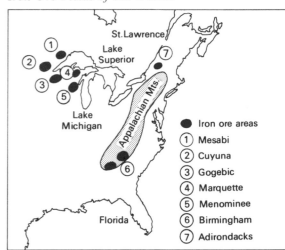

306

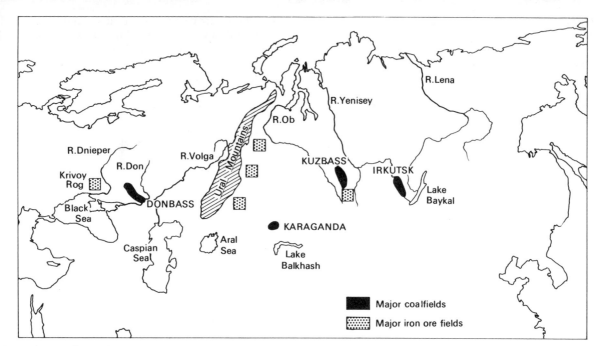

Iron Ore and Coal Deposits of the U.S.S.R.

Ships full of iron ore being unloaded near to a battery of blast furnaces in the industrial area of north-east U.S.A.

Europe

The most important iron ore deposits of Europe are in Lorraine (northern France), near Gallivare and Kiruna (Sweden), and near Scunthorpe and Corby, in eastern England.

Iron Ore Fields of Western Europe

Asia

China has considerable deposits of iron ore near to Anshan, in Heilungkiang, and deposits of high grade ore near Tayeh in the Yangtse Basin, and at Bayin Obo, near to Paotow, in inner Mongolia.

Iron Ore Deposits of Anshan, Bayin, Obo, and Tayeh, in China.

High-grade iron ores also occur in India, especiall at Singhbhum, near to Jameshedpur, in north east India. Japan has limited iron ore resource the most important of which are at Kamaishi, i northern Honshu and in parts of Hokkaido.

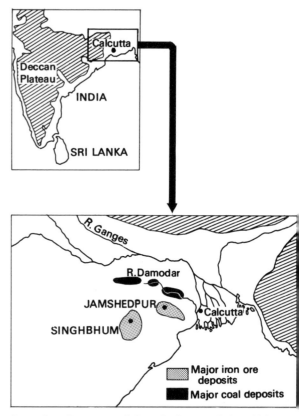

Iron Ore Deposits of Jamshedpur and Singhbhum i India

South America

Vast deposits of iron ore occur in the south-eas part of the Brazilian Plateau, just north of Rio d Janeiro.

Australia

Most of Australia's production of iron ore come from Western Australia, from the areas aroun Yampi Sound, Mt. Tom Price, Koolanooka, an Koolyanobbing. The only other important iron ore region of the continent is at Iron Knob, in Sout Australia.

Canada

A belt of iron ore extends for about 800 kilometre (about 500 miles) from Ungava Bay to the Hamilto river in Labrador. The iron content of the ores i about 60%, and these are worked mainly aroun Schefferville and Mt. Wright. The ore is taken b

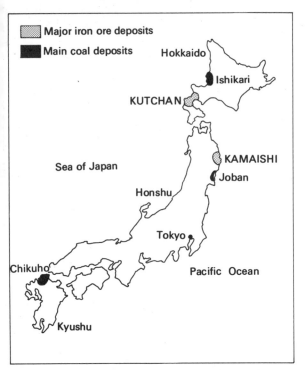

Iron Ore Deposits of Kamaishi and Kutchan in Japan

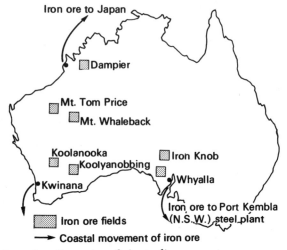

Iron Ore Deposits of Australia

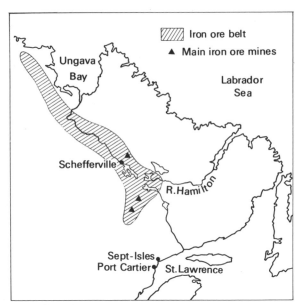

Iron Ore Areas of Quebec and Labrador

Mauritania

Iron ore was first mined in 1961 in Mauritania, which is to the west of Mali in North Africa. By 1970 the region was producing about 8 000 000 metric tons of ore a year and its production is expected to increase. The ore reserves are estimated to be about 200 million metric tons and the ore has a 60% iron content. The ore is taken by rail to the port of Nouadhibou, in Spanish Sahara, from where it is shipped to the U.K. and to other countries of the European Economic Community.

Iron ore area of Mauritania

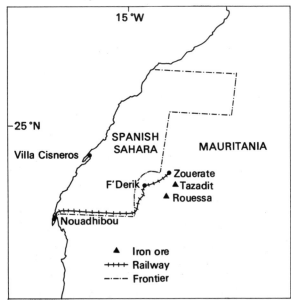

rail to Sept Isles and Port Cartier, on the St. Lawrence, from where they are shipped via the St. Lawrence Seaway to the iron and steel centres of the U.S.A. Mining conditions are difficult because the winters are long and cold. There are also important iron ore fields in Nova Scotia and Newfoundland. The ores of the latter are rich haematite, and in places it is mined from beneath the sea bed. The ore is shipped from Cape Breton Island and is sent to the iron and steel centres on the great lakes.

The making of iron

To extract iron from iron ore, the latter is mixed with coke and limestone which is then fired in a blast furnace. Coke is almost pure carbon and it burns with a fierce heat. It is also very hard, which enables it to support the heavy weight of the iron ore without it being crushed. This is very important because hot air which is blown into the base of the furnace, must be able to penetrate every part of the furnace to permit effective burning of the coke. The limestone acts as a flux and enables the impurities in the ore to be removed.

Temperatures in the blast furnace often reach 1650°C (about 3000°F). This is made possible by blowing hot air, enriched with oxygen, into the base of the furnace. The oxygen in the iron ore combines with the carbon in the coke to form carbon monoxide gas, while the impurities in the ore combine with the limestone to form *slag*. The molten iron is drawn off from the bottom of the furnace and is led into sand moulds where it solidifies. This iron is called *pig iron,* and from it are made *cast iron, wrought iron* and *steel.* Re-melted pig iron, when poured into moulds of the required shape, solidifies to give cast iron.

A large blast furnace, working continuously, produces about 500 000 metric tons of iron a year. This requires about 1 000 000 tons of ore, 250 000 tons of limestone, and a very large supply of water, which is needed for cooling. Blast furnaces are therefore sited near to abundant supplies of water. Pig iron contains some impurities, especially carbon, which makes it brittle. It can therefore be used only for making such items as pipes and drain covers. Wrought iron, which is made from pig iron, contains less impurities and it has many more uses. It is not brittle; it is very strong, and it does not rust as easily as pig iron.

The making of steel

Pig iron is made into steel by smelting it to remove all of its impurities, after which small amounts of other metals are added to make different types of steel, each of which has certain properties. Unlike cast iron, steel can be rolled into bars, plates and sheets. It can also be drawn into wire, and it can be forged. Steel can be made in several ways.

1 *The Bessemer Convertor Furnace* This is a furnace into which molten pig iron is poured, and through which air is blown to burn off the carbon and other impurities such as sulphur. When this is done, small amounts of anthracite, which is almost pure carbon, are added to harden the steel. Until 1878 pig iron made from phosphoric ores could not be made into steel because the Bessemer Convertor process was unable to get rid of the phosphorus. But in that year it was

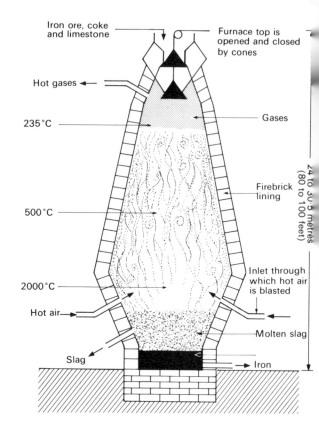

Diagram of a typical blast furnace

discovered that this could be done by lining the convertor with dolomite and by adding burnt lime to the molten pig iron. As a result of this discovery, phosphoric ores could be used to make steel. This led to the rapid development of the Lorraine iron ore fields of north-east France

Diagram of a Bessemer Converter

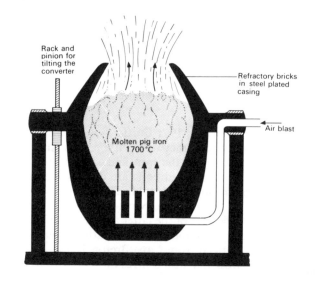

310

which contain phosphoric ores. Bessemer Convertors with this type of lining are known as basic convertors.

2 *The Open Hearth Furnace* This furnace uses a mixture of scrap iron and pig iron. Hot air and gases are blown onto the surface of the furnace.

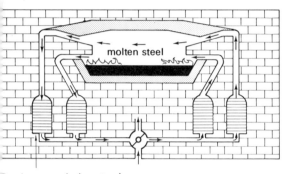

Coal gas and air enter here

Diagram of an Open Hearth Furnace

3 *The Basic Oxygen Furnace* This furnace was first used in Europe about 20 years ago. Molten pig iron, scrap and slag are poured into the cylindrical furnace and a jet of hot oxygen is forced, at supersonic speed, onto the surface of the pig iron. It penetrates deep into the molten metal, and impurities, such as carbon and sulphur, are rapidly burnt off. A high quality steel is produced in about one hour.

Integrated iron and steel plants

An integrated iron and steel plant consists of coke ovens, blast furnaces, steel furnaces and rolling mills, which means that all the raw materials necessary for the making of iron and steel are brought together at one site. By doing this, considerable savings in time and money can be made. For example, the gases from the coke ovens are used to heat the blast furnaces, and the molten pig iron from these is fed directly into the steel furnaces from which the molten steel passes to the rolling mills. These mills produce semi-finished products such as *sheet, pipes, bars, rods* and *wire*. These integrated plants occupy large areas of land and an important requirement for their construction is extensive cheap, flat land. They are often located on the coast to allow the easy import of raw materials and the export of finished or semi-finished products.

Factors influencing the location of the iron and steel industry

1 **Raw material**

At one time charcoal was used for smelting iron ores, but when coke replaced charcoal, blast furnaces were set up on the coalfields, especially those containing iron ores. When the ores of the coalfields became exhausted, iron ores had to be brought in, sometimes from abroad, to feed the blast furnaces. This gave rise to the construction of blast furnaces on coastal sites where the iron ore was unloaded. Two such sites in the U.K. were Cardiff and Swansea. When ores of low iron content began to be used, it was cheaper to take the coke to the iron ore field rather than take the ore to the coalfield. Blast furnaces were therefore set up on the ore fields. Two examples were Nancy and Longwy in Lorraine (France).

The raw materials of the iron and steel industry are very bulky and the location of the iron and steel plants is dependent on the relative costs of taking the ore to the coal or the coal to the ore.

2 **Power and Water**

Since coal is still the main source of power for the iron and steel industry, a location for the industry near to the coalfield is often chosen, usually provided that it is cheaper and easier to bring the ore to the coalfield than it is to take the coal to the ore. But if electricity is used for smelting steel, then the steel plant is often located near to the electric generating plant. A good example of this is the steel plant at Zaporozhe on the River Dnieper in the U.S.S.R. This steel plant is midway between the iron ore field at Krivoy Rog and the coalfield at Donbass.

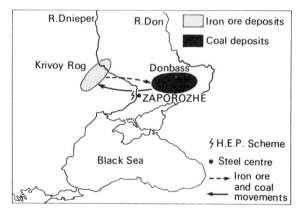

Location of Zaporozhe Steel Centre

Large amounts of water are used by the iron and steel industry which means that the location of the industry in arid regions is extremely difficult.

3 **Transportation costs**

The cost of transporting iron ore and coal by land is higher than it is by water, and iron ore deposits which are at considerable distances from coal deposits, are usually developed only if they occur near to the coast, or near to navigable waterways. The iron ore deposits of Western Australia,

Labrador, northern Sweden and Venezuela are remote from coal deposits but they are near to the coast. The Mesabi ore deposits of the U.S.A. are also remote from the coal deposits of the Pittsburg region but the navigable waterway of the Great Lakes, and adjoining rivers, enables both the ore and the coal to be brought together cheaply.

4 Flat land

Iron and steel plants, especially integrated plants, extend over large areas of land. An integrated plant may occupy an area as large as 15 square kilometres. Large areas of flat, cheap land are therefore required for establishing iron and steel plants. If the land is hilly, then it has to be levelled at considerable expense. This had to be done in the Pittsburg area of the U.S.A.

5 Political factors

As a result of the German penetration into the iron and steel industrial areas of the southern Ukraine, in the U.S.S.R., in World War II, the U.S.S.R. Government decided to develop new iron and steel centres in central Russia which are less vulnerable to attack from hostile countries. Iron and steel centres have been developed in the Urals using the local iron ones of Magnitogorsk and Sverdlovsk, and coal brought in by rail from the Kuzbass coalfield, over 2000 kilometres (about 1250 miles) to the east. To offset the high cost of transporting the coal, the trucks return from the Urals with iron ore to supply steel plants constructed in the Kuzbass. A similar situation exists in China where the government is developing numerous small iron and steel centres, in different parts of the country, to offset the dominance of the iron and steel complex of Anshan in Heilungkiang. Until the end of World War II, Heilungkiang was under Japanese control and although this region is now a part of China it is regarded by the government as being vulnerable to attack from hostile countries.

6 The Market

Because the products of the iron and steel industry are used by a great variety of other industries such as the shipbuilding, motor car, locomotive and engineering industries, these industries can influence the location of iron and steel plants. Wherever possible, iron and steel plants are set up near to the industry which uses its products.

Major iron and steel regions
The U.S.A.

The iron and steel industry of the U.S.A. is located in four main areas. These are:

(a) *The Pittsburg Region*

The Pittsburg iron and steel industry started by using iron ores from the Pittsburg area, and coal from the Pennsylvania Coalfield, but when the iron ores were exhausted, iron ore was brought in from the Mesabi Range, to the west of Lake Superior. Pittsburg and Youngstown are the main iron and steel centres. Most of the steel produced is used in the engineering industries of the region.

(b) *The Great Lakes Region*

When Pittsburg began to use the iron ores from Mesabi it was soon discovered that the ports along the southern shore of Lake Erie, where the iron ore was transferred from ships to rail wagons, were well located for the establishment of iron and steel plants. It was soon realised that it was more economical to

Main Iron Ore Regions of the World

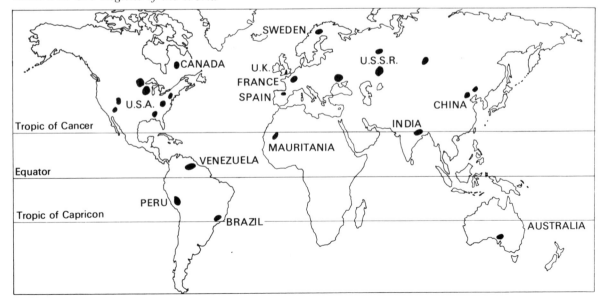

return the rail wagons filled with coal from Pittsburg rather than return them empty. Detroit, Cleveland and Buffalo are the main centres of this industry. Another important iron and steel area is located along the southern shore of Lake Michigan. The plants here use ore from Mesabi and coal from the Pennsylvania Coalfield. Cary and Chicago are the main centres and they produce steel which is used mainly by the engineering, shipbuilding and railway industries.

c) The Atlantic Coast

The iron and steel plants of this coastal region use iron ore imported from Labrador, Venezuela and Chile, and coal from the Pennsylvania Coalfield. The main centres are Sparrow's Point and Bethlehem.

d) The Birmingham Region

This industry uses iron ores, limestone and coal, all of which are mined in the region around Birmingham, at the southern end of the Appalachian Mountains. Birmingham is the centre of the iron and steel region.

The U.S.S.R.

The iron and steel industry of the U.S.S.R. has expanded enormously over the last 50 years, and today it produces about 30% of the world's annual output. Although it has vast deposits of iron ore and coal, these are often widely separated, as they are in the U.S.A. The industry is located in three main areas. These are:

a) The Donbass Region

The dissected plateau region of the southern Ukraine is called the *Donbass*. The industry here uses high grade coking coal from the coalfield, just north of the R. Don, and iron ore from Krivoy Rog, to the west of the R. Dnieper. The main iron and steel centres are on the coalfield especially at Donetsk, Kramatorsk and Makeyevka, but the empty ore wagons take coal, or coke back to the ore area where there are other iron and steel centres. The R. Dnieper flows between the coal and ore areas, and iron and steel plants have been set up at Zaporozhe and Dnepropetrovsk where there are large hydro-electric power stations. In addition, iron and steel plants are located along the coast of the Sea of Azov, at Zhdanov, where coke from the Donbass meets ore from the Kerch Peninsula. Iron and steel plants are also located at Kerch.

The movement of coal and iron in this region, especially between Donbass and the Kerch Peninsula, is similar to the movement between the Mesabi Range and the Pennsylvania Coalfield in the U.S.A.

b) The Ural Region

At one time the iron ores which occur in the Urals were smelted with charcoal made from the

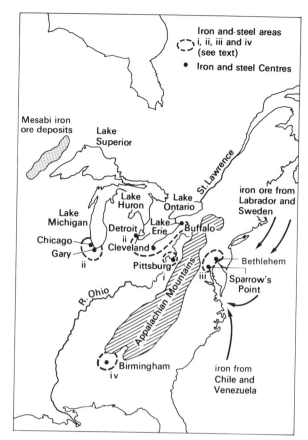

Iron and Steel Areas of U.S.A.

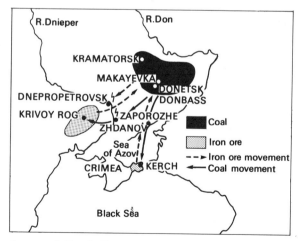

Iron and Steel Centres of the Southern Ukraine

nearby forests, but with the development of the huge coal deposits in the Kuzbass, about 2000 kilometres (about 1250 miles) to the east, and the building of the Trans-Siberian Railway, the whole

313

pattern of the industry changed. Coking coal is taken by rail from the Kuzbass to iron and steel centres such as, Magnitogorsk, Nizhni Tagil, Chelyabinsk and Sverdovsk, in the Ural Region. The rail wagons return with iron ore to the Kuzbass where there are similar steel centres.

The distance between the Urals and the Kuzbass is similar to that between the Mesabi Range and the Pennsylvania Coalfield in the U.S.A., but the Great Lakes which lie between the latter make iron ore and coal movement cheaper.

(c) The Kuzbass – Karaganda Region

The area between the Kuzbass and Karaganda is rapidly being developed as a major iron and steel producing belt. Coal is mined in both the Kuzbass and Karaganda regions and iron ore is brought in from the Urals, but new iron ore deposits in the southern part of the Kuzbass, and in other areas, are becoming increasingly important to the expanding iron and steel industry of this region. Novokuznetsk is an important iron and steel centre.

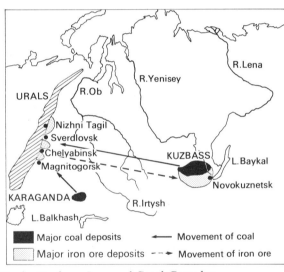

Urals–Kuzbass Iron and Steel Complex

European Economic Community (E.C.C.)

None of the countries of this Community is self-sufficient in both iron ore and coal deposits. For example, the U.K. imports iron ore from Sweden and North Africa, whilst West Germany imports iron ore from Lorraine, in France, and France, which has vast iron ore deposits in Lorraine, has to import coking coal from Germany. Prior to the formation of E.C.C., import regulations hindered the movement of raw materials among the countries of Western Europe, and it was partly because of this that the E.C.C. was formed. There are three main iron and steel areas in the E.C.C. and these are:

(a) The Ruhr Region (West Germany)

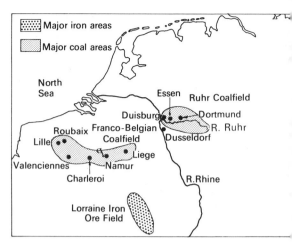

Iron and Steel Area of the European Economic Community

Iron ores from Lorraine and from Sweden are brought to the Ruhr coalfield which has excellent coking coal. Essen, Dusseldorf, Duisburg and Dortmund are the main centres.

(b) The Franco-Belgian Region

This region centres on the Franco-Belgian coalfield. The important iron and steel centres such as Lille, Roubaix and Valenciennes (all in France), and Charleroi, Namur and Liege (all in Belgium), use coal from this area, and iron ore from Lorraine.

(c) The U.K.

The main iron and steel centres in the U.K. are at Port Talbot, Cardiff, Margam, and Ebbw Vale (all is South Wales), at Consett and Middlesborough (in North-East England) and at Corby and Scunthorpe in central England. All of these centres except Ebbow Vale, Consett, Corby and Scunthorpe, have coastal locations and they rely largely on iron ore imported from North Africa, northern Spain, and Venezuela (for South Wales), and from Sweden and North Africa (for Consett and Middlesborough). The steel centres of Corby and Scunthorpe use iron ores that are mined nearby.

Japan

Japan has very limited iron ore resources which yield an annual production of about 2 500 000 metric tons. Her coal deposits are more extensive but they are not of good coking quality. In spite of these facts, Japan's iron and steel industry has expanded tremendously since the early 1950's, and today she is one of the world's major iron and steel producers. Most of the iron ore used in Japan iron and steel industry is imported from Australia, India and Chile, and most of the coal is imported from the U.S.A. and Australia, Because of this

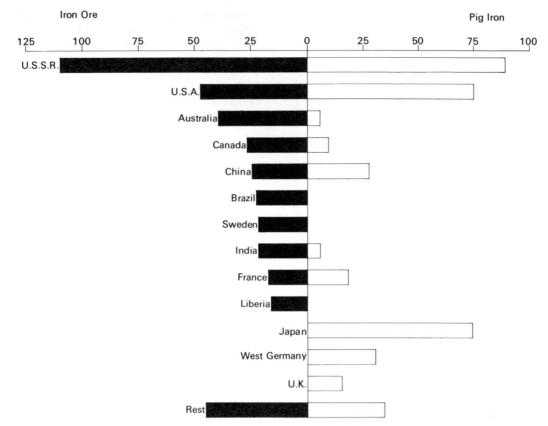

Annual World Production of Pig Iron (average for the last five years in millions of metric tons). Approximate annual production 400 000 000 metric tons.

Japan's major iron and steel centres are located near the coast. This is a considerable advantage because most of Japan's large urban centres also have a coastal location and these provide labour for the iron and steel industry, and a market for its products. The main iron and steel centres are between Osaka and Kobe, near to Tokyo, and Nagoya (all in Honshu), and at Yawata in Kyushu.

Australia

Australia's iron and steel industry is small compared with that of any one of the regions already examined, but it is playing an increasingly important role in the industrial development of the country. Important iron ore deposits occur in Western Australia, at Mount Tom Price, and these are mined by opencast methods. The ore is transported to Dampier, 290 kilometres (about 180 miles) away, mainly for export to Japan. The rich iron ores of Iron Knob are smelted at Whyalla using coal brought in from the Sydney–Newcastle field. These ores are also taken by coastal boats to the coalfield where they are smelted at Port Kembla. Another iron and steel centre is at Kwinana, near to Perth.

Iron and Steel Areas of Japan

315

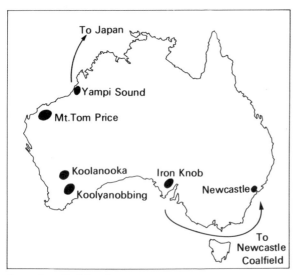

Iron Ore Fields of Australia

EXERCISES

1 (a) What raw materials are required for the manufacture of pig iron?
   (b) Draw a sketch map of one area of where iron ore is produced but which sends its ore outside the region for manufacture into iron and steel. Your example is to be chosen from Europe, or North America or Africa.
   (c) For the area you choose, explain why the ore is not processed in the area where it is mined.

6 The manufacture of iron and steel is one of the main industrial activities of industrial regions.
   (a) Carefully explain how iron and steel are made from iron ore.
   (b) Name *one* area where an important iron and steel industry is located near to iron ore deposits, and another area where the industry is located near to coal deposits.
   (c) Draw a map of *either* north-eastern U.S.A. *or* southern Russia to show the locations of iron and steel industries in relation to the locations of iron ore and coal.

   Your map should also show the routes taken by the coal and iron ore so that they can be brought together.

Objective Exercises

1 Metallic minerals do **not** usually form in
   A granitic rocks
   B plutonic rocks
   C organic rocks
   D metamorphic rocks
   E sedimentary rocks

   A B C D E

2 Metallic mineral lode deposits are associated with area composed of hard crystalline rocks. In which part o North America is this association **best** shown?
   A the Prairies
   B the Californian Valley
   C the Southern Appalachians
   D the Canadian Shield
   E the Mississippi Valley

   A B C D F

3 Some iron and steel industries are located on the coal fields, some on the iron ore fields, and some betweer the two. An example of the latter is the iron and stee industry of
   A Pittsburg (U.S.A.)
   B Nagoya (Japan)
   C the Ruhr (West Germany)
   D Scunthorpe (U.K.)
   E Zhdanov (U.S.S.R.)

   A B C D F

4 The two largest producers of iron and steel are
   A U.S.S.R. and Japan
   B U.S.S.R. and U.S.A.
   C U.S.A. and Germany
   D Germany and China
   E Japan and China

   A B C D E

5 Which of the following is **not** a type of iron?
   A slag
   B wrought
   C cast
   D pig

   A B C D

6 The first process in the manufacture of iron and steel, by refining iron ore, is done in
   A an iron refiner
   B an ore kiln
   C a steel convertor
   D a blast furnance

   A B C D

7 Which one of the following statements is **not** true with reference to the manufacture of steel?
   A The most modern way of making steel is the basic oxygen process.
   B Steel cannot be made from scrap metal.
   C Modern steel mills are often located near to coke ovens and blast furnaces.
   D The open hearth method of making steel is still used.

   A B C D

The Engineering, Chemical and Textile Industries

ENGINEERING

Engineering includes a wide range of industries from the manufacture of tools and machinery, to shipbuilding and the manufacture of locomotives and road vehicles. It also includes the manufacture of electrical and electronic goods. Practically all types of engineering involve the use of iron and steel, and other metals, such as copper for electrical engineering. Engineering can be divided into two basic groups. These are:

1 *Heavy Engineering*
This includes the manufacture of steel girders and plates, locomotives, ships, road vehicles and heavy machinery and engines. This type of engineering is usually located near to regions which produce iron and steel and near to the markets which use its products. Shipbuilding, and the manufacture of locomotives and road vehicles, fall into this category.

2 *Light Engineering*
This involves the manufacture of machine tools, electrical goods and domestic appliances. This type of engineering can often be more dispersed because the raw materials used, and the finished goods produced, are relatively cheap and easy to transport. Also, light engineering often uses electric power which enables it to become established in regions where there is skilled labour, or near to the markets for its products.

Shipbuilding

The location of the shipbuilding industry is greatly influenced by four factors. These are: (1) deep navigable water, (2) easy access to raw materials, such as steel girders and plates, marine engines and boilers, (3) cheap land, (4) access to skilled labour. Most shipbuilding yards are therefore located along the sides of sheltered estuaries where waters are deep and navigable, and which are near to port facilities, and near to iron and steel works supplying a large part of the raw materials. Up to World War I (1914), the U.K. was the greatest shipbuilding country in the world mainly because it was a great trading country and had an immense trading fleet. By the end of World War II (1945), the British shipbuilding industry had declined mainly because of the growth of shipbuilding in other countries, especially in Japan. Since 1945, Britain's relative decline as a leading builder of ships has continued, and today, Japan has the largest shipbuilding industry.

Shipbuilding in Japan

The expansion of Japan's shipbuilding industry after World War II has been phenomenal. The main reasons which enabled this to happen were:
1 Japan's enormous increase in external trade with the consequent increased demand for merchant ships.

Shipbuilding in Japan

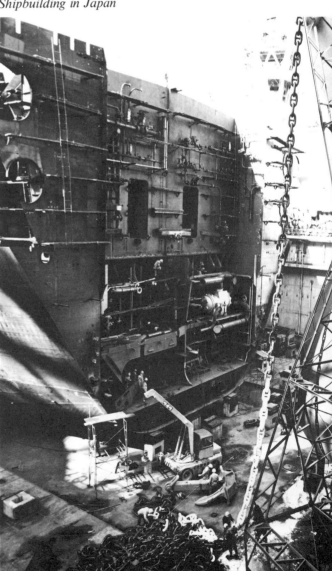

2 Japan's large capacity to produce heavy engineering products.

3 The absence of an old established shipbuilding industry which would have held back expansion because of the difficulties of introducing modernisation.

4 A large, skilled labour force.

5 A determination to survive the devastation caused by World War II, and, to become a successful, industrial and trading country.

6 The introduction of prefabricated shipbuilding.

Most of Japan's shipbuilding yards are located near to the ports of Kobe, Osaka, Chiba, Yokohama, Nagasaki and Hiroshima.

In comparison, the shipbuilding industry of the U.K. has been faced with many serious problems. Because the industry was established a long time ago, its modernisation proved very difficult and in this respect it is still well behind the shipbuilding industry of Japan.

Road Vehicles and Aircraft

Both these industries are basically assembly industries. They put together the various component parts manufactured by other industries. Most motor cars and lorries are now mass-produced on assembly

The manufacture of Volkswagen cars in W. Germany

lines, but considerable engineering skills are required in the assembly. In contrast, aircraft cannot be mass-produced in the same way. In some way these are like ships – they have to be built very much to individual schedules.

The U.S.A. is the largest producer of road vehicles and the industry is sited along the shores of the Great Lakes in such centres as Detroit, Cleveland and Buffalo. Another important centre is Los Angeles, in California. Other important road vehicle industries are in West Germany, centring on Nuremberg, Dusseldorf and Stuttgart, and in the U.K. centring on Coventry, Birmingham, Derby Oxford and Dagenham, and in Japan in the centres of Chukyo and Yokohama. The car industry requires large areas of flat land, and it must have good communications with the industries which supply the component parts.

Aeroplanes are often assembled in the areas where road vehicles are made. Important aircraft assembly plants are located in the U.S.S.R., in the Moscow-Gorki region, and at Kiev and Kharkov, in the Ukraine, and at Kuybyshev and Sverdovsk. In the U.S.A., which is the largest producer of aircraft, the assembly plants are located near to Los Angeles, at Seattle (on the Pacific Coast), and in the state of Washington. The aircraft industry of the U.K. is located at Bristol, Conventry and Manchester.

THE CHEMICAL INDUSTRY

The Chemical industry is concerned with the manufacture of a large range of substances for use in almost every type of industry. These substances are made mainly from raw materials such as petroleum, coal, wood and plants, by breaking them down into their constituent elements, and then putting these together again in different combinations and groups. These substances are usually called *plastics,* examples of which, are polythene, nylon and rayon. The chemical industry also manufactures large amounts of chemical raw materials, such as, acids, especially sulphuric acid, dyes, and fertilizers. As industrial expansion and technological developments continue, the demand for chemical substances will increase. Since World War II, tremendous achievements have been made by the chemical industry especially in the field of plastics. These synthetic materials usually have qualities of durability, strength and lightness which far exceed those of similar natural substances.

Plastics

The properties of materials called plastics varies very greatly. Some are hard and can be worked as if they were metal or wood, and others are flexible and soft. Some resist acids and other chemicals, and ome are transparent. Others can be made into fibres for the manufacture of textiles. Plastics can be easily moulded by applying heat and pressure, and because of this, they are extensively used for the production of shaped articles of many kinds. Here are some common examples of plastics:

Acrylic plastics These are transparent and they are used for making aircraft windows, street light fittings, basins, sinks and baths. *Perspex* is a common acrylic plastic.

2 *Polystyrene* This is used for making packaging materials, refrigerator parts, electrical components such as transistor and radio cases, and household utensils such as dishes and egg cups. A principal disadvantage of this plastic is that it is very brittle.

3 *Polythene* This plastic can be made into flexible sheets which are tough and waterproof and which are extensively used in food-packing industries. It is also made into pliable bottles and containers.

4 *Nylon* This plastic has several uses. It can be made into a fibre for manufacture into textiles such as cloth and rope. It can also be made into 'bristle' for making brushes. Because of its high mechanical strength, nylon is used for making lightweight gear wheels and other machine components, and curtain rails and runners, none of which need to be lubricated.

The main producers of plastics are the U.S.A., the U.S.S.R., the U.K., Japan, and West Germany. France and Italy also have growing plastic industries. The plastics industry is usually located near to the source of the raw materials, for example, near to oil refining centres.

Other substances produced by the chemical industry

These include such products as hydrochloric and sulphuric acids, and caustic soda. These are usually produced either near to the source of raw materials, or in areas where heavy industries are located. Chemical industries located on, or near to coalfields, which use coal-based chemicals, produce dyes, disinfectants and explosives. Synthetic rubber and plastics are usually made from petroleum by-products, and the factories making these chemical products are often sited near to oil refineries. Electrochemical industries which manufacture wood pulp and fertilizers are often sited near to abundant sources of cheap power, usually, hydro-electricity. In contrast, factories manufacturing drugs and medicines are often located near to large urban centres which form the market.

THE TEXTILE INDUSTRY

During the 19th and early 20th Centuries, natural fibres such as cotton, flax, wool and silk were the principal fibres used by the textile industry. But since the end of World War I, and more especially since the end of World War II, man-made fibres such as *nylon, rayon, terylene* and *orlon* have become increasingly important, and today, just under 50% of all textiles manufactured annually are made from these fibres. Textiles can be made from pure natural fibres or from pure man-made fibres, or from a mixture of the two. But of all the fibres used, cotton is still the most widely used.

Textiles made from natural fibres
Cotton Textiles

Although cotton cloth was probably made in some tropical countries before the Industrial Revolution, its manufacture first became important in the U.K. in the mid-17th Century, with the invention of the *cotton gin*. The spinning and weaving of cotton yarn increased rapidly, which led to a rapid increase in the production of raw cotton in America. The industry became established in Lancashire, in the U.K., using water to wash the cotton fibres and to drive the early spinning machines. The use of steam power towards the end of the 18th Century, resulted in the industry moving to the coalfields of Lancashire, and this region soon became the largest

Cotton Textile Area of Lancashire

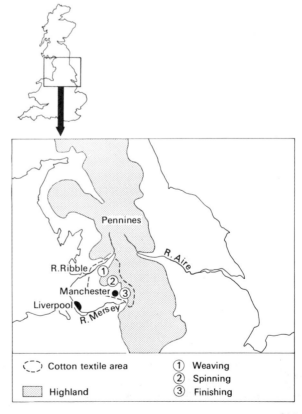

⌒ Cotton textile area	① Weaving
	② Spinning
▨ Highland	③ Finishing

A cotton textile factory in Lancashire (Great Britain). This factory makes heavy canvas for making cycle tyre covers and the tops of canvas shoes, etc.

cotton textile producer in the world. The industry involved three processes, each of which was located in its own area. Spinning became established in a group of towns around Manchester, cotton weaving was concentrated in towns to the north, especially in the Ribble Valley, while the bleaching, dyeing and printing of the cotton cloth became located between the spinning and weaving areas. This early specilization has now been largely replaced as a result of the modernisation of the industry.

During the 19th Century, cotton textile industries became established in the U.S.A., and in the 20th Century, they became established in Japan and in some of the countries producing cotton, e.g., India, Pakistan and Egypt. More recently, cotton textile manufacture has become important in the new industrial countries of Hong Kong and Singapore. Today, cotton textile manufacture is important in the U.S.A., at *Lowell* and *Lawrence;* in Japan at *Osaka, Nagoya* and *Tokyo;* in China at *Shanghai;* in India at *Bombay;* in the U.S.S.R. at *Moscow;* in France at *Lille,* and in Italy at *Milan.*

Woollen Textiles
The manufacture of woollen textiles took place much earlier than that of cotton textiles. It became established in the Middle Ages in the U.K., mainly in East Anglia, but with the use of steam power in the 18th Century, the industry moved to the coal-fields, especially those of Yorkshire.

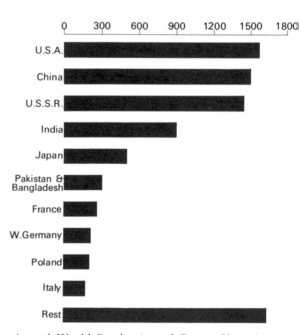

Annual World Production of Cotton Yarn (average for the last five years in thousands of metric tons) Approximate annual production 8 750 000 metric tons.

Wool is a greasy fibre, and the grease must be removed before the fibres can be spun. This is done by washing in soft water, and an abundant supply this is an important factor in the siting of woollen factories. The grease that is removed from wool is called *lanolin,* and this has many uses especially in the manufacture of medicinal creams. After it has been washed, it is *combed,* a process which separates the long fibres from the short. The fibres are then spun and woven into cloth. The long fibres produce a high quality textile called *worsteds* while the short fibres produce a lower quality textile called *woollens.* The various processes of the woollen industry, from spinning to the manufacture of the finished product, tends to take place in the same factories which contrasts sharply with the cotton textile industry.

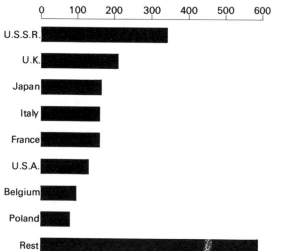

Annual World Production of Woollen Yarn (average for the last five years in thousands of metric tons). Approximate annual production 1 900 000 metric tons.

Textiles made from man-made fibres

The use of man-made fibres in the textile industry has increased enormously since the end of World War II. Rayon, which is made from wood pulp, was one of the first man-made fibres but in recent years several new fibres have been manufactured. Man-made fibres are of two types – 1 *synthetic,* 2 *regenerated.*

Synthetic fibres

These are made from chemicals by a process called *polymerisation* in which simple hydro-carbon molecules are made to combine to form larger molecules. Synthetic fibres are almost entirely made from the by-products of coal and petroleum, and the most common fibres now made are *nylon, terylene* and *crilon.* The main advantages of these fibres are:

1 They are strong.
2 They are light in weight.
3 They can be made into delicate fabrics.
4 They are crease-resisting.

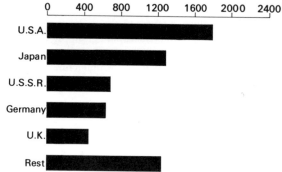

Annual World Production of Synthetic Fibres (average for the last five years in millions of US dollars). Approximate annual production US$6 000 000 000.

The principal disadvantages are:
1 They are expensive to make.
2 They do not absorb moisture.
3 They are sometimes difficult to dye.
4 They lack weight and warmth compared with wool.

Most of these disadvantages can be overcome by mixing synthetic fibres with natural fibres.

Regenerated fibres

These are made from plant and animal products. The most common of these fibres is rayon, which is made from wood cellulose. Rayon, which is often called artificial silk has largely replaced natural silk. As early as 1930, the annual production of rayon was three times that of silk. Other types of regenerated fibres are *fibrolane,* which is made from milk casein, *ardil,* made from groundnuts, and *vicara,* which is made from maize. These latter three fibres are *protein fibres* which contrast with rayon, which is a *cellulose fibre.*

EXERCISES

1 (a) On a world outline map, mark and name:
   (i) *two* areas where shipbuilding is important.
   (ii) *one* port in each area where ships are constructed.
  (b) Beneath your map:
   (i) Say whether the raw materials used by the two areas are imported.

321

2 Write a short essay on shipbuilding for *either* Japan *or* the British Isles. Illustrate your answer with relevant sketch-maps.

Objective Exercises

1 All of the following favoured the early growth of an industry in Lancashire: (i) cheap water supply, (ii) skilled labour, (iii) large market, (iv) damp climate. The industry is concerned with the manufacture of
   A motor cars
   B chemicals
   C machine tools
   D rubber goods
   E textiles

   A B C D E

2 Which are of the following factors need a country **not** possess for the successful development of a shipbuilding industry?
   A a market for the ships
   B a mercantile fleet
   C a plentiful supply of iron and marine engineering products
   D a skilled labour force
   E deep water harbours

   A B C D E

3 Synthetic fibres have all the following advantages **except**
   A they are strong
   B they are crease-resisting
   C they are light in weight
   D they can be made into delicate fabrics
   E they do not absorb moisture

   A B C D E

4 Japan revolutionised its shipbuilding by
   A building sections of ships in factories
   B employing more labour
   C introducing containerisation
   D building more shipyards in S. Honshu

   A B C D

5 Which one of the following statements is **not** true?
   A The U.S.A. is a leading producer of both cotton yarn and synthetic fibres.
   B Japan is a major producer of both woollen and cotton yarn.
   C Regenerated fibres, e.g. rayon, are made from such materials as wood cellulose.
   D The U.S.S.R. manufactures more woollen yarn than cotton yarn each year.

   A B C D

6 Which one pair of the following countries has developed an export-orientated textile industry though it produces no fibres?
   A Egypt and Japan
   B Hong Kong and Singapore
   C Hong Kong and Japan
   D Singapore and Egypt

   A B C D

7 All of the following are products of the chemical industry **except**
   A nylon
   B sulphuric acid
   C rubber
   D dyes

   A B C D

8 The largest part of the world's shipbuilding industry is located along the coasts of North-Western Europe, Japan and North-Eastern U.S.A. Which one of the following conditions has exerted the *most* powerful influence in bringing this about?
   A the development of advanced technology
   B the demand for ships by trading countries
   C the presence of good natural harbours
   D the presence of a skilled labour force

   A B C D

9 Heavy engineering industries, such as the shipbuilding and locomotive industries, are usually located
   A near to the markets
   B on the coalfields
   C close to the iron and steel making industries
   D on the iron ore fields

   A B C D

10 Many light engineering industries have a dispersed location and are usually located near to the market for their products, or near to areas where there is skilled labour. All of the following statements in respect of this are true **except**
   A Many such industries often use electric power.
   B Most of the raw materials used by these industries are easy to transport.
   C Large amounts of water are usually needed for such industries.
   D The goods these industries produce are often easy to transport.

   A B C D

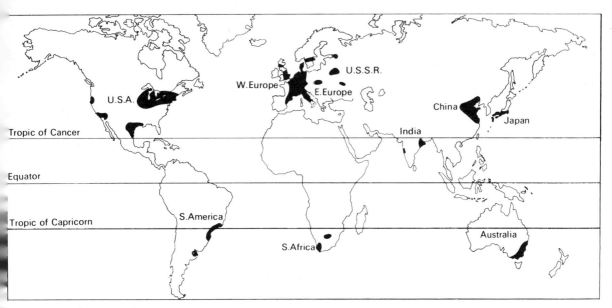

Major Manufacturing Regions

The turning of inorganic and organic raw materials, and refined materials, by mechanical or chemical means into new products, is known as *manufacturing*. For example, the palm oil factory takes in oil palm fruit at one end – this is the raw material, and turns out palm oil at the other. Palm oil is the end product of a palm oil factory but it is the raw material of soap and margarine factories. Thus, the end product of one manufacturing industry may be the raw material for another manufacturing industry. Manufacturing today usually means the employment of a large labour force in factories equipped with power-driven machines which mass produce goods. It also implies heavy financial investment by large-scale operators. This type of manufacturing first took place in the U.K. in the mid-18th Century, when improved farming techniques resulted in an increase in food production. This enabled the U.K. to support a large population whose increased demand for goods could not be satisfied by *domestic manufacture* which existed up to that time. This type of manufacture centred on individual craftsmen, such as the blacksmith, the silversmith and the carpenter, and on cottage manufactures. The main differences between these two types of manufacturing is that while the crafts-man usually buys his raw materials and sells the finished products made from these, the cottage industry operator is usually supplied his raw materials by a middle man, who pays the operator to turn them into finished goods which are then sold by the middle man.

From the end of the 18th Century, factory manufacture was extensively dependent on the steam engine and this resulted in many factories being located on the coalfields. As factory manufacture expanded, the demand for raw materials for the factories increased, and this in turn led to the opening up and development, by the U.K., of large parts of Africa, South-East Asia, Australia and the West Indies, as producers of raw materials such as cotton, rubber, wool and sugar. This led to the development of factories at those ports in the U.K. which imported these raw materials. The development of railways and canals in the U.K. made possible the cheap movement of bulky imported raw materials to the factories at the ports.

The development of agricultural and mineral resources in the overseas countries, which formed part of the old British Colonial Empire, in turn led to a demand for manufactured goods which further enhanced manufacturing in the U.K. The

323

port-site factories had a favourable location to manufacture for these new overseas markets. This early pattern of manufacturing was soon repeated in other West European countries, and in the U.S.A., and later still, especially after World War I, in Russia and Japan. Since World War II, extensive manufacturing has developed in many Asian countries as well as in most countries of the Southern Hemisphere. Manufacturing in the original industrial countries of Europe, and in the U.S.A., Japan and Russia, was at one time dominated by such industries as textiles, shipbuilding, iron and steel, and railway engineering. But with the invention of the internal combustion engine at the end of the 19th Century, followed by research in the fields of communications, and the structure of natural materials in the first part of the 20th Century, subsequently resulted in the development of a large number of new industries. The motor car industry, and related industries, developed rapidly in the early 1900's, and this was followed by the development of the aircraft industry. Since then, the discovery of how to make synthetic fibres and materials has given rise to numerous industries which make all types of semi-finished and finished goods. In addition, technical advances in communications have resulted in the development of electronics and related industries. Manufacturing industries can be put into two groups. These are:

1 *Heavy Industry* This includes the making of iron and steel, shipbuilding, railway engineering, the aircraft and car industries and the refining of metallic minerals such as copper and aluminium and the chemical industry.
2 *Light Industry* This includes the textile industries, light engineering, electronic industries, printing, the making of furniture and food processing.

A bicycle assembly plant in Singapore

Highly industrialized area of Sparrow's Point on the Coastal Plain near to New York (U.S.A.)

Factors affecting the location of an industry

The main factors which influence the location of an industry are:

Proximity to raw materials

Industries which use large quantities of bulky raw materials tend to be located near to the source of the materials. Thus, if iron ores and coking coals occur in close proximity, iron and steel industries will be located near to these raw materials. Also, iron and steel industries tend to be located near to worked iron ore deposits, especially if these have a low iron ore content, e.g. at Corby and Scunthorpe in the U.K., or near to, or on the coalfield, especially if the iron ore used has a high iron content which justifies it being transported to the coalfield. A good example is the iron and steel industry of the Pittsburgh area of the U.S.A. However, heavy industries are not always near to the source of the raw materials which they use, e.g. a high percentage of the iron ores mined in Western Australia are consumed by the iron and steel industry of Japan which is several thousands of kilometres distant from Australia.

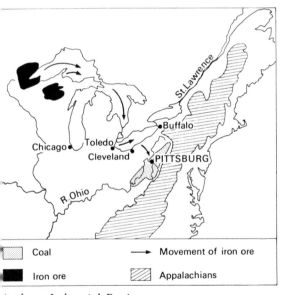

Coal

Iron ore

Movement of iron ore

Appalachians

Pittsburg Industrial Region

Raw materials which are not bulky, and which are of relative high value, do not exert the same influence in the location of industry. For example, refined metals such as tin and copper are mainly used in industries which are not located near to the copper-producing areas. A good example of an industry whose location is influenced by the location of the raw material it uses, is an industry which processes perishable materials. Thus, fruit and vegetable canning industries are nearly always in the areas which grow the fruit and vegetables, e.g. fruit and vegetable canning in California. Another example is the processing of oil palm in Malaysia. The fruit has to be processed within a few hours of picking and the oil processing factories are therefore sited near to the palm oil estates.

2 Proximity to power

All modern manufacturing processes require a regular supply of power. When steam power was extensively used, many industries were located on the coalfields, but with the development of electric power, factories using it had much greater freedom of location because electric power can be transmitted over great distances by overhead or underground cables. Industries which use moderate amounts of electric power, and which produce high-cost goods, e.g. light engineering and electronic industries, can be located at considerable distances away from the source of power. By contrast, industries which require vast amounts of cheap electric power, are usually located near to the source of power. The aluminium industry, the pulp and paper industries, and the electro-metallurgical and electro-chemical industries, all use large amounts of electric power and they tend to be located near to hydro-electric power stations because these produce the cheapest type of electric power. Examples of these industries are: the aluminium industry at Kitimat in Canada, which is near to the H.E.P. stations of the Coast Ranges, and which uses aluminium ore imported from Guyana (Chapter 29); the pulp and paper in-dusty of Corpach, near to the H.E.P. station at Fort William, in Scotland, which uses timber from Canada; and the electro-metallurgical industries near to Mount Lyell in Tasmania (Australia) which use electric power from nearby H.E.P. stations.

3 Water supply and flat land

Some industries require an abundant and a regular supply of water. The iron and steel industry requires it for various cooling processes, the textile industry requires it for washing the fibres, and the pulp and paper industry requires it for processing the timber. Other industries require smaller, though regular supplies of water. Flat and cheap land is essential to the development of many industries which cover large areas. That is why many industrial complexes are located on reclaimed marshlands, or on sandy wastelands.

4 Skilled labour force

Skilled labour is required for most types of industry, even those that are highly automated. New industries tend to be set up in areas which are already in-dustrialised. This is especially true of those areas where some established industries are closing down thus resulting in an unemployed labour force. This has happened in several coal-mining areas of the U.K. where coal mines closed down because it was uneconomical to continue working them. Similarly, new industries are sometimes developed in depressed agricultural areas. Since the end of World War II, extensive industrial development has taken place in many tropical developing countries, and this was centred mainly in those areas which had large populations engaged in trading or simple manufacturing activities. Industrial developments in Singapore and Hong Kong are good examples of this type of industrial development. Some industries, such as electronics and food processing industries, prefer to employ female workers and provided other factors are suitable, these industries are often established in industrial areas which are geared to the employment of male workers. For example, light industries using female workers have been established in North-East England where heavy industries employing male workers are dominant.

5 Access to good communications
All industries require a good system of communications to permit the import of raw materials and the export of finished products. Areas which have good communications therefore tend to attract industry. Ports which have established entrepôt trade, e.g. Singapore, Hongkong and Rotterdam, tend to become good locations for industrial development. Similarly, ports which handle the import of raw materials, especially bulky materials for industries in their hinterlands, often become industrial centres.

6 Access to markets
Industries which manufacture perishable goods, such as foodstuffs, and those which make goods which are relatively low-priced but for which transport costs are high, such as newspapers and soft-drinks, are usually located as near as possible to the consuming centres. Thus, confectionary manufacturers, newspaper printing works and breweries are usually located in, or near to, large urban markets.

7 Political influences
Government action, either direct or indirect, can influence the location of an industry. For example, protective tariffs are sometimes introduced to protect or foster the development of an industry which a government considers necessary to the well-being of its country. Most countries use tariffs to protect some of their industries. Economic planning by governments is now practised in many countries. This involves, among other things, the development of specific areas as industrial areas. Other types of political influence include special inducements in the form of low land rents and tax rebates, which are offered by governments to manufacturers who set up industries in depressed areas.

326

8 Access to management and capital
Good management and access to capital are essential for efficient industrial development, and in some regions industries are established mainly, but not solely, because of these two factors. Industrial expansion and the development of new industries in Singapore and Hongkong in the late 1960's and early 1970's was in part influenced by the movement of capital from other countries in South-East Asia, such as Taiwan, Indonesia and Malaysia.

THE AMERICAN AND EUROPEAN REGIONS
Together they form the greatest concentration of industry in the world. The significant points about the two regions are:

(i) together they use over 90 per cent of the world's annual output of energy;
(ii) together they produce over 90 per cent of the world's annual production of iron and steel;
(iii) the two regions are linked by one of the world's greatest sea routes, that of the North Atlantic;
(iv) together they form the greatest exporter of manufactured goods.

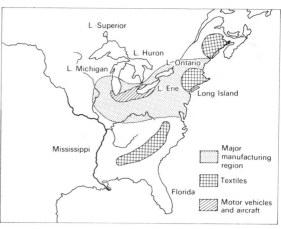

Manufacturing Areas of North-East America

The American Region
It extends from the western shores of Lake Michigan, through the Northern Appalachians and the St. Lawrence Lowlands to the Atlantic Coast. The main factors responsible for the development of this manufacturing region were:

(i) location of high quality coal in the Northern Appalachians;
(ii) location of high quality iron ores near Lake Superior;
(iii) excellent natural communications of the Great Lakes and the St. Lawrence River;
(iv) natural harbours;

(v) an enterprising immigrant population;
(vi) good timber resources;
(vii) a well established farming community.
(viii) a plentiful supply of water.

Types of manufacturing

The iron and steel industry, which is located in the Lakeside Region and around Pittsburg, is the main basic industry, but this has given rise to many other types of manufacturing industries. The chief of these are, transport equipment manufacture in the western part of the region, textiles and engineering in the north-east, chemicals and shipbuilding in the south-east, and the manufacture of cars in Detroit.

The European Region

This region extends from the west coast of England eastwards across the Low Countries and northern France into north-west Germany. Off-shoots spread from this region, south via the Rhone Valley to Switzerland and northern Italy, north through Denmark to central Sweden, and east into East Germany, southern Poland and northern Czechoslovakia (central Europe) and then further east and south-east into the Ukraine region of south-west Russia. Some of the chief factors which helped to give rise to this industrial belt were:

(i) important coal deposits in Great Britain, northern France, West Germany, East Germany and Poland.
(ii) important iron ore deposits in northern France and northern Sweden;
(iii) highly indented coastline on the Atlantic Ocean, and one which has many natural harbours;
(iv) excellent rivers which form natural routeways connecting north-west Europe with central and eastern Europe and the Mediterranean region;
(v) a long established prosperous farming community;
(vi) considerable timber resources;
(vii) a highly skilled craftsmen community.

Types of manufacturing

Manufacturing is both varied and extensive. The manufacture of motor vehicles, ships, textiles and electrical and chemical equipment form some of the more important industries, and all of these take place in Great Britain, the Low Countries, West Germany and France. There is, of course, regional specialisation such as textiles in Belgium, watches and clocks in Switzerland and cutlery in Britain.

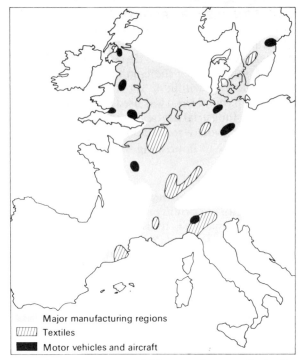

Major manufacturing regions
///// Textiles
■■■ Motor vehicles and aircraft

Manufacturing areas of Western Europe (including Great Britain)

Assembling watches in a factory in Switzerland

WESTERN U.S.S.R. REGION

Moscow is the centre of a major industrial region which began to develop before the introduction of the steam engine. The coal deposits near Tula were worked shortly after the coming of the industrial revolution and within a short time the locally-produced flax and wool was used in the manufacture of textiles. Industrial centres rapidly grew up around Moscow and today roads and railways radiate out from Moscow to flourishing towns such as Kalinin and Ivanovo.

Types of manufacturing

Moscow is an important producer of motor vehicles, machinery, machine tools and electrical goods, whilst in Ivanovo and Yaroslavl there are major textile-manufacturing centres. Although this central region has insufficient coal deposits, sufficient electric power for the factories is derived from burning peat, of which there are vast deposits, and from the H.E.P. stations on the Volga.

THE JAPANESE REGION

This is located in southern Honshu, in northern Kyushu and in Shikoku. The region has developed very recently because it started its industrial revolution under the Emperor Meiji, at the end of the 19th Century, and only established its heavy industrial base this century. Since 1945, when most of its industries were devastated, it has developed amazingly quickly, and it now has the third large *gross national product* (total wealth produced) i the world, after the U.S.A. and West Germany. used to produce cheap and shoddy products, b it has now established a reputation for high quali goods such as electronic equipment, cameras, mote cycles and cars, special textiles, and clothing.

OTHER REGIONS

The remaining manufacturing regions of the worl are of two types:

1. Those located in developing countries and usin local power and raw materials to industrialis speedily, e.g. China and India.
2. Those located in more advanced countries whic are remote from the large industrial regions e.g. Brazil, Argentina, South Africa and Australi

China The main manufacturing region is in north east China. The manufacture of cotton textiles i important, though an increasingly large range o consumer goods is being made.

India Bengal, Bihar and Bombay contain the mai manufacturing areas of India. Textiles and meta products are especially important.

Note India has considerable coal, iron, manganese cotton and H.E.P. resources.

Argentina Manufacturing is concentrated in th eastern areas near to Buenos Aires, and it dea mainly with food-processing and canning, and th manufacture of consumer-goods.

Brazil Similar types of industries occur in the area around Sao Paulo and Rio de Janeiro.

South Africa Industrial development is limited t the manufacture of light engineering goods, con sumer goods, motor vehicles and the processing o foods.

Australia Industrial development is concentrate in south-east Australia. Besides the processing o goods, the manufacture of motor vehicles, chemical and consumer goods is becoming increasingly im portant.

Assembling motor cycles in a large factory in Japan

EXERCISES

1. (a) Name *three* large manufacturing regions, one i Europe, one in Asia and one in North America.
   (b) Draw a large sketch-map of *one* of the regions name

and on it, mark and name:
(i) *two* important manufacturing centres.
(ii) the locations of *two* raw materials used by the manufacturing centres, *or* arrows showing where these raw materials come from if they do not occur within the manufacturing area.
(c) Beneath your map, state:
(i) the main goods manufactured.
(ii) whether these goods are exported, and if so, where.

Briefly describe the importance of manufacturing industries in *either* Japan *or* West Germany. Illustrate your answer with relevant sketch-maps.

The following are four important manufacturing areas in the world:
(i) The Pittsburg – Lake Erie region of the U.S.A.
(ii) The Birmingham region of Great Britain.
(iii) The area around the Inland Sea of Japan.
(iv) The Ruhr Valley of West Germany.

Choose *two* of these regions, and for each, explain how the development of their industries is related to:
(a) sources of power
(b) sources of raw materials
(c) transport facilities
(d) trade with overseas countries.

Objective Exercises

Which of the following is **not** a requirement for a large-scale manufacturing industry?
A a large labour supply
B plenty of capital
C easy access
D power and raw materials
E rapid transport to local market

A B C D E

2 The location of an industry is influenced by all of the following factors **except**
A proximity to power
B presence of a skilled labour force
C access to a dependable supply of water
D access to a good communication system
E close proximity to a coalfield

A B C D E

3 All of the following are classed as light industries **except**
A shipbuilding
B textiles
C food-processing
D electronics
E machine tools

A B C D E

4 Most of the old established manufacturing industries developed in regions which had
A large internal markets for the goods they manufactured
B easy access to raw materials and sources of power
C a large unskilled population
D a tropical location
E government financial assistance

A B C D E

5 Which of the following industries are *best* located near to an abundant supply of cheap electric power?
A shipbuilding
B aluminium industry
C newspaper printing
D motor car industry
E textile industry

A B C D E

6 All of the following statements are true **except**
A North-West Europe is a major manufacturing region.
B Central Africa lacks major manufacturing industries mainly because the people prefer farming.
C Australia's most important manufacturing industries are concentrated in South-East Australia.
D Manufacturing industries can develop successfully in regions which do not have mineral resources.

A B C D

7 Japan has developed an extensive range of industries which has resulted in the country becoming a major industrial power. No other country has become so industrialised within a framework of conditions broadly similar to that of Japan, which makes that country's industrial achievements unique. Which of the following conditions enabled Japan to occupy this unique position?
A The industrial belt is confined to a narrow coastal plain.
B Japan's limited agricultural resources caused it to develop industries.
C Japan has extremely limited coal, iron ore and power resources.
D Japan is a sea-faring country.

A B C D

8 Heavy industry requires
A a large area of flat land
B plentiful supplies of power
C plenty of water
D a large highly-skilled labour force

A B C D

329

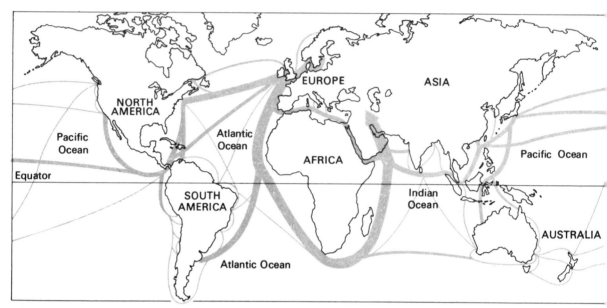

The World's Major Sea Routes (the width of the bands is in proportion to the tonnage and value)

The effective development of industry, commerce, and trade on a regional, an inter-regional, and an international basis, is mainly dependent upon efficient transport and communication systems. Considerable progress has been made to these systems over the last few decades which has resulted in the increased development of world economies and a corresponding increase in world trade. In the mid-19th Century it took about three months for ships from Europe to reach ports in South-East Asia, and about three weeks for ships from Europe to reach North America, whilst prior to the construction of the Asian Pacific Railway, the journey from the east to the west coast of North America took about three months. The introduction of the steam-powered ship, and the railway, later in the 19th Century, resulted in a remarkable reduction in these travel times. Indeed, it was only after the introduction of these quicker means of transport that development was able to take place in the *new lands* of the Americas and Africa. The invention of the internal combustion engine revolutionised land transport in the early 20th Century, and this was

soon followed by the development in air transport. At about the same time great improvements were made to communications, first by the invention of the telephone, and later by the invention of radio. More recently the launching of earth satellites has revolutionised international communications. This technological achievement will probably prove to be one of the most important of man's achievements because it has helped, more than any other of man's inventions, to bring the peoples of the world closer together.

Types of transport

There are three methods of transporting people and goods. These are:

1 *Land transport* (by roads and by railways).
2 *Water transport* (by inland waterways, such as rivers, lakes and canals, and by oceans).
3 *Air transport*

The type of transport used depends upon (i) the nature of the item transported, (ii) the cost of transporting it, and (iii) the speed at which it has

be transported. Perishable goods, such as fruit
nd vegetables, have to be moved quickly, and this
actor decides the type of transport used. Bulky
oods, like coal and iron ore, are usually transported
y the cheapest method, that is, by water, even
hough this may take a long time. In general, it is
nore economical to move a large quantity of goods
r people than it is to move a smaller quantity, and
ecause of this, super oil tankers and the jumbo
ts have been put into service.

AND TRANSPORT

and transport is either by rail or by road. To a
arge extent the two are complementary, though in
nany industrial regions they are now competitive
vith each other. Road transport has two main
dvantages. First, since roads can be constructed
lmost anywhere, for example over very steep
errain, they permit goods to be moved to almost
very part of a region. Second, roads often form
 direct link between producer and consumer. For
hese reasons the importance of road transport has
ncreased enormously over the last fifty years. There
re, of course, advantages of having rail transport.
irst, railways can transport very heavy goods and
aw materials. They are especially important in
egions where great distances are involved, for
xample, the U.S.S.R. and the U.S.A. Second,

railways can carry vast quantities of goods at any
one time. This is perhaps the greatest advantage
of the railway. The main disadvantage of railways
is that their flexibility is limited by gradient, that is,
railways usually follow level land or nearly level
land.

Transport by Road
Although roads have been in existence since man
first started to trade, the building of all-weather
roads did not really begin until the mid-18th Century
when Macadam and Metcalfe introduced new techniques in road construction. The invention of the
internal combustion engine and the subsequent
construction of motor cars, buses and lorries, led to
the building of extensive road networks in the industrial countries of Europe and in North America.
Since World War II, the number of vehicles on the
roads of industrial and advanced countries has
increased enormously, and this has resulted in the
construction of new roads extending across countries
and continents, and in the widening of old roads.
These new arterial roads by-pass urban centres but
connect with the road systems of them by take-off
roads. These arterial roads are called *autobahnen*
in Germany, *autostrada* in Italy, *highways* in the
U.S.A. and *motorways* in the U.K. Similar roads
have been built, and are being built across continents. Of these the Alaskan Highway linking

Highway in the U.S.A.

The Spadina Interchange in Toronto (Canada)

Alaska with Chile, and the Trans-African Highway from Mombassa to Lagos, in Nigeria, are good examples.

Transport by Rail
The invention of the steam engine, and the introduction of railways in the early 19th Century, resulted in railways being laid down to connect established settlements in the developing industrial areas of Europe. The more industrialised a region was, the greater was its network of railways. Contrasting with this was the development of railways in the *new lands* of North America, South America and Australia. In these continents it was the building of railways which encouraged emigration to the new lands and which later resulted in the extensive development of settlements. The introduction of the railways in industrial Europe enabled bulky new materials to be transported quickly and relatively cheaply to the industrial centres and this helped open up the process of industrialisation. Since World War II, the building of motorways and the use of articulated lorries, and more recently, containerised vehicles, has adversely affected transport by rail in many industrial countries. But in developing countries in Asia and Africa, the railway is still a very important means of transport. Whether rail transport is more important that road transport in a region, really depends on the economic conditions of that region. In highly developed countries, complex road networks have developed because

it is essential to move goods and people easily a quickly from place to place. In these regions, t importance of rail transport has declined. In exte sive regions where urban centres are separated l wide belts of rural settlement, these are usual linked by rail transport, though as we have see the construction of roads is altering this.

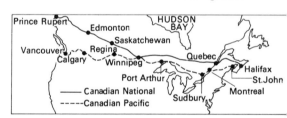

Trans-continental Railways of Canada

WATER TRANSPORT
The oldest method of moving goods over lon distances was by water transport, first by rive then by coastal waters, and later by sea route: Water transport today is of two types: (i) by inlan waterway (river, lake and canal), and (ii) by ocean.

Inland Waterway
Rivers have been used as a means of transport fo a very long time and many rivers are still used i this way today. Large rivers linking region which have important economies, and which d not contain obstructions, such as waterfalls an

332

pids, are invariably used for transporting raw materials, and sometimes, semi-finished and finished goods. The River Rhine is a good example. This river, which rises in Switzerland, flows a distance of about 1330 kilometres (825 miles) before entering the North Sea. It flows through some of the most densely populated parts of Western Europe; it is not obstructed by rapids or waterfalls, and it passes through rich agricultural and industrial regions.

Lakes, too, are important for transporting goods. The Great Lakes of North America are a particularly good example. These lakes enable bulky raw materials such as iron ore, coal, and wheat grain, to be transported considerable distances by large ocean barges.

One of the oldest canals in the world is that linking Peking with Central China. Since its construction, many canals have been dug in many parts of the world to improve communications. Many canals were built in the U.K. in the late 18th and early 19th Centuries for the purpose of transporting, easily and cheaply, coal, iron ore and other raw materials to feed the growing industrial centres. These canals flourished until the introduction of the railway, after which their importance steadily declined. Other canals were later built in North America to improve navigation on the Great Lakes, across the Isthmus of Panama to shorten the sea route between the east and west coasts of North America, and across the narrow neck of land in Egypt to enable ships to pass from the Mediterranean Sea to the Red Sea. These canals are discussed later in this chapter.

Ocean Transport

Ocean transport is cheaper than any other type of transport mainly because there are no construction costs, and because maintenance costs are minimal. The latter only arise where entrances to some harbours have to be dredged to maintain deep water channels. The principal disadvantages of ocean transport are (i) indirect routes usually have to be taken to get from one region to another, and (ii) storms, ice and fog sometimes interfere with the movements of ships in some oceans and seas. The two principal advantages of ocean transport are (i) the enormous quantities of goods that can be transported, and (ii) the relative low coast of transportation.

Types of ships used in ocean transport

There are several types of ships which use ocean routes. The main types are:

1 *Liners* These usually carry only passengers. The number of liners has declined appreciably over the last few decades.

A network of road, rail and water transportation which feeds raw materials to the factories of the Pittsburg region of North-east U.S.A.

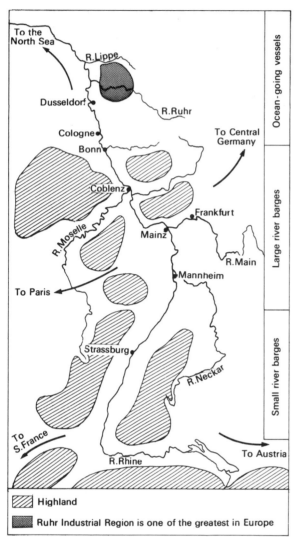

The Rhine Waterway. It is 1332 km (828 miles) long and it is the most important waterway in Europe flowing through, or near to, some of the most densely populated parts of the Continent.

The Great Lakes and St. Lawrence Seaway. Sinc[e] 1959, ocean-going ships with a draft up to 7.6 [m] (25 feet), have sailed from the Atlantic to Lak[e] Superior, a distance of about 3700 km (2300 miles) This was made possible by the construction of lock[s] and canals to by-pass rapids and waterfalls at variou[s] points along the seaway. A vast amount of ra[w] materials (iron ore, coal and wheat) and manu[-] factured goods (farming machinery) moves acros[s] the Waterway each year. Some of it is in respec[t] of trade between the U.S.A. and Canada, and som[e] is in respect of overseas trade.

2 *Cargo Liners* These ships carry both passengers and general cargo and they follow scheduled ocean routes.

3 *Bulk Cargo Ships* These ships carry large quantities of one type of goods such as oil, iron ore, and wheat grain. They are usually large and the loading and unloading of the cargo is mechanized. Oil tankers are an example of a bulk cargo ship. After the closing of the Suez Canal in 1967, ships carrying oil from the Persian Gulf regions to Europe and North America had to travel via the Cape of Good Hope which added greatly to cost and time. This, plus increased demand for

oil, and other factors, led to the construction o[f] much larger oil tankers, some of which no[w] exceed 450 000 metric tons. These large tanker[s] are called *super tankers,* and they are more econo[-] mical to operate relative to the small tankers. [A] major disadvantage is that their size prevent[s] them from using a number of oil terminal port[s] which means that either new ports have to b[e] built, or special unloading facilities have to b[e] constructed far out to sea. Mona Al Ahmadi i[s] one of these new ports which was built on th[e] Persian Gulf to facilitate the loading of oil int[o] the largest super tankers. It is, in fact, the larges[t] oil port in the world. Milford Haven, in the U.K. can take tankers up to 100 000 tons and its specia[l] unloading facilities enables the oil to be piped t[o] the refineries at Llandarcy, over 100 kilometre[s] (about 63 miles) away. The rapid increases in th[e] price of oil, which began in 1874, resulted in [a] fall in oil consumption by most countries. Thi[s] led to an almost complete halt to the constructio[n] of super oil tankers.

Ships carrying iron ore, and wheat grain, acros[s] the Great Lakes of North America, are anothe[r] example of bulk cargo ships. These ships are loaded and unloaded by highly mechanized equipment which reduces the *turn round time* to a minimum thus enabling the ships to be used very economically. Similarly, bulk cargo ships carry iron ore from Western Australia to Japan.

Container ships The use of containers for carrying goods started in the U.S.A., after World War II, but by the 1960's, trade in containerised cargo by land, sea and air, had grown enormously throughout the world. A container is a large metal box which has an average capacity of 33.3 cub m (1175 cub feet), able to carry goods weighing 20 300 kg (about 44 800 lbs) and into which goods are packed either on pallets or in cartons. Loading is usually done by fork lift truck. The largest container can be filled and sealed within a few hours. Containers are filled at the point of production and are sent sealed to the customer who receives them unopened unless customs inspection takes place. The principal advantages of container trade are:

(i) The handling of containers by mechanical means which means that less manual labour is required. In industrial countries, where labour costs are high, the saving is very appreciable.

(ii) Containers minimise breakage and theft.

(iii) Specially designed loading and unloading port facilities enable containers to be loaded and unloaded quickly thereby reducing the turn round time at the ports.

(iv) Goods carried in containers can either be palletised (strapped onto wooden pallets of uniform size which can be moved by fork lift truck) or packed into cartons. The former method is the most common and this enables containers to be filled and emptied by one man operating a fork lift truck. This effects a great saving in labour costs.

Container trade involves the use of specially designed lorries, trains and ships, in addition to specially designed crane equipment and port storage goods.

The main container sea routes are across the Atlantic Ocean (between North America and Europe), across the Pacific Ocean (between North America and Japan), and across the Atlantic and Indian Oceans (between Europe and South-East Asia and Australia). Hong Kong and Singapore are important container ports of South-East Asia. Trade in containers will continue to develop as more ports become equipped to handle them.

5 *Tramp cargo ships* These ships carry assorted cargo, but they do not sail along regular sea routes. They call at any port either to unload or load cargo, and because of this they do not have regular sailing dates. They are also much

Ship-to-shore operations at the Container Terminal of Singapore Port

slower than cargo liners and bulk cargo ships.

6 *Other types of ships* These include fishing vessels and lighters, tugs and pilot vessels. Fishing vessels operate in both coastal waters and in the open seas, whilst vessels which attend to ships entering and leaving ports operate mainly in coastal waters.

A straddle carrier carrying a 40 foot container at the Container Terminal Yard of the Port of Singapore Authority

A container ship at sea. Container transportation provides a door-to-door service between manufacturer and consumer.

World Shipping Routes

There are five major shipping routes. These are:
1 The route connecting North-West Europe and North-East North America.
2 The Panama route, connecting the east and west coasts of North America, the east coast of North America and the west coast of South America, Europe and the west coast of South America, and the east coast of North America with Asia and Australia.
3 The Cape of Good Hope route. This was extensively used by ships trading between Europe and Australasia when the Suez Canal was closed, but its importance is likely to decline with the re-opening of the Canal in June, 1975.
4 The South Atlantic route connecting North-West Europe with South America.
5 The Suez Canal route connecting Europe with Asia and Australia. This route will resume its former importance with the re-opening of the Suez Canal.

Important Canals

Several important canals have been built across narrow strips of land separating adjacent seas and oceans to shorten the transit time between trading countries. The most important of these are the Suez and Panama Canals which were built in 1869 and 1914 respectively. These canals enormously reduce sailing distances between the east coast of the U.S.A. and Japan (via the Panama Canal), and between North-West Europe and Asia (via the Suez Canal).

1 *The Suez Canal* This canal is 160 kilometres long (about 100 miles) and it connects the Mediterranean and Red Seas. It has no locks and the movement of ships through it is therefore very easy. At the time of its closure in 1967, almost 20 000 ships used it annually. An important cargo transported through this canal was oil from the Persian Gulf region to North-West Europe. With the re-opening of the Canal, oil may once again move through it but not in super tankers which are too large for the Canal.

The Suez Canal

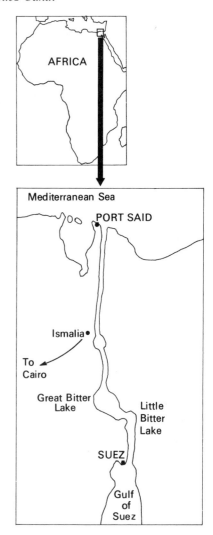

The Panama Canal This canal joins the Caribbean Sea (and hence the Atlantic) with the Pacific Ocean. It is 80 kilometres long (about 50 miles) and it has three sets of locks which makes the movement of ships through it slower than through the Suez Canal.

3 *The St. Lawrence Seaway* The construction of a series of deep water canals and locks on the St. Lawrence was completed in 1960 to enable ships of up to 10 000 tons to reach the Great Lakes from the Atlantic Ocean. Two of the main cargoes transported in bulk are wheat, from the Canadian Prairies, and iron ore, from Labrador, for the steel mills on the shores of the Great Lakes. This seaway is virtually closed to shipping for about four months a year because of the freezing of the St. Lawrence.

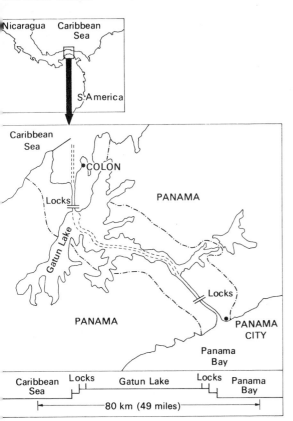

he Panama Canal extends from Colon to Panama ity

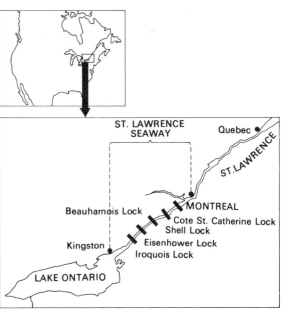

The St. Lawrence Seaway bypasses the section of the St. Lawrence River which contains numerous lakes to the west of Montreal

 ship passing through the locks of the Panama Canal on its way to the Pacific.

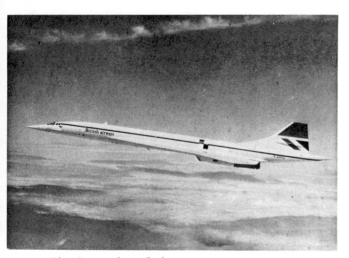

The Concorde in flight

AIR TRANSPORT

Air transport, unlike land and sea transport, is mainly concerned with carrying passengers. However, the transport of goods is becoming of increasing importance. The two main advantages of air transport over other types of transport are (i) the tremendous saving in transit time, and (ii) direct routes (great circle routes) can be followed except where political factors operate which prevent aircraft from flying over certain regions.

Air transport is of recent development, but since the invention of the jet engine, it has assumed very considerable importance. The jet engine has enabled bigger aircraft to be constructed and it has also enabled aircraft to fly faster. The jumbo jet, which is now in regular service with many airlines, carries about 350 passengers, whilst the Concorde, which came into service in 1976, will fly at supersonic speeds thus reducing the travel time still further.

The increase in the number and size of aircraft has resulted in enlarging existing airports, in improving handling facilities at airports, and in the construction of new airports. Since most of the passengers travelling by air are urban dwellers, airports have to be sited close to urban centres. This means that new airports, which are built outside of, but as near to urban centres as is permissible, must have excellent communication systems with the urban centres.

Although cargo still plays only a small part in air transport, as planes become larger and as the need to get goods more quickly from producer to consumer increases, there is every possibility that air freight planes carrying containerised cargo will play an increasingly important role in international and national trade. Other important advantages of air transport include the carrying of food, clothing and medical supplies to areas stricken by floods, earthquakes and other natural disasters.

WORLD TRADE

Most raw materials are of little, if any use, un they have been processed, e.g. bauxite is of no u until it has been smelted and its aluminium conte released. When this is achieved the aluminiu becomes a useful and valuable material for man facture into a variety of goods. Over the last 2 years the manufacture of consumer goods h increased greatly to meet increased world deman This increase was made possible by a correspondi increase in both the production and mining of ra materials, and also in the production of foo Since raw materials, food production, manufa turing and markets rarely occur in the same regio raw materials have to be transported to the man facturing regions, consumer goods to the marke both internal and external, and food to the regio which do not produce it or which cannot produ enough of it. This movement of raw materia manufactured goods and food, forms the basis trade. An examination of the pattern of world tra reveals:

1 The industrial countries of the Northern Hem sphere are the main importers and exporters a they account for about 90% of the world's trad

2 A large percentage of world trade involves t movement of primary products such as mineral wool, cotton and food, most of which is directe to the industrial countries, especially those of t Northern Hemisphere.

3 The trade in raw materials and foods among t producers of primary products is relatively sma

4 The trade in primary products and manufacture goods among industrial countries is relative large.

Factors influencing trade

Several factors influence the type and amount trade that a region has. Of these, the most importa are:

1 **Natural resources**
   Countries which possess and exploit the followi resources usually carry on active trade wi external regions:
   (a) petroleum, e.g. The Middle East
   (b) timber, and wheat, e.g. Canada
   (c) rubber, e.g. Malaysia
   (d) gold, e.g. South Africa
   (e) wool, wheat, meat and dairy products, e. New Zealand
   (f) iron ore, e.g. France

2 **The degree of industrial development**
   High industrialised regions such as, Japan, We Germany and the U.K., produce considerab quantities of manufactured goods for export.

3 **Geographical position**
   Some countries are well placed to take part

trade with countries having contrasting economies, e.g. Singapore handles part of the trade for South-East Asian countries, especially Malaysia; and Holland, through Rotterdam, handles a large part of the trade of several West European countries. This type of trade is called *entrepôt trade*.

4 Tariffs and import duties

Heavy duties often reduce trade among countries. Similarly, low tariffs, or preferential tariffs, often increase trade among countries. Since World War II, large trading groups have dominated world trade by operating preferential tariffs. The important trading groups are:

(a) *The European Economic Community (E.E.C.)* consisting of West Germany, France, Italy, the Benelux countries, and the U.K., Denmark and Eire.

(b) *The European Free Trade Association (E.F.T.A.)* consisting of Sweden, Norway, Austria, Portugal, Spain, Switzerland, and Finland.

(c) *The Council for Mutual Economic Aid (C.O.M.E.C.O.N.)* consisting of the U.S.S.R., Poland, Rumania, Bulgaria, Czechoslovakia, Hungary, East Germany and Mongolia.

(d) The Latin America Free Trade Association (L.A.F.T.A.) consisting of *Mexico, Venezuela, Chile, Ecuador, Colombia, Peru, Brazil, Uruguay, Paraguay and Argentina*.

(e) The Central American Common Market (C.A.C.M.) consisting of *Costa Rica, Honduras, Nicaragua, El Salvador and Guatemala*.

(f) The Organisation of Economic Cooperation and Development (O.E.C.D.) consisting of the *U.S.A., Canada,* the *U.K., Iceland, Denmark, France, West Germany, Belgium,* the *Irish Republic, Italy, Greece, Luxembourg, Netherlands, Norway, Turkey, Sweden, Switzerland, Spain, Portugal, Austria* and *Japan*.

Distribution of world trade

More than one half of the world's trade is handled by North America and the E.E.C. In comparison, the trade handled by Asia, South America, Africa and Australia, is relatively small. For most of these regions, the value of imports is about balanced by the value of exports, but for North America the value of exports exceeds that for imports whilst for the E.E.C. the value of imports exceeds that for exports.

Balance of Trade

Most countries try to ensure that they earn as much as they spend. This is achieved when the value of exports balances the value of imports. This is called the *Balance of Trade*. However, some countries whose value of imports is greater than the value of exports, achieve a balance of trade by having invisible earnings (payment received by carrying goods for other countries, the interest on loans to overseas countries, tourism, and remittances, i.e., money sent home by people working abroad).

EXERCISES

1 (a) Name *three* ways in which goods are transported from a producing region to a consuming region.
  (b) Name *any three* important sea routes, and for *one* of these:
    (i) make a list of goods carried along this route.
    (ii) name two countries which either export or import these goods.

2 Explain *three* of the following statements:
  (a) It is faster but more expensive to carry goods by aeroplane than by land or sea transport.
  (b) A large percentage of the world's trade is still in basic raw materials.
  (c) The development of trade may become easier for a country if it has easy access to the sea.
  (d) Few countries have balanced trade.
  (e) Trade results in better communications.

3 Make a list of the more important factors which influence the trade of a region.

4 Carefully explain the meanings of *three* of the following terms which are used in connection with trade.
  (a) cargo-liner
  (b) tanker
  (c) refrigerated ship
  (d) barge
  (e) import duty
  (f) invisible exports.

Objective Exercises

1 All of the following statements about the Panama Canal are true **except**
  A It has 3 sets of locks.
  B It extends from the Gulf of Panama to the Caribbean Sea.
  C Ships carrying raw materials and goods from the east coast to the west coast of the U.S.A. all use the Canal.
  D The locks restrict the number of ships that can use the Canal daily.
  E The Canal is extensively used by ships operating between Japan and eastern U.S.A.
  A B C D E
  ·· ·· ·· ·· ··

2 Which one of the following types of goods would almost certainly be transported from one place to another by air, where the two places are far apart?

A electronic equipment
B copper
C diamonds
D photographic equipment
E citrus fruits

$$\overset{A}{\underset{..}{}} \quad \overset{B}{\underset{..}{}} \quad \overset{C}{\underset{..}{}} \quad \overset{D}{\underset{..}{}} \quad \overset{E}{\underset{..}{}}$$

3 In which of the following types of natural landscape would the construction of railways be the easiest?
A a lowland jungle-covered plain
B a gently undulating grassland
C an upland region crossed by a rejuvenated river
D an indented coastal plain
E a densely populated urban lowland

$$\overset{A}{\underset{..}{}} \quad \overset{B}{\underset{..}{}} \quad \overset{C}{\underset{..}{}} \quad \overset{D}{\underset{..}{}} \quad \overset{E}{\underset{..}{}}$$

4 Which of the following statements is **not** true?
A The exports and imports of the main industrial countries of the Northern Hemisphere account for about 80% of the annual value of world trade.
B Trade in raw materials and foods among the producers of primary goods is fairly small.
C The import and export of primary products form a small part world trade.
D Trade in primary products and manufactured goods among industrial countries is considerable.

$$\overset{A}{\underset{..}{}} \quad \overset{B}{\underset{..}{}} \quad \overset{C}{\underset{..}{}} \quad \overset{D}{\underset{..}{}}$$

5 Which of the following is a disadvantage for containerisation, compared with traditional means of transporting goods?
A it needs less capital
B it needs less labour
C goods are damaged less
D goods can easily be transferred from land to sea transport

$$\overset{A}{\underset{..}{}} \quad \overset{B}{\underset{..}{}} \quad \overset{C}{\underset{..}{}} \quad \overset{D}{\underset{..}{}}$$

5 Which type of the following goods is suitable for air transport?
A high value, high bulk, non-perishable
B low value, low bulk, non-perishable
C low value, high bulk, perishable
D high value, low bulk, perishable

$$\overset{A}{\underset{..}{}} \quad \overset{B}{\underset{..}{}} \quad \overset{C}{\underset{..}{}} \quad \overset{D}{\underset{..}{}}$$

7 90% of the world's trade is carried on in
A developing countries of the Tropics
B industrial countries of N. Hemisphere
C agricultural producers of the S. Hemisphere
D developing countries of the S. Hemisphere

$$\overset{A}{\underset{..}{}} \quad \overset{B}{\underset{..}{}} \quad \overset{C}{\underset{..}{}} \quad \overset{D}{\underset{..}{}}$$

8 Which of the following statements is **not** true about th Suez Canal?
A The canal was constructed to shorten the distance for ships travelling between Europe and Asia.
B The canal joins the Red Sea to the Mediterranea Sea.
C The canal has been in use continuously since it wa constructed in the 19th Century.
D The canal is approximately 100 miles long.

$$\overset{A}{\underset{..}{}} \quad \overset{B}{\underset{}{}} \quad \overset{C}{\underset{..}{}} \quad \overset{D}{}$$

9 Of the several types of ships operating on internationa sea lanes, one type keeps to definite routes betwee regular destinations. Which type of ship does this?
A fishing vessels
B tramp steamers
C naval ships
D ocean liners

$$\overset{A}{\underset{..}{}} \quad \overset{B}{\underset{..}{}} \quad \overset{C}{\underset{}{}} \quad \overset{D}{}$$

10 Although road and rail communications are comple mentary, roads have one considerable advantage ove railways. This is:
A the comparative ease with which they are built
B they can be built almost anywhere and over almos every type of terrain
C many different types of vehicles can use them
D they provide communications among settlements

$$\overset{A}{\underset{..}{}} \quad \overset{B}{\underset{..}{}} \quad \overset{C}{\underset{}{}} \quad \overset{D}{}$$

11 Which one of the following regions exports and import less goods in terms of value than any of the others'
A Africa
B Europe
C North America
D South America

$$\overset{A}{\underset{..}{}} \quad \overset{B}{\underset{..}{}} \quad \overset{C}{\underset{}{}} \quad \overset{D}{}$$

A settlement is a place where people live. It may consist of a farm or homestead, or a few houses, or it may cover several square kilometres and contain not only houses but shops, schools, offices, factories, government buildings, places of entertainment, and other buildings. Each settlement has a *site* and a *location*. Site refers to the land on which the settlement is built whereas location, or situation, refers to the position of a settlement in relation to other places in the region. For example, Singapore is sited on the south side of Singapore Island but it is located in the heart of South-East Asia.

Settlements can be divided into two main groups:

A   rural settlements
B   urban settlements

RURAL SETTLEMENT

Rural settlement is the name given to all villages and dispersed settlements. Their inhabitants are mainly engaged in agriculture. There are three types of rural settlement:

1   *village settlement* (nucleated or linear)
2   *dispersed settlement*
3   *isolated settlement*

Village settlement A village contains many buildings, such as houses, shops and often schools. It is usually situated so that it has access to the region around it, and it tends to cater for the needs of the region. It is often the shopping centre, the social centre and the religious centre for the region. Villages are often, though not always, associated with intensive farming.

Villages are, perhaps, the most common type of settlement in the lowlands of Asia where rice is grown, especially the rice regions of India, China and Java, and in the mixed farming regions of Europe and North America. The market gardening region of Holland also has a well developed village settlement pattern.

The shape of villages

A village usually has one of three basic shapes. (i) compact type, (ii) linear type, and (iii) cross type.

Compact (nucleated) type The shape may be round or square but in all cases the buildings are close together and they are connected by roads, or footpaths, or both. These link all buildings together. The village has a definite 'boundary' (line beyond which it does not extend).

An aerial view of a nucleated settlement in Nigeria

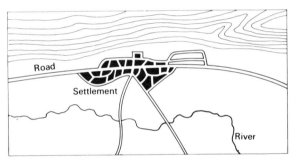

Nucleated settlement

Linear type Buildings form a line, straight or curved, which may follow a line of movement, e.g. a road, a river, a relief feature, for example a coast, the base of a ridge, or a zone where water is near the surface. Linear settlement does not follow railways except near to stations which are relevant for transport.

Linear settlement

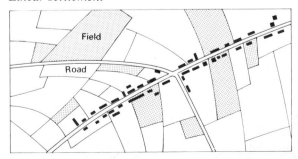

An aerial view of a line settlement in the Netherlands

Cross type These form where two lines of movement cross, e.g. the crossing of two roads.

Cross type of settlement (nucleated and linear)

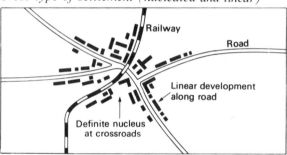

Railway

Road

Linear development along road

Definite nucleus at crossroads

Road

House

River

Dispersed settlement

Dispersed settlement An area of dispersed settlement is quite different. It usually consists of houses and is often some distance, often several kilometres from the nearest village. Dispersed settlements are often associated with extensive farming, e.g. the wheat and sheep regions of South-East Australia.

Dispersed settlements in a Yorkshire Valley in Great Britain.

Isolated type A single house or farm, very remote from any other settlement, constitutes an isolated settlement.

URBAN SETTLEMENT

Urban settlements are called *towns, cities* or *conurbations.* The first towns probably developed in the river lowlands of the Tigris, Euphrates, Indus and other sub-tropical rivers, also in the Hwang Ho Valley in China. These towns grew from the need of the people to live together for security and trade. Also, the area was productive enough to free some people for commerce and administration. A town today is often a centre for administration, banking, commerce and education, and it is intimately connected with the region surrounding it. A town is the product of man's activities. A city is a large town, and a conurbation is several towns joined together.

Urban settlements can be classified in several ways, but the three most useful are:
A according to location
B according to function
C according to size (hierachy)

Classification of settlements according to location

On a river bend The bank of a river meander is a frequent site for the growth of a settlement and there are many examples of large urban settlements of this type. Some of the best known are *Dnepropetrovsk* (on the R. Dnieper, U.S.S.R.), *Orleans* (on the R. Loire, France), *Kaifeng* (on the Hwang-ho, China), and *Telok Anson* (on the R. Perak, Malaysia). This site has a long waterfront for mooring, water supply and effluent disposal.

2 *At a river confluence* The volume of water increases at this point and navigability usually increases. Several trade routes by river may meet at a river confluence. Because of this many large settlements have arisen, such as *St. Louis* (confluence of the Mississippi and Missouri, U.S.A.), *Khartoum* (confluence of the Blue and White Nile, Sudan), *Wuhan* (confluence of the Yangtse-kiang and Han, China) and *Kuala Lumpur* (confluence of the Gombak and Klang, Malaysia).

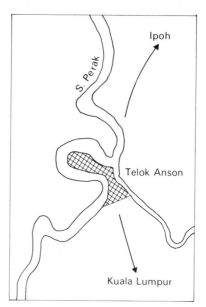

River bend town – Telok Anson (Malaysia)

Confluence town – Khartoum

 *Where a river can be forded* (or easily crossed or bridged) e.g., *Oxford, Rome,* and *Paris.* Paris makes use of an island.

4 *At the entrance or exit of a gorge* The valley at either end of a gorge is wider than it is in the gorge and settlements often develop at each end of a gorge. Examples are *Bingen,* where the R. Rhine enters the Rhine Gorge, and *Bonn* where it leaves the Rhine Gorge (both in West

Germany), *Ichang* and *Wanhsien* on the R. Yangtse-kiang (China), and *Vienna* on the R. Danube (Austria).

Gorge towns – Bingen and Bonn

5 *At the head of a river plain,* for example, *Dresden* (East Germany) and *Taskhent* (the U.S.S.R.).

6 *Where a river leaves or enters a lake* Good examples are *Geneva,* where the R. Rhone leaves Lake Geneva (Switzerland), *Duluth,* where the St. Louis River enters Lakes Superior, and *Detroit,* where the St. Clair River enters Lake Erie (both in the U.S.A.).

7 *On an island in a river, or on a prominent hill, which affords a defensive site* Perhaps the best example of the former is *Paris,* on the R. Seine (France). *Edinburgh* castle and the related old

Where a river leaves a lake – Detroit

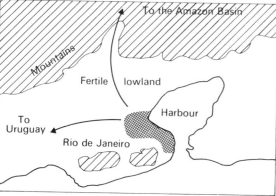

Natural harbour settlement – Rio de Janeiro

settlement is located on a steep-sided volcanic outcrop and it is a good example of the latter. *Durham* (England) is on a defensive site on the hilly core inside an incised meander, while *St Malo* (France) is on a rocky tidal island off the coast.

8 *Near the limit of navigation on a river* At such a point ships are loaded and unloaded, and hence river ports may develop. Some examples are *Wuchang* on the Yangste-kiang (China), *Rouen* on the R. Seine (France), *Bremen* on the R. Weser (West Germany), and *York* on the R. Ouse (the U.K.).

9 *At the lowest bridging point on a river* This often gives rise to trans-shipment ports, and route centres. Some examples are *London* on the R. Thames (the U.K.), *Hamburg* on the R. Elbe (West Germany), *Nanking* on the Yangtze (China). The 'lowest bridging point' moves downstream as technology advances and wider river widths can be spanned, but the former trans-shipment ports usually remain, once established.

10 *On a river where falls occur* Waterfalls limit navigation but they sometimes are utilised first for developing water power, and later hydroelectric power. When this happens, settlements arise at, or near the site of the fall. Good examples are *Sault St. Marie*, at the eastern end of Lake Superior (the U.S.A.), *Schaffausen* on the R. Rhine (West Germany), and *Buffalo* near the Niagara Falls (the U.S.A.)

11 *On an indented coast where there are sheltered anchorages* for example, *Singapore*, whose harbour is sheltered by the two islands of Sentosa

London, capital of the U.K., flanks the River Thames for several kilometres

and Pulau Brani, *Tokyo* (Japan), *Bergen* (Norway) and *Rio de Janeiro* (Brazil).

2 *In the centre of a fertile plain* Good examples include *Harbin* (Heilungkiang) and *Paris* (France).

Fertile plain town – Harbin

Lakeside town – Entebbe

13 *Where routeways meet* Two good examples are *Chicago*, which is the meeting point of road and rail routes going from east to west and from south to north, as well as being at the end of the Great Lake water route across Lake Michigan (the U.S.A.), and *Singapore* where east to west sea routes via the Strait of Malacca cross north to south air routes from Europe to Australia, and where both of these meet the southern end of land routes from Malaysia, Thailand and neighbouring countries.

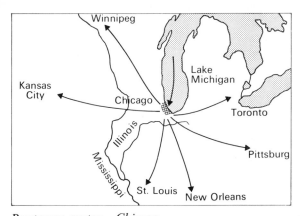

Routeway centre – Chicago

14 *By the side of a lake.* Lakes may offer inland transport, fishing, or tourist attractions, or all three. There are plenty of examples of lake-side settlements, such as, *Entebbe* on Lake Victoria (Uganda), *Como,* on Lake Como (Italy) and *Geneva,* on Lake Geneva (Switzerland).

15 *Near to mineral wealth* There are many examples of towns which have grown at the site where minerals are mined. Some of these are located in environments conducive to settlement, for example, *Pittsburgh* (the U.S.A.), *Essen* on the Ruhr Coalfield (West Germany), *Sudbury* (Canada) and *Johannesburg,* the gold-mining town (S. Africa). Other mining towns have grown up in regions which are certainly not conducive to settlement. Examples are, *Kalgoorlie,* the gold-mining town of Australia, *Potosi,* the once important silver-mining town of Bolivia and *Chuquicamata* the copper-mining town of Chile.

Mining towns – Johannesburg and Witbank

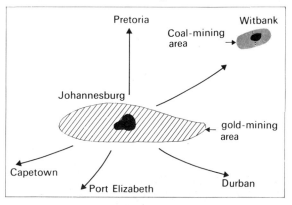

345

16 *Coastal plain towns* Some coastal plains are backed by steep-sided highland on the landward side. If a pass occurs across the highland and opens onto the plain, a town may develop near the entrance of the pass. *Bombay* is located on the western coastal plain of India at a point where it commands the gap across the W. Ghats. Some coasts have very little coastal plain, with the mountains coming down to the sea. Such limited areas of coastal plain, especially if they can be cultivated, or if the nearby sea provides good fishing, may see the development of scattered settlements focussing on one large central settlement. *Trondheim,* on the west coast of Norway is such a town.

Coastal town – Bombay

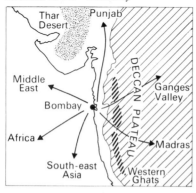

17 *On levées,* or *on scattered patches of raised, dry soils in a region which floods seasonally or periodically* Some flood plains, such as the Mississippi flood plain, have levees which serve as useful sites for settlements. Other lowland areas which often flood have low mounds which are just above flood level. Such mounds often become the sites for settlements. *Ely,* in eastern England is an example. Settlements of this type also occur in parts of the rice-growing regions of Asia, for example, parts of Kelantan and Kedah in West Malaysia. *Rangoon,* on the eastern side of the Irrawaddy Delta (Burma), is another a good example of such a settlement.

B Classification of settlements according to function

Most settlements are not new. They have grown up over a long period of time and as a result their present-day functions may not be the same as they were when the settlements first began to develop. Although most towns can be classified according to definite functions, the function of a town does not always explain the origin of the town. For example, one of *Oxford's* main functions is motor engineering but the town originated as a university centre. *Benares,* which originated as a religious centre, now has industrial and commercial function as important as its religious functions. Although most towns have several functions it is possible t define the main function of most of them.

The main types of town according to function ar as follows.

1 *Ports* These are settlements which are site either on the coast, on a river, on a lake or o a canal, whose main function is the handling o goods brought into, or sent out of the settlement It handles goods for an area which may be a part of a country, or a whole country, or severa countries. The area which a port serves is callec its *hinterland.* The main types of port are a follows:

(a) *Seaports* These handle the imports and exports of a region which may be a part of a country, or a whole country, or several countries Ports are large and are designed to allow large ships to berth alongside wharves, or to ancho in nearby sheltered harbours. They also have facilities for handling large quantities of goods and sometimes passengers. Good examples include *Glasgow* (the U.K.), *Hong Kong, Singapore, Hamburg* (West Germany), *Yokohama* (Japan), *Sydney* (Australia), and *Auckland* (New Zealand).

(b) *Outports* The trade of a port is rarely static. In general the trade of most major ports has increased enormously over the past 50 years, and some ports, because of their physical setting, are unable to expand to handle the increased trade. When this happens a new port is sometimes constructed to handle a part of the trade of the original port. The new port is called an *outport* and it is often situated near to the original port. Whether an outport is constructed or not, depends on whether deep water wharves can be constructed. Examples of outports are, *Avonmouth* for Bristol (the U.K.), *Cuxhaven* for Hamburg (West Germany), *Tanjong Priok* for Jakarta (Indonesia), and *Freemantle* for Perth (Australia). These are related to the size of ships and the depth of water.

Outport – Tanjong Priok

ort handling facilities at New Liverpool Dock (U.K.)

(c) *Entrepôts* These ports handle the imports and exports of other countries. *Singapore, Colombo* (Sri Lanka), and *Rotterdam* (Holland) are examples.

(d) *Packet stations* These are ports whose main function is to handle mail and passengers moving across narrow stretches of water. Some examples are *Harwich* and *Dover* (the U.K.), *Calais* (France) and *Ostend* (Belgium).

(e) *Inland ports* These are located far from the sea. They are connected to the sea either by river or canal. *Duisburg-Ruhrort* (West Germany), *Manchester* (the U.K.), and *Duluth* (the U.S.A.) and *Wuhan* are examples of inland ports.

(f) *Fishing ports* Good examples are *Grimsby* and *Aberdeen* (the U.K.), *Bergen* (Norway), and *Halifax* (Nova Scotia).

(g) *Naval ports* Perhaps the best known of these are *Plymouth* and *Portsmouth* (the U.K.), and *Toulon* (France).

2 *Market Towns* Some towns occupying a fairly central position in a well-settled region, become

he inland port of Duisburg-Ruhrort on the R. Rhine (W. Germany)

A fishing port in Tasmania (Australia)

route centres to which the regions send their local produce for sale, *Norwich* (the U.K.), *Kumasi* (Ghana), *Kano* (Nigeria), and *Seremban* (Malaysia) are good examples of market towns. They are often situated between two regions with different products to exchange at the route centre.

Market Town – Kano

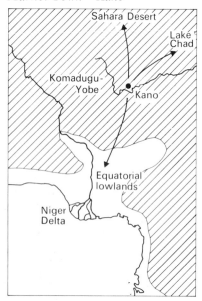

3 *Mining towns* We have already touched on these towns in the classification of settlements according to location. There are plenty of examples of mining towns. Some are locate in well-populated regions, for example, *Merth Tydfil* – coal-mining (the U.K.), *Krivoy Rog* iron ore mining (the U.S.S.R.) and *Anshan* coal-mining (Heilungkiang). Others are locate in sparsely populated regions, for exampl *Kalgoorlie* – gold-mining (Australia), *Chuquic mata* – copper-mining (Chile) and *Sudbury* nickel-mining (Canada).

Mining Town – Pittsburg

4 *Industrial towns* These towns are concerne with manufacturing, either the processing o raw materials, or the production of finishe goods. Examples are *Manchester* and *Leeds* textile manufacture (the U.K.), *Birmingham* engineering (the U.K.), *Pittsburg* – steel manu facture (the U.S.A.), and *Wuppertal* and *Dui burg* – chemical manufacture (West Germany

5 *Banking towns* Some towns have developed into important financial centres. *London, New York* and *Zurich* (Switzerland) are examples of this type of town.

6 *Capital towns* These towns are the headquarters of governments which means that they are administrative centres. Some examples are *Canberra* (Australia), *Washington* D.C. (the U.S.A.), *Brasilia* (Brazil) and *Moscow* (the U.S.S.R.).

7 *Religious towns* These towns have developed in several parts of the world. Many of them are old, and some of them are visited by pilgrims on specific occasions. Examples of religious towns are *Jerusalem* (Israel), *Mecca* (Saudi Arabia), *Benares* (India), *Canterbury* (the U.K.), *Lourdes* (France), and *Lhasa* (Tibet).

8 *Cultural towns* These towns are often very closely associated with educational activites, for example, *Oxford* and *Cambridge* (the U.K.), *Heidelberg* (West Germany), and *Leiden* (Holland).

9 *Resort towns* Some towns owe their origins to beautiful scenery or to mineral waters which are supposed to be of medicinal value. *Innsbruck* (Austria), *Grindelwald* (Switzerland) and *Banff* (Canada) are mountain towns which attract tourists who come to admire the scenery or to take part in local sports, such as skiing. *Bath* (the U.K.) and *Baden* (West Germany) are resort towns which have mineral waters. There are also coastal resorts to which tourists come to enjoy sailing, swimming or just to be near the sea. Examples of this type of resort town are *Brighton* (the U.K.) *Miami* (the U.S.A.) and *Beirut* (Lebanon).

10 *Satellite towns* These are new towns which have been deliberately built either as new manufacturing centres to relieve the congestion of old centres, or to re-house people who work in over-crowded urban centres. There are numerous examples of such towns in the industrial countries for example, *Crawley* (the U,K,). Other examples are *Petaling Jaya* (Malaysia) and *Jurong Town* (Singapore).

349

Classification of settlements according to size

The following table shows how settlements can be classified according to size. Some of these types have already been discussed under rural settlement.

TYPE (SIZE)	DESCRIPTION
Isolated settlement	One house only in a remote area.
Hamlet	Two or three houses without shops, schools or services, in most cases.
Village	Up to several thousand people, usually with small shops, a school, cafe or public house, often with a petrol station and a post office.
Town	Over 'several thousand' people. This varies with the population density: in Australia a settlement with about 5000 people is regarded as a town while in Java, S. Japan, and the densely populated parts of China, this would be called a village. A town has most types of shops some of which are large chain stores and department stores. It provides most types of services.
City	Main towns in a country e.g. large county towns.
Conurbation	Several towns joined together.
Megalopolis	Many cities joined together as in E. U.S.A., and in S. E. Japan.

EXERCISES

1 With the aid of sketch-maps and naming specific examples, carefully describe *each* of the following:
  (a) Nucleated settlement
  (b) Line settlement
  (c) River confluence town
  (d) An outport.

2 (a) Briefly describe the more important factors which influence the location of rural settlement.
  (b) Make a simple classification of the main types of rural settlement.

3 Towns can be classified according to their function.
  (a) Name five such types of towns and give a specific example of each.
  (b) Draw a sketch-map of *two* of the towns named to show their locations in relation to physical features and human activities.

Objective Exercises

1 Which of the following is **not** a main type of rural settlement pattern?
  A nucleated
  B isolated
  C dispersed
  D linear
  E gap

  A B C D

2 Which of these is **not** a type of settlement classified function?
  A at lowest bridging point
  B port town
  C capital city
  D serving a religious centre
  E satellite town

  A B C D

3 The principal difference between a nucleated settlement and a dispersed settlement is that
  A it contains more buildings and more people
  B it is compact, that is, the buildings are clustered together
  C it often occurs in regions of intensive farming
  D it is usually at the meeting points of roads

  A B C

4 The lowest bridging point is often a trans-shipment point because
  A the lowest bridging point moves downstream with time
  B the highest point for ocean-ships moves downstream
  C the large ocean ships today cannot pass under most bridges
  D ships unload at the first point near to land transport

  A B C

350

35

Population Density

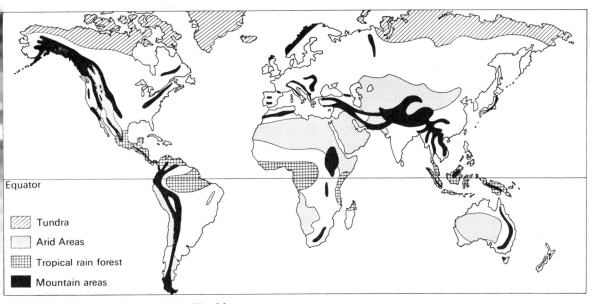

The Major Negative Areas of the World

In 1975 the population of the world was just over 3 000 000 000. Every hour 10 000 babies are born and 5000 people die which means that the world's population increases by 2 000 000 people, or the equivalent of Singapore's population, every 17 days. At this rate the world's population will be about 7000 million people in 2000 A.D. The world's population is not evenly distributed over the earth's surface: some parts are densely peopled and other parts are almost empty.

The land surface of the earth occupies only about 30 per cent of the total surface, and of this about 10 per cent contains very few people either because it is too cold, or too dry, or too hot and wet, or too mountainous. The remaining 70 per cent of the surface is occupied by water which means that 3 000 000 000 people live on 20 per cent of the total earth's surface. This gives an average of just over 50 people per 2·5 square kilometres (1 square mile). Most of the people live on this part of the earth's surface because of suitable climates, fertile soils, mineral deposits and other natural resources.

The density of population, which is the number of people per unit area, varies considerably from region to region. This is because the degree of suitability for human settlement varies from one region to the next.

Population Density			
Country or Area	*Size (sq. km)*	*Population*	*Density per sq. km*
Java	128 000	85 million	664
Sumatra	422 000	22 million	52
Singapore	580	2.4 million	4137
Australia	7 500 000	12 million	2

This table shows the areas and populations of four regions. You can see from the table that the population densities are much higher for Singapore and Java than they are for Australia and Sumatra. This may lead us to think that Singapore and Java are seriously over-populated, and that Australia and Sumatra are equally seriously under-populated.

What we have to examine is not the total surface area but the area that is habitable, and the degree to which it can be used by man. Java has extremely fertile volcanic soils which have been cultivated for a long time. The rich soils give high crop yields which have resulted in the development of a dense settlement pattern. The high yields also resulted in a surplus which enabled some people to take part in non-agricultural activities, and because of this, towns developed. In contrast, the soils of Sumatra are not as fertile and the population pattern is not as dense as in Java. The high population density of Singapore is not based on agricultural productivity. This island state has very little land which is suitable for agriculture, but its geographical position, astride the major shipping routes linking eastern Asia and Australia with Europe, has resulted in its becoming a trading centre. In recent years it has developed industries which have broadened its economic base, which has resulted in the country being able to support a dense urban population. Australia, on the other hand, is largely desert and its population is concentrated in the south-east, and in a few areas in Western and Northern Australia, mainly near to the coast.

Factors affecting population density

We have already seen that there are several factors which affect this, but for most countries the main factors responsible for population density are as follows:

1 **Agriculture** Extensive agriculture which gives low crop yields per hectare will support a smaller population than intensive agriculture which has high crop yields per hectare. Intensive market gardening in Holland, and intensive padi cultivation in Java are good examples.

2 **Industry** Industrial activities are much more productive than agricultural activities and the population density of an industrial region is nearly always higher than that of an agricultural region. And some countries are highly industralised, for example, Great Britain where over 50 per cent of the population live in industrial regions.

3 **Commerce** Centres of trade are usually centres of high population density. This is especially true of ports, for example, New York, London and Singapore.

Some regions have a higher population density than others

The natural environment of most regions usually enables the inhabitants to carry out several types of work. If a region contains forests, good soils

and valuable mineral deposits, it is possible tha lumbering, agriculture and mining may all tak

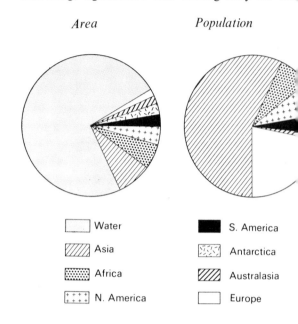

Area *Population*

☐ Water   ■ S. America

▨ Asia   ▨ Antarctica

▨ Africa   ▨ Australasia

⊹ N. America ☐ Europe

place. But whether they take place depends on th suitablity of the climate (for agriculture) and th accessibility of, and demand for, minerals (fo mining). Some regions can be more easily utilise than others. Lowlands with warm, moist fertil soils can be more easily cultivated than the wet cold soils of temperate plateaus.

Over the centuries, man has picked out the fertil regions and has cultivated them intensively, an these regions, because of abundant harvests, hav become densely populated.

The coal and iron deposits of western Europe, plu the ingenuity of the people, gave rise to the Industria Revolution, which resulted in this region becomin densely populated in the 19th Century. It als resulted in its peoples having a high standard o living today. There are regions in Asia with a hig density of population, for example, the Gange Valley and Java, but most of the people in thes regions are poor because they are either estat workers or subsistence farmers. Estate worker usually receive lower wages than factory or offic workers. Subsistence farmers may produce a sma surplus which can be sold, but they do not get goo prices for this surplus. This situation is commo usually in most parts of Asia where half the world population lives. Other regions, like the tundra land and the Amazon lowlands, are difficult climati regions and they have a low density of populatio But although the natural environment influence the population density, the educational level of th people also plays an important role.

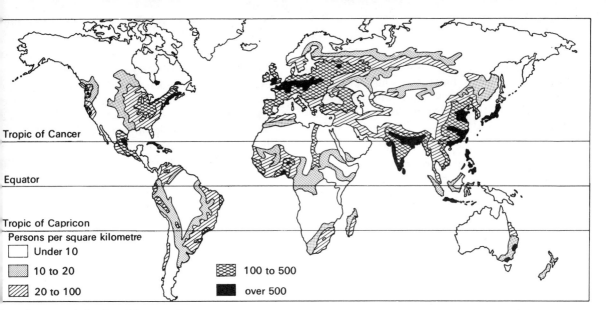

Distribution of the World's Population

DISTRIBUTION PATTERN OF THE WORLD'S POPULATION

Nearly one half of the world's population lives in monsoon Asia, chiefly on the river and coastal lowlands. There are two main regions:

China and Japan.

India, Pakistan and Bangladesh and Sri Lanka. These two regions are separated by the high, rugged mountain country of the Himalayas, and its off-shoots into Burma, northern Thailand and south-west China, and they together occupy the south-east part of the great land mass called Eurasia (Europe and Asia). Almost on the opposite side of this land mass is another region of great population. This is western Europe which has a population of almost 700 000 000. These Asian and European regions are separated by a vast 'empty belt' which extends from the Sahara Desert eastwards through Arabia to the dry temperate desert lands and mountains and plateaus of Siberia, and central and western Asia. Finally, the eastern part of North America forms a fourth region of dense population (approximately 130 000 000). In addition to these regions there are many other regions which have large populations, for example, Java, the lower Nile Valley, parts of west Africa and the south-east coastlands of Brazil.

There are three main population belts:

South, South-East and East Asia.

This region has a population of 2000 million of which there are 750 million people in China, 540 million people in India, 112 million in Pakistan and Bangladesh and 105 million in Japan. Within each of these countries there are regions of out-standingly high population density, for example, the fertile valley lowlands of the Hwang-ho, Yangtze kiang and Si-kiang (all in China) and the Indo-Gangetic Plain of India, Pakistan and Bangladesh.

In the rest of this part of Asia there are other similarly densely populated regions such as, the Irrawaddy Delta of Burma, the Chao-Phraya Basin of Thailand, and the Red and Mekong Deltas of North and South Vietnam. And on the islands of South-East Asia there are densely-populated lowlands in Java (Indonesia) and the Philippines.

2 *Peninsular Europe*

The most advanced countries, in terms of in-dustrial and technological development, occur here. The total population of this part of Europe is about 500 million people who live in the densely populated countries of the U.K., West Germany, France, Benelux (Holland, Belgium and Luxemburg) Denmark and Italy.

The principal reasons for the high density of population in this region are (i) the temperate climate which is conducive to human settlement and to men working hard, (ii) the large variety of natural resources such as good soil, coal, iron ore and other metals, timber and fertile sur-rounding seas, (iii) the 'pull' of the sea which has encouraged men to explore and thus develop trading links with other parts of the world.

3 *North-East America*

The population of this region is almost 140 million, most of whom live in the extensive industrial belt extending from the Atlantic seaboard, south of the St. Lawrence, south-westwards along the coast to New York and westwards through the Pittsburg region to the shores of Lakes Ontario and Erie.

A STUDY OF TWO REGIONS WHICH HAVE CONTRASTING POPULATION DENSITIES

The two regions to be studied are Java, which has a population density of over 400 people per square kilometre, and Saudi Arabia which has a density of 3 people per square kilometre.

1 Java lies just within the equatorial latitudes but because of its position it has an equatorial climate which is modified by monsoon winds. Most of the island receives rain throughout the year, though this has a seasonal pattern. Monthly temperatures are high. In comparison Saudi Arabia, which lies in the tropical latitudes, has an arid climate consisting of a hot and a warm season. The climate of Java is more favourable to human settlement than is the climate of Saudi Arabia.

2 Most of the lowlands and the lower slopes of mountains in Java have fertile soils. This is especially true of the areas containing volcanoes. In contrast, the only regions in Saudi Arabia which contain usable soils are those of parts of the coastal plains and the interior oases.

3 The warm, moist climate and fertile soils of Java have resulted in a luxuriant growth of natural vegetation, mainly of forests. Most of these have been cut down so that the land could be used for growing padi, sugar cane, spices and other crops. Agriculture has been established for a long time and all the land which could be cultivated, is intensively cultivated. Indeed, sloping land has been carefully terraced for padi cultivation and the total production of food crops is enormous: sufficient to feed the island's large population. Large parts of Saudi Arabia in contrast have no vegetation at all and the land is either stony, or rocky, or sandy. Agriculture can take place only in those regions with soils and with access to water, and the total production of food crops is very limited. Some areas have a steppe-like vegetation which provides food for the sheep, camels, and goats of nomadic herdsmen.

4 Although both Java and Saudi Arabia contain some minerals, the mineral oil deposits of Saudi Arabia are of greater value than those of Java. However, in neither case have the mineral deposits had any important effect upon the population density in either country.

5 Agricultural products have been an important export of Java for a long time and this has given rise to trade between Java and her neighbours as well as with countries in Europe. This in turn has helped various types of industry to be established in Java. In comparison, Saudi Arabia's trade is only in oil and this has not given rise to the development of manufacturing industries.

POPULATION GROWTH AND SOME OF ITS PROBLEMS

The world's population is increasing at an alarming rate. Each year another 50 000 000 people are added all of whom have to be fed, housed and clothed. Statistics show that for the world as a whole, the availability of food, adequate housing and basic medical and educational facilities are never sufficient to meet the demands of the growing population.

The annual world population increase is not evenly spread over all countries. In some countries the rate of increase is higher than in others. The increase for India is higher than that for Malaysia which in turn is higher than that for Japan. In general agricultural populations have a higher rate of increase than do industrial populations. Also, countries with a low standard of living, for example, Indonesia and India have a higher rate of increase than do countries with a high standard of living, e.g. the U.S.A., the U.K. and West Germany. Urbanisation causes a drop in the birth rate as people can see the advantages of small families in housing, education and material wealth. Most of the countries where populations are increasing at a rate higher than the average for the world, are in monsoon Asia and Africa. Countries such as India, Bangladesh, parts of Indonesia and China are by far the worst affected, and it is these countries, especially India which produce insufficient food (or money to buy it), for their peoples. This state of affairs is clearly seen in times of inadequate rainfall or excessive rainfall when crops fail and widespread famine occurs. Besides having insufficient food, countries of this type also have insufficient housing and clothing and insufficient medical and educational facilities for their people. Such countries can be said to be over-populated because they are unable to provide all members of their populations with a standard of living which gives a proper diet and effective housing, medical and educational facilities. That is, such countries would seem to be better off with less people. However, this is, in many borderline cases, a matter of opinion.

Over-Population

When a country has more people than its resources can support, then that country can be said to be over-populated. We have seen that the world's population is increasing at an alarming rate and that by the end of this century it will have risen from the present level of about 3600 million to over 7000 million. Practically all countries are faced with major problems arising out of the rapid increase in their populations over the past 50 years, and those that do not yet have these problems will probably have them within the next 20 years. Several countries are already over-populated, but their populations are still increasing. This is true of India, Pakistan and Bangladesh. Since these countries have agricultral economies, their population problem is one of rural over-population. Rural over-population also occurs in Java, parts of Sumatra, and in the valleys of several South-East Asian rivers, especially that of the Mekong. Rural over-population also occurs in parts of Africa, S. America, China and the Philippines, but positive action has already been taken in some countries, and as a result, the over-population problem has greatly diminished. Real and imminent over-population in other parts of the world such as, Japan and Hongkong, is urban over-population because the economies of these countries are industrial and commercial. This diagram shows the expected increases in urban and rural populations for developed countries (mainly western countries) and for developing countries (mainly Asian countries) for the period 1960 to 2000.

The Prevention of over-population

The greatest single challenge facing nearly all countries is the need to control the growth of population. Improvements in sanitation, hygiene, and medical science, over the past 100 years, have resulted in a sharp decrease in the death rate of babies and children, and in consequence an overall increase in births over deaths.

In some countries the natural rate of increase of population is as high as 25%. The only way to control population increase is to introduce effective birth control. But before this can be achieved, religious, and sometimes political objections have to be overcome, and in addition, money has to be made available to enable poor people to obtain contraceptives cheaply or to receive medical treatment free-of-charge. The idea of family planning is based on the fact that if two parents have an average of two children, then the population will not increase. Action of this type has already been taken by many countries, of which, in Asia, China, Japan, India, and Singapore are outstanding examples. But urgent action is still required in many regions, to

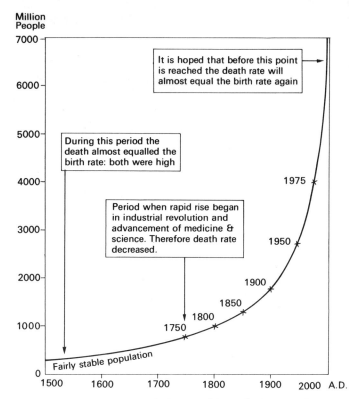

World Population Growth from 1500, with Projected Growth to the end of 20th Century.

ensure that effective measures are taken, to prevent the population of a country from increasing to a point which will threaten, and eventually erode the standard of living of those regions.

Possible Solutions for Over-Population

A *Rural Over-population*

We have seen that rural over-population is very common in several Asian countries where population densities are often between 400 and 1000 persons per sq km. In South-East Asia, and in most of the tropical regions of the rest of Asia, these high population densities are dependent completely on rice cultivation. If the rice crop fails, that is, if the yield per unit area is below average, then famine inevitably follows. The late arrival of monsoon rains or insufficient rain during the growing season, or the ravaging of the crop by insect pests, always results in the crop failing. If the rural population of a country is greater than its agricultural resources can support, then it will have to introduce measures to balance its population to its resources. One of the most important actions is to increase agricultural productivity. This can be done by:

(a) Introducing new farming techniques. These are intended to increase crop yield:

355

2000							
1990							
1980							
1970							
1960							
0	1000	2000	3000	4000	5000	6000	7000

Population in million

Developed Countries

☐ Urban population

▨ Rural population

Developing countries

■ Urban population

▦ Rural population

A Comparison of Rural and Urban Populations of Developed and Developing Countries.

1 By making extensive use of chemical fertilizers.

2 By improving soil fertility by ensuring that all trace elements, necessary to proper plant growth, are maintained in the soil, Such trace elements are cobalt and copper.

3 By using better quality seed, especially those resistant to disease.

4 By using new food crops especially in those areas which are not suitable for the cultivation of basic crops such as rice.

5 By improving the breed of animals, and, if necessary, by introducing animal farming to supplement crop farming.

6 By increasing the area under crops through land drainage and irrigation.

7 By using farming machinery. This will result in less people being employed directly in farming. Those no longer employed in farming can be trained to manufacture farming equipment, to process agricultural crops and to extend agricultural areas by developing drainage and irrigation works.

(b) By land reform.

In many parts of Asia farmers do not own the land they cultivate, and even those that do usually have areas of land which are too small to cultivate economically. This problem can be overcome:

1 By making it possible for as many farmers as possible to own the land they cultivate which will result in the farmer having a real incentive to make his land highly productive.

2 By the abolition of land division based on inheritance. Laws will have to be introduced to prevent the land of a farmer being divided on his death, into areas which are too small for effective cultivation. The creation of large units also makes mechanisation possible. This has been achieved in China over the past twenty years.

B *Urban Over-Population*

Although urban over-population is not as serious as rural over-population in Asia, evidence suggests that it can become a major problem over the next 25 years unless action is taken now to ensure that urban job opportunities increase and that housing, education, and transport facilities, are expanded. The movement of people from the rural to the urban areas took place in many European countries during the 19th and early 20th Centuries, but in Asia it has only taken place over the last 25 years. This has been caused partly by crop failures, and partly by the general poverty in farming communities. This movement has resulted in the over crowding of towns and cities, and because industrial and commercial activities have not expanded to meet this increase in urban population, unemployment in many urban centres is high because these centres contain more people than there are jobs available for them. Some of the measures which can be taken to reduce these problems are:

(a) The building of *satellite* or new towns. This results in:

356

1 A reduction in over-crowding.
2 A reduction in pollution.
3 The creation of work in the building and construction industries.
4 An expansion in commercial and industrial activities.

) The building of high-rise blocks of flats. This results in:
1 The creation of open spaces which can be used for parks and for recreation areas.
2 A reduction in over-crowding.

) The expansion of existing commercial and industrial activites and the creation of new ones.

onculsion

he various suggestions which have been made for lieving over-population, both rural and urban, e only short term measures. Large areas of Asia e still in a developing stage and the problem of ver-population is becoming increasingly serious. drastic reduction of the birth rate, coupled with considerable increase in agricultural and industrial roductivity will have to take place if this problem to be resolved. But this requires both finance and sources on a vast scale, and since most of the eveloping areas in Asia are lacking both of these, iormous help will have to be given by the eveloped countries.

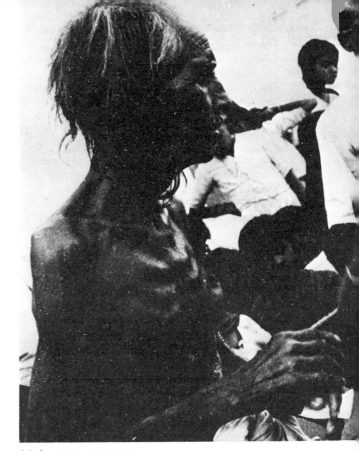

Malnutrition in India

ERCISES

(a) On an outline world map shade:
    (i) *two* areas which have high population densities.
    (ii) *two* areas which have low population densities.
(b) Choose *one* area from (i) and another from (ii), and from each, describe the conditions which are responsible for the population density.

Explain *any three* of the following statements:
(a) Although the middle and lower Ganges Valley are not industrialised they have very high population densities.
(b) Some areas which have a population density as low as 10 persons per square kilometre are said to be over-populated.
(c) Singapore Island grows only limited quantities of food yet it supports a population of 9 000 persons per square kilometre.
(d) Areas with no material resources do not always have low population densities.

bjective Exercises

The world's population, in millions, in 2000 A.D., is expected to be about

A 4000
B 5000
C 7000
D 10 000
E 15 000

A B C D E

2 Monsoon Asia now contains nearly . . . . . . of the world's population

A $\frac{1}{4}$
B $\frac{1}{3}$
C $\frac{1}{5}$
D $\frac{1}{2}$
E $\frac{2}{3}$

A B C D E

3 The population density of a country depends on
A the rate of population increase
B the size of territory and the size of population
C the birth rate and the death rate
D the standard of living and urbanisation

A B C D

4 Which one of the following refers to *Population Density?*
  A The total population of a region in respect to the region's productivity.
  B The number of people per unit area of land.
  C The average standard of living for a given region.
  D The total population of a country in relation to inhabited land area of that country.

  A   B   C   D

5 Which of the following countries has the highest population density?
  A Australia
  B India
  C Singapore
  D Indonesia

  A   B   C   D

6 Most of the people in Australia live within fifty to eighty kilometres of the coast. This may be explained by all of the following **except**
  A the economy in the 19th Century was based on the export of pastoral farming products
  B mineral ore deposits are concentrated in coastal areas
  C rainfall reliability is high in coastal areas
  D manufacturing industries developed in the 20th Century required imported raw materials

  A   B   C   D

7 Countries which have very large populations and a very high population growth are usually those countries which
  A possess very varied natural resources in considerable quantities
  B produce inadequate food supplies
  C are heavily industrialised
  D have an enervating climate

  A   B   C   D

Answer the following three questions by using the accompanying map.

8 In which of the following areas is the population density very low because of adverse dry climatic conditions
  A *1*
  B *2*
  C *3*
  D *4*

  A   B   C   D

9 A high population density based on intensive agriculture occurs in the area numbered
  A *1*
  B *2*
  C *3*
  D *4*

  A   B   C   D

10 The area numbered *4*
  A is thinly populated
  B has a high population density based on agriculture
  C is moderately populated
  D has a high population density based on industry

  A   B   C   D

11 The world's population in the last one hundred years has increased enormously mainly because
  A families increased in size
  B fewer natural catastrophes occurred than in previous centuries
  C medical science improved
  D people married at earlier ages

  A   B   C   D

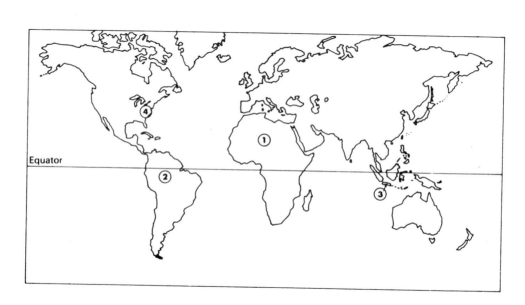

358

36

Conservation and Pollution

The exploitation of the natural environment has resulted in the creation of numerous problems which have adversely affected the *ecosystems* (plant and animal communities) of many parts of the world. In some instances, natural resources such as soils, have been destroyed, and fresh water supplies have been heavily polluted with the result that our survival is now being threatened. Fortunately, the governments of many countries are aware of these problems and attempts are being made to try to develop ways and means by which the environment can be utilised more meaningfully, without causing major changes to the balance of plant and animal species in relation to the natural environment. The problem facing man is how best can he utilise, or develop, his natural environment, and, at the same time, conserve it. In every environment where man's activities are minimal, there is a natural balance among animal and plant species, and between these and the physical environment. If this balance is greatly disturbed, as it is for example when an animal or plant species is destroyed by man, then a chain reaction sets in which may ultimately result in the partial or complete destruction of the original ecosystem.

Conservation

We have seen that the population of the world is increasing at an alarming rate, and that by the end of this century the population will reach the enormous figure of about 7000 million. As the population expands, ever increasing demands will be made on land which has been little touched by man. Some of this land will be needed for agriculture, some for industry, and some for building new settlements. In addition, increased demands will be made on all types of natural resources such as, mineral resources, forests and products of the sea. When the remaining lands and resources of the world are developed, it will be essential to ensure that the methods used do not result in either the destruction of the balance of nature, or the over-exploitation of soil, mineral, animal and plant resources. In other words, the utilisation of all these resources must be accompained by the conservation of them.

Conservation of natural resources

Some natural resources such as oil, gas, coal and metallic minerals are finite and irreplaceable, and the utilisation of these must be carefully controlled. Other natural resources, such as trees, are replaceable, and the utilisation of these should be accompained by the replacement of them.

Forests

As mature trees are felled, young trees should be planted to ensure that the forest reserves do not diminish. In other words, the annual cut of timber should not exceed the net annual growth of trees. *Tree farming* is practised in many parts of the temperate latitudes. This involves the planting of trees on large areas of land for future use. As the trees of the plantations are cut, new trees are planted. However, the ruthless exploitation of natural forests is still going on in many countries. For example, in the U.S.A. the annual cut of timber exceeds the annual growth in all major forest areas. In California, the annual cut is about 7000 million board feet while the net annual growth is estimated to be only 4000 million board feet. This means that in California alone, the forest areas are decreasing by about 3000 million board feet per year.

Marine animals

Fishing is an activity which often results in the over-exploitation of this natural resource. This resource is also replaceable and provided the number of fish caught does not exceed the number replaced either by natural means, or by man-controlled fish breeding, the seas and rivers of the world can continue to provide man with a valuable source of food in the future. Most countries which have large fishing industries stipulate that all small fish which are caught must be returned to the sea immediately to enable them to reproduce themselves. But whether this is done or not depends on the fishermen. Sometimes the small fish are not thrown back into the sea immediately and large numbers of them die. Another factor which has adversely affected the natural balance among marine animals is the preference for certain types of fish in many European countries, as well as in North America. This preference has led to the over-fishing of some fish species which has resulted in a dramatic reduction in the population of these species and certain other inter-dependent species.

Fish farming is an important activity in many Asian countries and it has been practised for a very long time in countries like China and Japan. It is usually integrated with general farming activities, such as rice and vegetable farming. In fish farming, the type and number of fishes bred are carefully controlled. Fish farming usually takes place in fish ponds and rivers which means that it can often be located near to the consuming centre.

Water

Water is one of the most important natural resources in that all forms of animal and plant life are dependent on it. Although the amount of rain falling onto the earth's surface varies both regionally and annually, the total amount received each year remains fairly constant. However, some parts of the world are receiving less rain each year and these areas are becoming increasingly arid. This is happening in north Africa where the Sahara Desert is moving southwards by several thousands metres annually. Such increasing aridity is sometimes the result of bad farming practices, such as the over-grazing of semi-arid grassland. The loss of fresh water to man and his animals through practices of this type is serious only on a regional basis, that is, it is confined to a few areas. Far more important is the loss of fresh water caused by pollution. This loss is much more widespread and it occurs in most industrial regions.

Conservation of the earth's fresh water supplies is of pressing importance and it is made all the more urgent by the rapidly expanding world population and the increased demand for fresh water that this will bring. Cities of the future may well have to be planned so that all the rain falling on the roofs of buildings is collected and led away to nearby reservoirs by underground pipes. Such water would require only limited treatment to make it fit for human consumption. The rain falling on the roads and other areas of ground, in these cities of the future, will probably be led away by another network of underground pipes to reservoirs whose water is for industrial use only.

The anticipated acute shortage of fresh water in many of the heavily populated industrial regions of the world, especially in Europe and North America, has led to serious discussions by scientists on the possibility, and feasibility, of towing large masses of ice from the polar regions to the temperate regions to provide water for specially constructed reservoirs.

Whilst this is still only theoretical, the water shortage problem will be so real by the end of this century that a means will have to be found to bring the fresh water, at present locked up in the polar ice masses, to those parts of the world where the problem is most acute.

Conservation also has other meanings

Conservation is sometimes used to refer to the protection of plant and animal species in their natural environments. Areas which are protected for this purpose are often called *nature reserves*. Such reserves sometimes have one or more other functions such as collecting surface run-off which is directed into reservoirs and providing recreational areas for urban dwellers.

When a nature reserve has been established, it has to be carefully controlled by man to ensure that no one species of animal or plant becomes dominant over other species. In some parts of East Africa, the protection of the elephant in reserves from the activities of poachers resulted in an increase in the elephant population which led to the destruction of trees at a faster rate than they were replenished by natural means. In order to prevent the destruction of their tree cover, which would have led to severe soil erosion, some of the elephants had to be killed to keep the elephant population at a controllable level. Similar action has to be taken in seal sanctuaries to prevent the seal population from growing too rapidly. If this is not done, the number of seals will increase at the expense of the fish on which they feed.

Pollution

Pollution of the natural environment by man has been going on for centuries, but prior to the Industrial Revolution, the damage caused was negligible. The use of coal, and later oil, as a source of energy, plus the development of chemical, engineering and other industries, have caused pollution to become one of the major problems facing man today. Pollution has become increasingly alarming over the last five decades with the result that today most countries are taking active measures not only to curb the rate of pollution but also to reduce it.

Causes of Pollution

Pollution arises in several ways.

1 In the mining of coal and minerals, large quantities of mineral waste were dumped, and in some regions are still being dumped, on surrounding land which makes that land of little, if any, use for a long time.

2 The discharge of waste gases, and often poisonous fumes, into the air by industries such as the iron

and steel industry, chemical and engineering industries. This type of pollution, often called *atmospheric pollution,* is usually dangerous to plants, animals and man.

The discharge of waste materials from industry and urban settlements into rivers, lakes and seas. Some of this waste contains toxic matter which is poisonous to many plant and animal species, both fresh water and marine. In addition, fresh water which is polluted in this way, has to be very carefully treated before it can be used for drinking. Sometimes, it can never be used for drinking, as for example, when it contains chemicals such as mercury which is poisonous to man. This type of pollution is called *water pollution.*

Problems created by Pollution

The dumping of industrial and urban waste onto the land, the tipping of industrial fluid waste into the rivers and the sea, and the discharge of waste industrial gases into the air, all endanger the health and well-being of plants, animals, fish and man. Some years ago, *Minimata Disease* occurred in Japan as a result of mercury waste being tipped into the coastal waters near to Minimata in Japan. The waste came from a large chemical factory which has been producing chemicals since about 1920. As early as 1927, it was noticed that the catch of fish around Minimata declined annually, and it was soon recognised that this was caused by the deadly methylmercury waste that was being dumped into the sea. Although the chemical company paid compensation to the fishermen, the dumping continued. Then, in the 1940's, it was noticed that cats around Minimata were developing a strange disease which caused many of them to go mad: At about the same time, some people in the same region developed paralysis of the limbs. This illness was particularly common in the fishing community.

Perhaps the greatest health hazard created by pollution results from the discharge of radio-active materials in the seas and air. The increased development of nuclear power stations will inevitably result in increased radio-active pollution. This type of pollution is difficult to control. A small leakage of radio-active materials can cause surrounding ground waters to be contaminated which results in the vegetation becoming contaminated. This happened a few years ago, at Winscale in Cumbria (England) and it resulted in cow's milk being contaminated through the grass they ate. To ensure that the radio-active materials did not affect people, all the milk in the affected areas had to be destroyed for a period of several weeks.

The only known way of safely disposing radio-active waste material, is to seal it in reinforced lead containers and to dump this onto the ocean bed. However, the waste materials remain radio-active for a very long time, possibly for several thousands of years, which clearly means that the containers in which the materials are sealed, must be able to remain intact for an equally long period of time.

Summary

We have seen that the Earth is finite, with mostly non-renewable resources. As the world's population rapidly increases, ever-greater demands are being made by man on his environment, which is his home. Nobody wants a deterioration in his quality of life, and we must therefore conserve our environment as we develop it. The environment must be properly managed, and this, because of the very immensity of the operation, can only be undertaken on a cooperative international basis.

GENERAL EXERCISES

1 Explain, with the help of sketch-maps or diagrams, the meanings of three of the following statements:
   (a) Days are shorter in Peking than in Singapore during December.
   (b) Great Circles are frequently used in navigation.
   (c) The length in miles of 1° of longitude increases from latitude 80°N to the Equator.
   (d) When flying from Bombay to Manila, passengers are told to put their watches forward.

2 With the help of well-annotated diagrams, explain why:
   (a) The U.S.S.R. has a large number of time zones.
   (b) The sun never rises above the horizon at the South Pole during June.
   (c) The average temperatures are lower in regions near the Poles than in regions near the Equator.

3 Briefly explain the causes of the winter and summer seasons. Illustrate your answer with diagrams.

4 Explain the following:
   (a) The number of hours of daylight equals the number of hours of darkness along the Equator throughout the year.
   (b) Day equals night all over the earth's surface twice a year.
   (c) The sun is visible in the sky for several weeks during June at the North Pole.
   (d) Winds and ocean currents are deflected to the left in the Southern Hemisphere.

5 Write brief notes on *three* of the following, and illustrate with diagrams where appropriate: sedimentary rocks, continental shelf, earthquakes, joints, igneous rocks.

6 Choose *two* of the following features: fold mountain, rift valley, lava plateau, block mountain. For each one you choose:
   (a) Describe its appearance and explain its formation with the help of diagrams.
   (b) Name an area where an actual example could be seen.

7 (a) Draw a large, annotated diagram to show the structure of a typical volcano.
   (b) Describe how a volcano has been formed and some of the ways in which volcanic eruptions have influenced Man's activities.
   (c) Name one region which has active volcanoes.

Carefully explain the following processes which help in the formation of physical features, and describe the results of these:
   (a) Chemical weathering of rocks in an equatorial climate.
   (b) Mechanical weathering of rocks in a cold climate.
   (c) Weathering of limestones.

9 Write short notes on *three* of the following, and illustrate your answer with clear diagrams: artesian basin, water-table, spring, well, landslide.

10 Describe *three* of the following terms which are as-sociated with a river's course: waterfall, delta, gor[ge], meander. Your answer must contain relevant diagra[ms] or sketch-maps.

11 Choose *two* of the following valley features: ox-b[ow] lake, flood plain, rapids, terrace. For each one y[ou] choose:
   (a) With the aid of diagrams, (i) describe its appearan[ce] and (ii) explain its formation.
   (b) Locate an area where an example could be see[n].

12 The following features sometimes occur in a rive[r] course: delta, flood plain, gorge, waterfall.
   Select *two* of these and for each one:
   (a) Locate an example by means of a sketch-map.
   (b) Describe the example and explain its formation.

13 Draw two contour maps, one of a young river valle[y] and the other of a mature valley, and then describe t[he] characteristic features of each. You should illustra[te] your answer with diagrams.

14 With the aid of clearly labelled diagrams, describe *thre[e]* of the following: spit, delta, hanging valley, caldera.

15 Illustrating your answer with diagrams and sketc[h] maps, describe:
   (a) The appearance and formation of a flood plain,
   (b) The importance of flood plains to agriculture in [a] specific Asian country.

16 You are required to make a geographical study of [a] small river in your country. Describe the appearanc[e] of features which you would examine and draw diagram[s] to illustrate these. You should give specific examples o[f] such features from your knowledge of an actual regio[n].

17 Name *two* features produced by wind erosion and *on[e]* produced by wind deposition, which frequently occu[r] in tropical deserts. For each, describe: (a) its appearanc[e] and (b) its formation. Illustrate your answer wit[h] diagrams.

18 Choose *three* of the following rift valley, wadi, matur[e] river valley, canyon.
   For each one you choose: (a) explain how it has bee[n] formed, and (b) give an example.

19 With the aid of well-labelled diagrams or sketch-maps
   (a) Describe the function,
   (b) Name an actual example of *three* of the following crater lake, delta, U-shaped valley, coral reef, loess plain

20 Choose any *three* of the following features: ria, stack lagoon, spit, sand-dunes, fiord. For each *one* chosen:
   (a) Describe its physical appearance.
   (b) Describe how it has been formed.
   (c) Name a region where an actual example could b[e] seen.

21 Explain, with the aid of well-labelled diagrams, why
   (a) Rias often provide better harbours than fiords.
   (b) Headlands have cliffs but bays have beaches.
   (c) Spits sometimes form across the mouth of a river

22 Illustrating your answer with diagrams, describe three

ways in which mountains are formed.

3 With the aid of sketch-maps, locate examples of *two* of the following:
(a) Coral reef,
(b) A ria coast,
(c) A raised coast.
For each one you choose, describe its chief features and explain how they have been formed.

4 Name four physical features which have been formed by glacial action and for each locate a region where it could be seen. Choose *three* of these and for each:
(a) Draw a well-labelled diagram to show its chief characteristics.
(b) Describe how it has been formed.

5 In what ways does the appearance of a glaciated valley differ from that of a river valley? Draw a large contour map for each type of valley and for each, name one region where an example could be seen.

6 With the aid of annotated diagrams or sketch-maps, describe:
(a) The appearance.
(b) The function of *three* characteristic features of a glaciated highland.

7 Write a short essay on the ways in which a glaciated region can be of value to Man. Illustrate your answer with reference to specific regions where the utilisation of glacial features has already taken place.

8 Draw a large map of Africa and on it:
(i) Draw and name the Equator.
(ii) Shade one region which contains rift valleys and print on it the letters R.V.
(iii) Mark and name two important ocean currents, indicating clearly the direction in which they flow and whether they are warm or cold.
(iv) Mark (by arrows), and name, *one* wind which brings rain to South Africa and *one* wind which brings rain to North Africa. Alongside the arrows write the months during which they blow.
(v) Draw a small circle where the sun will be overhead on December 21st.
(iv) Indicate, by printing the capital letter, where you would expect to find:
(a) Basin of inland drainage (capital B)
(b) Young fold mountains (capital Y)
(c) Delta (capital D).

9 Name the fishing grounds of the world, and for any one of them describe its chief conditions which account for its abundance of fish.

10 Concisely explain why some ocean currents are warm and others are cold. Name two warm currents of the North Atlantic and two cold currents of the North Pacific. For each:
(a) State the time of year when it is most pronounced.
(b) Describe any effect which it has on the climate of a specific region.

31 Illustrating your answer with sketch-maps or diagrams, describe how a lake can be formed (a) *by erosion*, and (b) *by deposition*. For each, name a region where an actual example could be seen.

32 For any *two* of the following features: lagoon, mountain lake, delta, artesian basin, (a) describe how it has been formed, (b) make a list of its possible uses, (c) name a particular region where an example could be found.

33 (a) (i) Name *two* temperature readings which are taken at weather stations.
(ii) Name and describe the instruments used for these measurements.
(b) Clearly state what you understand by the terms: humidity, precipitation, atmospheric pressure.

34 All weather stations have a Stevenson Screen. Make a large drawing of this and then:
(a) (i) Describe its structure.
(ii) Name the instruments it contains.
(iii) Briefly describe the nature of its location.
(b) Name any two other instruments in a weather station.

35 Carefully explain the meanings of the following terms: mean daily temperature, diurnal temperature range, relative humidity.
(a) Name the instruments with which each is associated.
(b) Choose one of these instruments and briefly describe how it works.

36 A weather station contains the following instruments: wind vane, aneroid barometer, maximum and minimum thermometer. Choose two of these, and for each:
(a) describe its appearance, (b) explain how it is used to obtain weather records.

37 With the aid of diagrams explain the effect of:
(a) longitude on temperature, (b) longitude on time, (c) earth rotation on wind direction.

38 (a) Briefly describe three ways in which rainfall may be caused.
(b) Name the instrument which is used for measuring rainfall and explain how rainfall is shown on a distribution map.
(c) Name (i) a region where rain falls all the year, (ii) a region where rain falls from June to September only, (iii) a region where it rarely falls.

39 Using well-labelled diagrams, explain the meaning of *three* of the following geographical terms:
Convection Rainfall, Land and Sea Breezes, International Date Line, Tropical Cyclone, Midnight Sun.

40 (a) Draw an outline map of South America and on it shade and name (i) *one* region having an Equatorial Climate, (ii) *one* region having a Mediterranean Climate, (iii) *one* region having a Savana Climate.
(b) Describe briefly each of these climates.

41 With the aid of diagrams, explain *three* of the following: tropical cyclones, monsoon winds, prevailing winds, rain shadow. Give an actual example for each one you choose.

363

42 Draw a simple sketch-map of either western North America or western Europe and on it shade *three* regions, each of which has a different type of climate. Name the climatic types and then describe *two* of them.

43 Describe and account for the main features of the climates of *two* of the following regions: Burma; Northern China; the Amazon Basin. Illustrate your answer with sketch-maps or diagrams.

44 (a) (i) Name *four* types of natural vegetation which occur in Africa, and (ii) briefly describe the features of each type.
   (b) Choose *two* of the four types you have named, and for each, explain how the features you have described show the influence of climate.

45 With the aid of sketch-maps locate examples of *two* of the following:
   (a) A tropical grassland
   (b) An evergreen tropical forest
   (c) A deciduous temperate forest
   (d) A temperate grassland.
   For each one you choose, describe the characteristic features of the vegetation and show how they are related to the climate of the region.

46 Select *three* of the following: laterite, loess, lowland alluvium, desert sands, boulder-clay. For each: (a) locate and name an actual example, (b) briefly describe how it has been formed.

47 Choose *two* of the following: Market Gardening, Sericulture, Shifting Cultivation, Plantation Farming. For each of the two chosen;
   (a) Name and give the location of an area where it is practised.
   (b) Write a short account of the farming methods used using the headings
      (i) crops or animals,
      (ii) work through the year,
      (iii) the reasons why the type of farming is practised in the area you have located in (a).

48 (a) Name *one* hydro-electric power scheme and explain how its location is influenced by
      (i) rainfall,
      (ii) relief,
      (iii) distribution of population.
   (b) What other advantages, other than the provision of power, can a hydro-electric power scheme have? For *each* of the advantages you give, name a specific example of a scheme which has this.

49 Select *one* of the following: cacao, coffee, tea.
   (a) Draw sketch-maps to show the locations of *two* important producing areas.
   (b) Briefly state the natural conditions required for the cultivation of the plant.
   (c) State to what extent the plant is grown to supply overseas demands, and name *two* countries which import the product of the plant.

50 Annual coal production shows that some continen produce more than others.
   A High production continents: Europe, North Americ Asia.
   B Low production continents: Africa, Australia, Sou America.
   (a) Choose *one* continent from group A and for it:
      (i) Draw a sketch-map to show the locations ar names of *three* important coalfields, and tw large mining towns.
      (ii) Briefly describe the various ways in which co is used.
   (b) Choose *one* continent from group B, and for write a short account of the location and importan of other sources of power.

51 Choose *two* of the following: maize, wheat, rice, oats.
   (a) For *each* one chosen, draw a sketch-map to sho *one* area which grows and exports the commodit
   (b) For *each* area chosen, briefly describe
      (i) the relief and soils,
      (ii) the climate,
      (iii) other factors which affect the production of th commodity.
   (c) Describe any *two* processes which are carried ou after the crop is harvested.

52 Give reasons to explain why:
   (i) the Panama Canal is important to North America
   (ii) Australia is not a large producer of hydro-electri power;
   (iii) Western Europe has a very high population density
   (iv) air transport is very important to South America

53 On an outline map of the world;
   (a) Mark and name
      (i) *one* important coalfield in the U.S.A. and on in Europe.
      (ii) *one* iron-mining town in China and *one* i Canada.
      (iii) *one* oil-producing area in South America an *one* in Asia.
   (b) Beneath your map
      (i) Briefly describe how steel is made and nam *one* large steel centre which uses coal and iror ore mined within its area,
      (ii) Name *two* large oil-refining centres which ar not on or near the oilfields, and say from where they obtain their crude oil.

54 The following countries can be divided into two groups one having a high population density, and the other a low population density: Great Britain, Brazil, Australia Holland, Arabia and Singapore.
   (a) Divide these countries into two groups, arranging the countries in order of population density.
   (b) Name the capital city for *each* country.
   (c) Briefly account for the population density for *one* country from *each* group.

Write a brief account of the factors influencing the location of industry in any *two* of the following:
(a) The iron and steel industry of the Ruhr.
(b) Shipbuilding on Clydeside.
(c) The manufacture of woollen textiles in Yorkshire.
(d) The making of paper pulp in Finland.

Choose *two* of the following: cotton, rubber, hemp, silk, jute.
(a) For *each* one chosen, draw a sketch-map to show *one* area which produces and exports the commodity.
(b) For *each* area selected, describe
  (i) the relief and soils,
  (ii) the climate,
  (iii) any other factors which encourage production.
(c) Briefly describe how each product is treated before it can be used.

(a) By using sketch-maps, name and locate *two* large-scale manufacturing industries in Asia, *or* in North America *or* in Europe.
(b) Name the raw materials used by *each* and state the source of supply.
(c) Explain how the industries have grown in the areas selected.

3 Carefully explain why:
(a) A large part of Australia's population is located near the coasts.
(b) A large part of Brazil does have a close network of roads and railways.
(c) River transport in Western Europe is important.
(d) Large parts of central Asia have a low population density.

9 Choose *three* of the following: entrepôt, hinterland, fuelling port, outport, and for *each*:
(a) Name *one* example from any country south of the Tropic of Cancer.
(b) Draw a sketch-map to show its location.
(c) Carefully explain its meaning in terms of its position and function.

0 Choose:
(a) *Either* rubber *or* oil palm products.
(b) *Either* coal *or* iron ore.
For *each* commodity selected:
  (i) draw a sketch-map to show an important area of production outside Europe and North America,
  (ii) briefly describe the conditions of production in the area located.

1 Select *three* of the following: softwood lumbering, truck farming, collective farming, cattle ranching, shifting cultivation. For *each* one chosen:
  (i) draw a sketch-map to show the location of an area where it is important,
  (ii) describe the conditions which have resulted in this activity in the area selected.

62 (a) From what part of the world are the following obtained: sisal, citrus fruits, cacao beans, jute, copra?
(b) Choose *three* of these commodities and for *each* describe what geographical conditions are favourable for its production.
(c) For *each* of the commodities selected, draw a sketch-map of *one* area which is important for its production.

63 (a) Carefully explain why irrigation is necessary for crop cultivation in some parts of the world.
(b) Choose *two* areas where irrigation is practised on a large scale, and for *each*:
  (i) draw a large sketch-map to show the location of the area and the source of irrigation water,
  (ii) briefly describe the system of irrigation used and name the crops grown.
(c) Write a short account of the advantages and disadvantages of traditional and modern irrigation methods.

64 (a) Name *two* areas where extensive drainage schemes are carried out.
(b) For *each* area chosen:
  (i) draw a sketch-map to show its location,
  (ii) describe how drainage is effected and say for what purpose.

65 (a) Describe the natural conditions under which petroleum occurs.
(b) Name *three* countries which are major producers, and, for *one* of these:
  (i) draw a sketch-map to show the main areas of production and mark and name *one* town which handles the petroleum,
  (ii) carefully explain how the petroleum is taken to the refining centres, which you should name.

66 Select *three* of the following: dried fruit, canned fish, paper-pulp, wool yarn, pig-iron, and for *each*:
  (i) draw a sketch-map to show the location of an area where it is produced on a large scale,
  (ii) describe the conditions under which it is produced.

67 Choose any *three* of the following: meat, wool, hides and skins, dairy products.
(a) Name *one* important area of production for *each* one chosen and show this on a sketch-map.
(b) For *two* of the three chosen, briefly describe the method of production.

68 Choose *two* of the following: Dry Farming, Crop Rotation, Terrace Cultivation, Mixed Farming. For *each* of the two chosen;
(a) Name and give the location of an area where it is practised.
(b) Briefly describe the farming methods using the headings
  (i) crops or animals,
  (ii) work through the year,
  (iii) the reasons why the type of farming is practised in the area you have located in (a).

HINTS ON ANSWERING EXAMINATION QUESTIONS

At School Certificate level, students are expected to illustrate their answers with relevant sketch-maps and diagrams. When students are ready to take the School Certificate they should be able to draw simple diagrams of the principal landforms discussed in their physical geography course, and they should also be able to give specific examples of the more important of these. Examiners are always impressed with reference to features or processes that the students have studied at first-hand in the field.

All diagrams and sketch-maps should be bold, large and clear in outline. They should contain no irrelevant detail and any writing on the map should be in printed lettering. Coloured pencils, used in moderation, can improve a diagram and can help to make the more important pieces of information stand out clearly. All diagrams should be given a title and, where necessary, a key. The latter must always be kept as simple as possible. A common tendency of many students is to include too much information in their diagrams with the result that they become overcrowded and difficult to interpret.

Before answering any questions in the examination a student should:

1 Read the instructions at the head of the paper very carefully.
2 Read through all the questions in the relevant section and mark those which he decides to answer.
3 Answer, first of all, the compulsory question(s) which usually carry more marks, and more time should allocated to them.
4 Calculate how much time should be allowed for each question, making the necessary allowance for compulsory question(s). When the time limit expires, work on the answer should stop and the next question should tackled. Any time which is left over at the end of the examination can then be used for completing unfinished answers.
5 Make sure that all answers are to the point and do not include information which has no direct bearing on the questions.

A study of past examination questions will show that a definite terminology is often used by examiners. Words and phrases in common use are *development, locate, factor, significant*, etc. The student must know the precise meaning of each of these.

EXAMINATION QUESTIONS

1 **Either,** (a) (i) Describe and account for **three** of the landforms associated with hot deserts.
(ii) Show how physical conditions and human activities can lead to the establishment of settlements even in the hot deserts.
or, (b) (i) Describe and account for the landforms associated with the floodplains of large rivers.
(ii) Where are human settlements most likely to be found in floodplain areas, and why?
(Oxford G.C.E. 1972)

2 Give an account of the mode of formation of two of the following:
(a) swallow holes, (b) eskers, (c) springs.
(Oxford/Cambridge Alt 'O' 1970)

3 Describe the characteristics of **either** a Mediterranean **or** a hot desert climate and account for the global distribution of the climatic type you chose.
(Oxford/Cambridge Alt 'O' 1970)

4 **Either** Discuss the causes and effects of river capture.
Or Explain the development of superimposed drainage. Illustrate your answer with sketch-maps of **named** examples.
(Oxford/Cambridge Alt 'O' 1971)

5 With the aid of diagrams explain fully the nature of and the reasons for.
(i) 'Chinook' winds in Canada.
(ii) The general decrease in temperature polewards with increasing latitude from the equator.
(iii) The wide diurnal variations in temperature that occur in Hot Deserts
(Welsh Joint Education Committee G.E.C. 1972)

6 **Either**
(a) For **each** of **three** of the following draw a simple contoured sketch-map, numbering the contours: a glaciated mountain valley, a ria coastline, a dissected plateau, a volcano.
(b) For any **one** of those chosen in (a), (i) describe the feature, and (ii) illustrating your answer with diagrams, suggest how it may have been formed.
Or
Headland and bay, anticline and syncline, rift valley and horst (block mountain), dyke and sill.
(a) For each of **two** of the above pairs of features:
(i) draw a labelled diagram to show the rock formations,
(ii) name and locate an example of **each** feature.
(b) For any **one** pair you have chosen in (a), suggest how **each** feature may have been formed.
(Cambridge G.C.E. 1970)

Either What features characterise temperate maritime and temperate interior climates? Why do they differ?
Or Briefly describe and account for the principal regional variations of climate in Europe.

(Oxford/Cambridge Alt 'O' 1972)

Study the climatic maps of North America in your atlas and the following figures, then:
(a) name the state or province and the climatic region in which each place lies;
(b) explain the pattern of temperatures and rainfalls revealed by the figures;
(c) write a note on the natural vegetation to be seen round these places.

ma 32° 35′N   114° 40′W

	J	F	M	A	M	J	J	A	S	O	N	D
	54·5	59·0	64·5	70·0	76·5	85·0	91·5	90·5	85·0	73·0	62·5	55·5
	12·5	15·0	18·1	21·1	24·7	29·5	33·1	32·5	29·5	22·8	16·9	13·1
	0·4	0·4	0·3	0·1	< 0·1	< 0·1	0·2	0·6	0·4	0·3	0·2	0·5
	1·0	1·0	0·8	0·3	< 0·3	< 0·3	0·5	1·5	1·0	0·8	0·5	1·2

ppermine 67° 0′N   115° 0′W

	J	F	M	A	M	J	J	A	S	O	N	D
	−19·0	−19·5	−15·5	0·5	22·5	38·5	50·0	46·0	36·5	17·5	−6·0	−15·5
	−28·3	−28·6	−26·4	−17·5	−5·3	3·6	10·0	7·8	2·5	−8·0	−21·1	−26·4
	0·6	0·4	0·6	0·8	0·5	0·8	1·3	1·9	1·0	1·2	0·8	0·6
	1·5	1·0	1·5	2·0	1·3	2·0	3·3	4·8	2·5	3·6	2·0	1·5

(Southern Universities Board G.C.E. 1970)

Quoting examples and illustrating your answer with diagrams explain how two of the following physical features are formed:
(a) canyons,
(b) rift valleys,
(c) fiords,
(d) deltas

(Southern Universities Board G.C.E. 1971)

(a) Outline the chief factors which affect the amount and nature of the weathering of rocks.
(b) Describe, with the aid of diagrams, the formation of three of the following: *rounded boulders of hot desert areas, screes, earth pillars, clints and grykes.*

(Cambridge G.C.E. 1972)

(a) With reference to one type of forest:
(i) draw a sketch map to locate an extensive area where it is found,
(ii) show how the vegetation is adapted to the climatic conditions.
(b) Describe, with examples, how man has influenced the world's natural vegetation cover.

(Cambridge G.C.E. 1972)

12 Either Describe, mentioning named examples, the ways in which natural lakes may be formed.
Or Write a short essay on springs and underground water.

(Oxford/Cambridge Alt 'O' 1972)

13 Explain why:
(a) Evergreen woodland and scrub occur in areas of hot dry summers and warm wet winters.
(b) Grassland vegetation occurs in areas of hot moist summers and very cold dry winters.
(c) Very dense evergreen forest occurs in areas of very hot and very wet conditions throughout the year.

(Cambridge G.C.E. 1970)

14 With the aid of diagrams, describe and give reasons for three of the following:
(a) Daily land and sea breezes.
(b) Chinook (föhn) winds.
(c) The Trade Winds.
(d) The frequent changes in the direction of the wind over Britain.

(Cambridge G.C.E. 1970)

15 (a) Outline briefly the principal causes of the surface currents of the oceans.
(b) Draw a sketch-map to show the arrangement of the ocean currents north of the equator in either the Atlantic or the Pacific. Mark and name two warm and two cold currents.
(c) Explain with examples the part played by ocean currents in
(i) the formation of fog, and
(ii) the development of fishing grounds.

(Cambridge G.C.E. 1971)

16 (a) With the aid of diagrams explain why it is nearly midday in New Zealand when it is midnight in England.
(b) When it is midnight in England, what is the time in Canberra, Australia?
(c) With the aid of diagrams explain why all areas north of the Arctic Circle have 24 hours daylight on 21st June, and 12 hours on 21st March.
(d) For 21st December give the number of hours of daylight at
(i) the Arctic Circle
(ii) the Equator
(iii) the Antarctic Circle.

(Southern Universities Board G.C.E. 1972)

17 (i) Explain, with the aid of diagrams, what is meant by latitude and longitude.
(ii) Calculate the time at Greenwich when it is 10.00 p.m. in Buenos Aires (Longitude 60°W)
(iii) Explain why at the Poles the sun does not set for many weeks on end at certain times of the year.

(Welsh Joint Education Committee G.E.C. 1972)

18 The rainfall diagrams below refer to places which are about 30°N latitude and on or not far from the coast.
  (a) Describe the similarities and differences among the diagrams, and suggest where each place might be.
  (b) Taking Stations **B** and **C** in turn, suggest reasons for the seasonal variation in rainfall.

(Oxford G.C.E. 1972)

19 (a) Name the instruments used in a school weath station to record temperature, humidity, a rainfall.
  (b) With the aid of diagrams, describe any **two** of the
  (c) Explain how the following are calculated:
      (i)   the daily average temperature;
      (ii) the monthly average temperature;
      (iii) the monthly average rainfall.

(Southern Universities Board G.C.E. 197

20 Study the weather charts on page 369 and then answ the following questions:
  (a) On the maps:
      (i) in the spaces provided name the weather syste shown on **each** chart,
      (ii) name the warm sector.
  (b) On your answer paper:
      (i)   With reference to Scotland and Norther Ireland compare the weather experienced o 27 February with that experienced on 2 October.
      (ii) Suggest reasons for the differences you hav described in (i) above.
      (iii) Name **six** conditions of the atmosphere whic are recorded by the meteorologist but whic are not shown on the weather maps provided

(Cambridge G.C.E. 1972

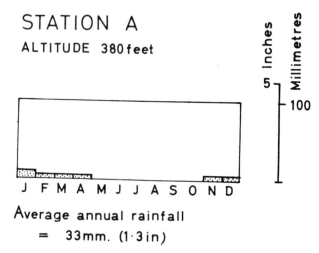

STATION A
ALTITUDE 380 feet

Average annual rainfall
   = 33mm. (1·3 in)

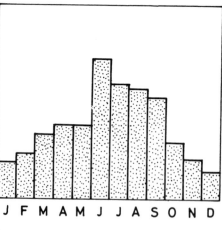

STATION B
ALTITUDE 33 feet

Average annual rainfall
   = 1135mm. (44·7 in)

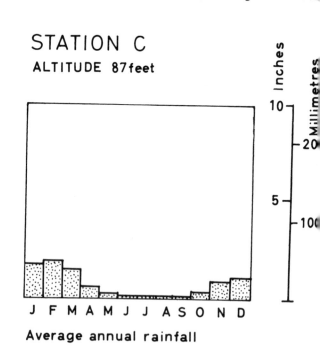

STATION C
ALTITUDE 87 feet

Average annual rainfall
   = 244 mm. (9·6 in)

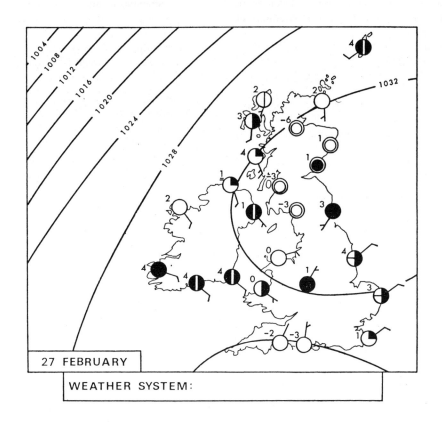

27 FEBRUARY

WEATHER SYSTEM:

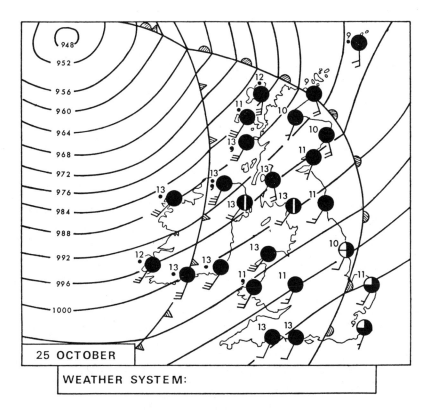

25 OCTOBER

WEATHER SYSTEM:

21 **Either** Illustrating your answer with a sketch map, briefly describe and account for major variations in climate in North America or Africa.
Or Describe and account for the monsoon climate of the Indian subcontinent.

(Oxford/Cambridge Alt 'O' 1971)

22 How are waves formed in the sea? How does wave and current action influence the form of coastlines.

(Oxford/Cambridge Alt 'O' 1972)

23 Choose an area where igneous rock occurs at the surface.
(a) Draw a simple sketch-map to locate the area.
(b) Describe the relief features of the area.
(c) Show how the rock has influenced physical features and human activities in the area.

(Oxford G.C.E. 1971)

24 State briefly what is meant by *peasant* (subsistence) *farming*. Describe the main features of this type of farming by reference to examples chosen from **two** continents.

(Cambridge 1971)

25 Choose **two** of the following: *Shifting Agriculture, Terrace Cultivation, Pastoral Farming. Tropical Plantation Farming*. For **each** of the **two** you have chosen:

(a) name and locate an area where it is practised,
(b) write an account of the farming under the headin
 (i) crops and or animals, (ii) work through the yea
 (iii) the reasons why the type of farming is practis
 in the area you have located in (a).

(Cambridge 19'

26 World map below shows a selection of the worl
major routes by sea, land and air. On the map:
(a) in the spaces provided name **eight** of the tow
 numbered 1-10.
(b) Indicate, by using the letters given in brackets.
 (i) where **one** of the railway routes crosses a h
 desert (H);
 (ii) the part of a sea route where ships may c
 counter icebergs (I);
 (iii) the part of an air route where a typhoon m
 cause a diversion of the flight (T);
 (iv) where **one** of the railway routes crosses a h
 range of mountains (M).
On your answer paper:
(c) With the aid of examples explain why some cargo
 between South Africa and London are now carri
 by air while others are still carried by sea.
(d) Name **two** products which are sent to the Unite
 Kingdom along route B.
(e) Explain why much wheat is transported alo
 railway route D at certain times of the year.

(Cambridge 197

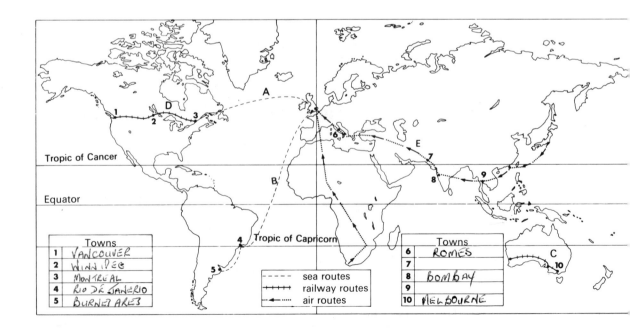

	Towns
1	VANCOUVER
2	WINNIPEG
3	MONTREAL
4	RIO DE JANERIO
5	BURNEI ARES

	Towns
6	ROMES
7	
8	BOMBAY
9	
10	MELBOURNE

- - - - sea routes
++++ railway routes
····◄···· air routes

Tropic of Cancer

Equator

Tropic of Capricorn

World's fuel and power production

	Approximate percentages	
	1937	1967
Coal and other solid fuels	74	40
Oil and natural gas	25	50
Hydro-electric power	1	10

Although the world's total production of coal has increased, its relative importance has declined.

(a) State briefly the difficulties of the coal mining industry.

(b) Give reasons for the increased importance of the forms of power other than coal.

Illustrate your answer by reference to named and located examples.

(Cambridge 1971)

28

	Estimated Population (millions)	Area millions sq km (millions sq miles)
Chinese Republic	700	9·736 (3·759)
Brazil	71	8·516 (3·288)
Europe (including European U.S.S.R.)	624	9·8 (3·8)

(a) Study the information given in the table and state which area is
  (i) the most densely populated,
  (ii) the most sparsely populated.

(b) For **one** of the **three** areas:
  (i) Locate on a sketch map the densely and sparsely populated regions,
  (ii) State the geographical factors which have influenced the distribution you have shown in (b) (i).

(Cambridge 1969)

29 Answer **one** of the alternatives (a) and (b) below.

Either (a) Explain the chief causes of the decay of parent rock by weathering in (i) tropical deserts, (ii) equatorial lands, (iii) cool temperate lands.

Or (b) (i) Explain how a glacier may be formed in a mountain valley.
  (ii) Name and describe the features you would expect to find in the valley and the surrounding highland after it has been eroded by the glacier.

Draw diagrams to illustrate your answers.

(Cambridge 1973)

30 (a) (i) Describe the main features of the natural vegetation of 'Mediterranean' lands and explain how it is adapted to the climate.

(ii) Name and locate by means of a sketch-map an area, other than the lands around the Mediterranean sea, where this type of natural vegetation may be found.

(b) Describe the main features of the natural vegetation known as the Taiga (Northern Coniferous Forests). State briefly why this type is not widely distributed in the Southern Hemisphere.

(Cambridge 1973)

31 Explain the following:

(a) The sun rises in an easterly direction and sets in a westerly direction.

(b) In the United Kingdom during the month of December the sun rises in the south-east and sets the south-west.

(c) In Europe the noonday sun is always due south while in New Zealand it is always due north.

(d) In Sri Lanka (Ceylon) the noonday sun is sometimes overhead, sometimes due north and at other times due south.

(e) The local time in Calcutta 23°N, 90°E, is four hours ahead of the local time in Leningrad 60°N, 30°E.

(Cambridge 1973)

32 (a) For **each** of the following: *Europe, Asia, Australia, North America,* name and locate **one** town or area which manufactures large amounts of iron and steel. For **one** of the towns or areas you have named in (a)

(b) Explain how the location of the iron and steel industry is influenced by (i) the location of raw materials, (ii) transport, (iii) labour, (iv) water supply and (v) markets.

(c) Describe the developments of the steel industry in either South America or Africa.

(Cambridge 1973)

33 (a) Describe the methods which are being used to increase agricultural output in tropical countries.

(b) (i) State with reasons and examples the advantages of the large plantation system of cultivation in tropical lands.
  (ii) State one of the possible disadvantages of this system.

(Cambridge 1973)

34 (a) By reference to **named** examples,
  (i) explain why many seaports are among the world's largest towns of over one million people;
  (ii) state the factors which have led to the growth of very large inland towns of over one million people.

(b) Give examples of and write on two major problems caused by the growth of extremely large towns.

(Cambridge 1973)

Index